Elasticity with MATHEMATICA®

This book gives an introduction to the key ideas and principles in the theory of elasticity with the help of symbolic computation. Differential and integral operators on vector and tensor fields of displacements, strains, and stresses are considered on a consistent and rigorous basis with respect to curvilinear orthogonal coordinate systems. As a consequence, vector and tensor objects can be manipulated readily, and fundamental concepts can be illustrated and problems solved with ease. The method is illustrated using a variety of plane and three-dimensional elastic problems. General theorems, fundamental solutions, displacements, and stress potentials are presented and discussed. The Rayleigh-Ritz method for obtaining approximate solutions is introduced for elastostatic and spectral analysis problems. The book contains more than 60 exercises and solutions in the form of MATHEMATICA notebooks that accompany every chapter. Once the reader learns and masters the techniques, they can be applied to a large range of practical and fundamental problems.

Andrei Constantinescu is currently Directeur de Recherche at CNRS: the French National Center for Scientific Research in the Laboratoire de Mecanique des Solides, and Associated Professor at École Polytechnique, Palaiseau, near Paris. He teaches courses on continuum mechanics, elasticity, fatigue, and inverse problems at engineering schools from the ParisTech Consortium. His research is in applied mechanics and covers areas ranging from inverse problems and the identification of defects and constitutive laws to fatigue and lifetime prediction of structures. The results have applied through collaboration and consulting for companies such as the car manufacturer Peugeot-Citroen, energy providers Électricité de France and Gaz de France, and the aeroengine manufacturer MTU.

Alexander Korsunsky is currently Professor in the Department of Engineering Science, University of Oxford. He is also a Fellow and Dean at Trinity College, Oxford. He teaches courses in England and France on engineering alloys, fracture mechanics, applied elasticity, advanced stress analysis, and residual stresses. His research interests are in the field of experimental characterization and theoretical analysis of deformation and fracture of metals, polymers, and concrete, with emphasis on thermo-mechanical fatigue and damage. He is particularly interested in residual stress effects and their measurement by advanced diffraction techniques using neutrons and high-energy X-rays at synchrotron sources and in the laboratory. He is a member of the Science Advisory Committee of the European Synchrotron Radiation Facility in Grenoble, and he leads the development of the new engineering instrument (JEEP) at Diamond Light Source near Oxford.

Elasticity with MATHEMATICA®

AN INTRODUCTION TO CONTINUUM MECHANICS AND LINEAR ELASTICITY

Andrei Constantinescu
CNRS and Ecole Polytechnique

Alexander Korsunsky
Trinity College, University of Oxford

CAMBRIDGE
UNIVERSITY PRESS

CAMBRIDGE UNIVERSITY PRESS
Cambridge, New York, Melbourne, Madrid, Cape Town,
Singapore, São Paulo, Delhi, Mexico City

Cambridge University Press
The Edinburgh Building, Cambridge CB2 8RU, UK

Published in the United States of America by Cambridge University Press, New York

www.cambridge.org
Information on this title: www.cambridge.org/9781107406131

First published 2007
First paperback edition 2012

A catalogue record for this publication is available from the British Library

Library of Congress Cataloguing in Publication Data

Constantinescu, Andrei.
Elasticity with mathematica : an introduction to continuum mechanics
and linear elasticity / Andrei Constantinescu, Alexander Korsunsky.
p. cm.
Includes bibliographical references and index.
ISBN-978-0-521-84201-3 (hardback)
1. Elasticity – Mathematical models. I. Korsunsky, Alexander. II. Title.
TA418.C66 2007
620.1′1232015118 – dc22 2007015321

ISBN 978-0-521-84201-3 Hardback
ISBN 978-1-107-40613-1 Paperback

Contents

Acknowledgments

The authors would like to thank colleagues and research organisations for their support.

We are grateful to our respective labs and teaching institutions, the Laboratoire de Mécanique des Solides at École Polytechnique and CNRS in Paris and the Department of Engineering Science, University of Oxford, for providing the space, environment, and support for our research.

We were particularly lucky to be able to enjoy the opportunities for meeting and working together at Trinity College, Oxford, in an atmosphere that is both stylish and stimulating, and would like to thank the President and Fellows for their generosity.

We are indebted to the funding bodies that supported our collaboration, including CNRS in France and EPSRC and the Royal Society in the United Kingdom. We are also grateful to industrial organisations that contributed support for the projects that both motivated and informed the research results reported in this book, including Peugeot, GDF, and Rolls-Royce plc.

A.C. would like to thank Professor Patrick Ballard for the opportunity to share thoughts about teaching and exercises and for numerous discussions that helped to clarify some of the more intricate aspects of elastic problems.

A.M.K. would like to express his special thanks to Professor Jim Barber for his sharp wit and acutely discerning mind, which made conversations with him so enjoyable and made it possible to unravel many an apparent mystery in elasticity.

Introduction

MOTIVATION

The idea for this book arose when the authors discovered, working together on a particular problem in elastic contact mechanics, that they were making extensive and repeated use of MATHEMATICA^{TM} as a powerful, convenient, and versatile tool. Critically, the usefulness of this tool was not limited to its ability to compute and display complex two- and three-dimensional fields, but rather it helped in understanding the relationships between different vector and tensor quantities and the way these quantities transformed with changes of coordinate systems, orientation of surfaces, and representation.

We could still remember our own experiences of learning about classical elasticity and tensor analysis, in which grasping the complex nature of the objects being manipulated was only part of the challenge, the other part being the ability to carry out rather long, laborious, and therefore error-prone algebraic manipulations.

It was then natural to ask the question: Would it be possible to develop a set of algebraic instruments, within MATHEMATICA, that would carry out these laborious manipulations in a way that was transparent, invariant of the coordinate system, and error-free? We started the project by reviewing the existing MATHEMATICA packages, in particular the `VectorAnalysis` package, to assess what tools had been already developed by others before us, and what additions and modifications would be required to enable the manipulation of second-rank tensor field quantities, which are of central importance in classical elasticity. In this book we present our readers with the result of our effort, in the form of MATHEMATICA packages, notebooks, and worked examples.

In the course of building up this body of methods and solutions, we were forced to review much of the well-established body of classical elasticity, looking for areas of application where our instrumentarium would be most effective. After a while it became apparently necessary for us to include this review in the text, in order to preserve the logic and consistency of approach and to achieve a level of completeness – although we did not aim to reach every region of the vast domain of continuum mechanics, or elasticity in particular.

This book is intended as a text and reference for those wishing to realise more fully the benefit of studying and using classical elasticity. The approaches presented here are not aimed at replacing various other computational techniques that have become successful and widespread in modern engineering practice. Finite element methods, in particular, through decades of application and development, have acquired tremendous versatility and the ability to deliver numerical solutions of complex problems. However, the power of analytical treatments possible within the framework of elasticity should not be underestimated: true understanding of physical systems often consists of the ability to identify the relationships and interdependencies between different quantities, and nothing serves this objective more elegantly and efficiently than concise analytical solutions.

It is our hope that any readers who have previous experience of courses in engineering mechanics and strength of materials will find something useful for themselves in this book. This might be just a practical tool, such as a symbolic manipulation module; or it might be an explanation that helps readers to make sense of a more or less sophisticated concept in elasticity theory, or in the broader context of continuum mechanics. In particular, we sought to use consistently, insofar as it was possible, the invariant form of operations with tensor fields. It is of course true that for practical purposes the results always need to be expressed in some specific coordinate system, to make them understandable to computer algebra systems and humans. Natural phenomena, however, do not require coordinate systems to happen – in fact, some of the most successful theories in the natural sciences are built on the basis of invariance with respect to transformations of spatial and temporal coordinates. The great benefit of the symbolic manipulation ability of MATHEMATICA is that it allows the (sometimes heavy) machinery of tensor manipulation in index notation to be hidden from the user. It is indeed our hope that providing readers with coordinate-invariant analytical instruments will allow them to concentrate on the intriguing underlying natural relationships that are the reason many people choose to study this subject in the first place.

Many books exist that are devoted to similar topics, and many of them are remarkably good. Some of them show readers in detail how important results in elasticity are derived, often frightening away beginners with lengthy derivations and numerous indices. Others select some of the most elegant solutions that can be obtained in a surprisingly concise way, if the right path to the answer is judiciously chosen, usually on the basis of many years of practice in algebraic manipulation. This work is unique in that it attempts to place the focus firmly on the analysis of the mechanics of deformation in terms of tensor fields, but to take away the fear of 'long lines,' freeing the reader to explore, verify, visualise, and compute.

As in any classical subject (and there are not many fields in hard natural science more classically established than elasticity), a great body of knowledge has been accumulated over decades and centuries of research. Detailed description of all of these areas could fill many volumes. Topics covered in this book were selected because they represent the common core of concepts and methods that will be useful to any practitioner, whether on the research or application side of the subject. They also lend themselves well to being implemented in the form of symbolic manipulation packages and illustrate key principles that could be applied elsewhere within the broader subject. We made a deliberate effort to make this book rather concise, aiming to illustrate an approach that can be successfully applied also to numerous other examples found in the excellent literature on the subject.

The authors' experience is primarily of teaching continuum mechanics and elasticity to European students in France and the United Kingdom. Some of the material included in this book was used to teach advanced mechanics and stress analysis courses. However, it is also the authors' belief that, in the context of the U.S. graduate teaching system, the scope covered in this work would be particularly appropriate for a one-semester course at the graduate level in departments of engineering mechanics, engineering science, and mechanical, aerospace, and civil engineering. It will equip the listeners with valuable analytical skills applicable in many contexts of applied research and advanced industrial development work.

The subject of the book is of particular interest to the authors because both of them have been involved, for a number of years, in the capacity of researchers, graduate supervisors, research project leaders, and consultants, in the application of classical methods of continuum mechanics to modern engineering problems in the aerospace and automotive industry, power generation, manufacturing optimisation and process modelling, systems design, and structural integrity assessment, etc.

Classical elasticity is one of the oldest and most complete theories in modern science. Its development was driven by engineering demands in both civil and military construction and manufacture and required the invention and refinement of analytical tools that made crucial contributions to the broader subject of applied mathematics.

In an old and thoroughly researched subject such as elasticity, why does one need yet another textbook? Elasticity theory has not experienced the kind of revolution brought about by quantum theory in physics or the discovery of the gene in biology. Development of elasticity theory largely followed the paradigm established by Cauchy and Kelvin, Lagrange and Love, without significant revisions. Certainly, one ought not to overlook the advent of powerful computational techniques such as the boundary element method and the finite element method. Yet these techniques are entirely numerical in their nature and cannot be used directly to establish fundamental analytical relations between various problem parameters.

For the first time in perhaps over 200 years, the practice of performing analytical manipulations in elasticity is changing from the pen and paper paradigm to something entirely different: analytical elasticity by computer.

The origins of elasticity are often traced to Hooke's statement of elasticity in 1679 in the form of the anagram CEIIINSSSTTUVO containing the coded Latin message 'ut tensio, sic vis,' or 'as the extension, so the force.' Development of elasticity theory required generalisation of the concepts of extension or deformation and of stress to three dimensions. The necessity of describing elastic fields promoted the development of vector analysis, matrix methods, and particularly tensor calculus. The modern notation used in tensor calculus is largely due to Ricci and Levi-Civita, but the term 'tensor' itself was first introduced by Voigt in 1903, possibly in reference to Hooke's 'tensio.'

The subject of tensor analysis is thus particularly closely related to elasticity theory. In this book we devote particular attention to the manipulation of second rank tensors in arbitrary orthogonal curvilinear coordinate systems to derive elastic solutions. Differential operations with second rank tensors are considered in detail in an appendix. Most importantly from the practical viewpoint, convenient tools for tensor manipulation are written as modules or commends and organized in the form of a MATHEMATICA package supplied with this book.

The theory of potential is another branch of mathematics that stands in a close symbiotic relationship with elasticity theory, in that it both was driven by and benefited from the search for solutions of practical elasticity problems. We devote particular attention to potential representations of elastic fields, in terms of both stress and displacement functions.

Fundamental theorems of elasticity are indispensible tools needed to establish uniqueness of solutions and also to develop the techniques for finding approximate solutions. These are presented in a concise form, and their use is illustrated using MATHEMATICA examples. Particular attention is given to the development of approximate solution techniques based on rigorous variational arguments.

Appendices contain some reference information, which we hope readers will find useful, on tensor calculus and MATHEMATICA commands employed throughout the text.

WHAT WILL AND WILL NOT BE FOUND IN THIS BOOK

The particular emphasis in this text is placed on developing a MATHEMATICA instrumentarium for manipulating vector and tensor fields in invariant form, but also allowing the user to inspect and dissect the expressions for tensor components in explicit, coordinate-system-specific forms. To this end, at relevant points in the presentation, the appropriate modules are constructed. This includes the definition of differential operators (`Grad, Div, Curl, Laplacian, Biharmonic, Inc`) applicable to scalars, vectors, and tensors. Importantly, in the case of tensor fields, definitions of right (post-) and left (pre-) forms of the `Grad` operator are made available. Analysis of biharmonic functions is addressed in some detail, and tools for the reduction of differential operators in arbitrary orthogonal curvilinear coordinate systems are provided to help the reader reveal and appreciate their nature. The modules `IntegrateGrad` and `IntegrateStrain` have particular significance in the context of linear elastic theory and are explained in some detail, together with their connection with the Saint Venant strain compatibility conditions. All packages, example notebooks, and solutions to exercises can be downloaded freely from the publisher's web site at www.cambridge.org/9780521842013

The development of MATHEMATICA tools happens against the backdrop of the presentation of the classical linear elastic theory. To keep the presentation concise, some care was taken to select the topics included in this treatment.

Chapter 1 is devoted to the kinematics of motion and serves as a vehicle for introducing the concept of deformation as a transformation map, leading naturally to the concept of deformation gradient and its polar decomposition into rotation and translation. The definition of strain then follows, and particular attention is focused on the concept of small strain. The procedure for reconstituting the displacement field from a given distribution of small strains is constructed based on rigorous arguments and implemented in the form of an efficient MATHEMATICA module. In the process of developing this constructive approach, the conditions for small strain integrability are identified (also known as the Saint Venant strain compatibility conditions).

The significance of some differential operators applied to tensor fields becomes immediately apparent from the analysis of Chapter 1. In particular, the second-order incompatibility operator, inc, is introduced, allowing the Saint Venant condition for compatibility of small strain ϵ to be written concisely:

$$\operatorname{inc}\boldsymbol{\varepsilon} = \mathbf{0}.$$

This operator has particular significance in the theory of elasticity, and further attention is devoted to it in subsequent chapters, as well as to its relationship with the laplacian and biharmonic operators.

Chapter 2 is devoted to the analysis of forces. Particular attention is given to elastostatics, that is, the study of stresses and the conditions of their equilibrium. We show how the principle of virtual power offers a rational starting point for the analysis of equilibria of continua. The concept of stress appears naturally in this approach as dual to small strain in a continuum solid. Furthermore, the equations of stress equilibrium, together with the traction boundary conditions, follow from this variational formulation in the most convenient invariant form. An interesting aside here is the discussion of the expressions for virtual power arising within different kinematical descriptions of deformation (e.g., inviscid fluid, beams under bending) and the modifications of the concept of stress that are appropriate for these cases.

The classical stress definition according to Cauchy is also presented, and its equivalence to the definition arising from the principle of virtual power is noted. The Cauchy–Poisson theorem then establishes the form of equilibrium equations and traction boundary conditions. (Discussion of the index form of equilibrium equations and boundary conditions that is specific to coordinate systems is addressed by demonstration in the exercises at the end of this chapter.) Some elementary stress states are considered in detail.

Having established the fact that equlibrium stress states in continuum solids in the absence of body forces are represented by divergence-free tensors, we address the question of efficient representation of such tensor fields. The Beltrami potential representation is introduced, in which the operator inc once again makes its appearance. Donati's theorem is then quoted, which establishes a certain duality between the conditions of stress equilibrium and strain compatibility.

Chapter 3 is devoted to the discussion of general anisotropic elasticity tensors. Important properties of elastic tensors are introduced, and the relationships between tensor and matrix representations are rigorously considered, together with efficient MATHEMATICA implementations of conversion between different forms. Next, classes of material elastic sysmmetry are considered, and the implications for the form of elastic stiffness matrices are clarified. Elastic isotropy is discussed in detail as a particularly important case that is treated in more detail in subsequent chapters.

MATHEMATICA tools for displaying elastic symmetry planes are presented, along with ways of visualising the results of extension experiments on anisotropic materials.

The methods of solution of elasticity problems for anisotropic materials are not considered in the present treatment, as the authors felt that this important subject deserved special treatment.

Modifications and perturbations to the liniear elastic theory are briefly discussed, including thermal strain effects and residual stresses. The chapter is concluded with a brief discussion of the limitations of the linear elastic theory and the formulation of Tresca and von Mises yield criteria.

Chapter 4 is devoted to the formulation of the complete problem of elasticity and the discussion of general theorems and principles. First, the formulation of a well-posed, or regular problem of thermoelasticity is introduced. Next, the displacement formulation (Navier equation) and the stress formulation (Beltrami–Michell equations) are introduced. As a demonstration of the application of elasticity problem formulation, the problem of the spherical vessel is solved directly by considering the radial displacement field in the spherical coordinate system, computing strains and stresses, and satisfying the equilibrium and boundary conditions.

Next, the principle of superposition is introduced, followed by the virtual work theorem. This allows the nature and the conditions for the uniqueness of elastic solution to be established. This is followed by the proof of existence of the strain energy potential and the complementary energy potential and of reciprocity theorems. Saint Venant torsion is next considered in detail with the help of MATHEMATICA implementation, serving as the vehicle for the introduction of the more general Saint Venant principle. The counterexample due to Hoff is given as an illustration, and a rigorous formulation of the principle, following von Mises and Sternberg, is given.

Chapter 5 is devoted to the solution of elastic problems using the stress function approach. The Beltrami potential introduced previously provides a convenient representation of self-equilibrated stress fields. The Airy stress function corresponds to a particular case of this representation and is of special importance in the context of plane elasticity due to its simplicity, and for historical reasons. Particular care is therefore taken to introduce this approach and to discuss the precise nature of strain compatibility conditions that must be imposed in this formulation to complement stress equilibrium. This allows the elucidation of the strain incompatibility that arises in the plane stress approximation. In passing, an important issue of verifying the biharmonic property of expressions in an arbitrary coordinate system is addressed symbolically through the analysis of reducibility of differential operators. It is then demonstrated how strain compatibility in plane stress can be enforced through the introduction of a corrective term. Plane strain is also considered, and the simple relationship with plane stress is pointed out.

The properties of Airy stress functions in cylindrical polar coordinates are addressed next. The general form of biharmonic functions of two coordinates, due to Goursat, serves as the basis for obtaining various forms of Airy stress functions as suitable candidate solutions of the plane elasticity problem. The Michell solution, although originally incomplete and amplified with additional terms by various contributors, is introduced and discussed due to its historical importance. Furthermore, it allows the identification of some important fundamental solutions that serve as *nuclei of strain* within the elasticity theory. In this way the solutions for disclination, dislocation, and other associated problems are analysed.

The Airy stress function solution is derived next for a concentrated force applied at the apex of an infinitely extended wedge. This important solution serves to introduce the Flamant solution for the concentrated force at the surface of an elastic half-plane. The combination of the appropriate wedge solution with the dislocation solution allows the Kelvin solution for a concentrated force acting in an infinite elastic plane to be derived by enforcing displacement continuity. The derivation makes use of the strain integration procedure presented earlier.

Williams eigenfunction analysis of the stress state in an elastic wedge under homogeneous loading is presented next. On the basis of this solution, the elastic stress fields can be found around the tip of a sharp crack subjected either to opening or to shear mode loading. Finally, two further important problems are treated, namely the Kirsch problem of remote loading of a circular hole in an infinite plate and the Inglis problem of remote loading of an elliptical hole in an infinite plate.

Chapter 6 is devoted to the introduction and use of the method of displacement potential. First, the harmonic scalar and vector Papkovich–Neuber potentials are introduced and the representations of simple deformation states in terms of these potentials are found. Next, the fundamental solution of three-dimensional elasticity is derived, the Kelvin solution for a force concentrated at a point within an infinitely extended isotropic elastic solid. The Kelvin solution serves as the basis for deriving solutions for force doublets, or dipoles, with or without moment, and also for centres of dilatation and rotation. These are further examples of *strain nuclei*, already introduced earlier in the context of plane elastic problems.

Solutions presented next are for the Boussinesq and Cerruti problems about concentrated forces applied normally or tangentially to the surface of an elastic half-space. The solution for a concentrated force applied at the tip of an elastic cone is given next. General solutions in spherical and cylindrical coordinates are discussed, and the use of spherical harmonics illustrated. The Galerkin vector is introduced as an equivalent displacement potential formulation, and Love strain function presented as a particular case. The chapter is concluded with a brief note on the integral transform methods and contact problems.

Chapter 7 deals with the subject of energy principles and variational formulations, which are of particular importance for many applications, because they provide the basis for most numerical methods of approximate solutions for problems in continuum solid mechanics. Using strain energy and complementary energy potentials introduced earlier, a suite of extremum theorems is introduced. On this basis approximate solutions (bounds) in the theory of elasticity are introduced, using the notions of kinematically and statically admissible fields. The problem of the compression of a cylinder between rigid platens provides an example of application of the method.

Next, extremal properties of free vibrations and approximate spectra are considered. Analysis of vibration of a cantilever beam serves as an example.

Appendices contain some background information on linear differential operators, particularly in application to tensor fields studied with respect to general orthogonal curvilinear coordinate systems. Also explained is the implementation of these operators within the `Tensor2Analysis` package. Some important MATHEMATICA constructs used in the text, such as the `IntegrateGrad` module, are also explained, along with other MATHEMATICA tricks and utilities developed by the authors for the visualisation of results.

This book does not dwell in any detail on many important problems in elasticity and continuum solid mechanics. Anisotropic elasticity problems are not addressed here in any detail, nor are the complex variable methods in plane elasticity. Contact mechanics forms another large section of elasticity that is not treated here. Elastic waves, dispersion, and interaction with boundaries are not addressed in this text, again due to the fact that the authors thought it impossible to give a fair exposition of this subject within the limited space available.

It is the authors' hope, however, that many of the methods and approaches developed and presented in this book will provide the reader with transferable techniques that can be applied to many other interesting and complex problems in continuum mechanics. To help achieve this purpose, the book contains over 60 exercises that are most efficiently solved using MATHEMATICA tools developed in the corresponding chapters. Many of these exercises are not original, and, whenever possible, explicit reference is made to the source. The authors' hope is, however, that in solving all of these exercises readers will be able to appreciate the advantages offered by symbolic manipulation.

1 Kinematics: displacements and strains

OUTLINE

This chapter is devoted to the introduction of the fundamental concepts used to describe continuum deformation. This is probably most naturally done using examples from fluid dynamics, by considering the description of particle motion either with reference to the initial particle positions, or with reference to the current (actual) configuration. The relationship between the two approaches is illustrated using examples, and further illustrations are provided in the exercises at the end of the chapter. Some methods of flow visualisation (streamlines and streaklines) are described and are illustrated using simple examples. The concepts are then clarified further using the example of inviscid potential flow.

Placing the focus on the description of deformation, the fundamental concept of deformation gradient is introduced. The polar decomposition theorem is used to separate deformation into rotation and stretch using appropriate tensor forms, with particular attention being devoted to the analysis of the stretch tensor and the principal stretches, using pure shear as an illustrative example. Trigonometric representation of stretch and rotation is discussed briefly.

Discussion is further specialised to the consideration of small strains. Analysis of integrability of strain fields then leads to the identification of the invariant form of compatibility conditions. This subject is important for many applications within elastic theory and is therefore dwelt on in some detail. Strain integration is implemented as a generic module in MATHEMATICA, allowing displacement field reconstruction within any properly defined orthogonal curvilinear system.

1.1 PARTICLE MOTION: TRAJECTORIES AND STREAMLINES

Lagrangian description

Let us suppose that the material body under observation occupies the domain $\Omega \in \mathbb{R}^3$ in a reference configuration $\mathcal{C}$. Each material point is identified by its spatial position $\boldsymbol{X}$ in the reference configuration.

Let us assume that the motion of a particle is described by a function

$$\boldsymbol{x} = \mathcal{F}(\boldsymbol{X}, t) \tag{1.1}$$

which maps each point $\boldsymbol{X}$ of the reference configuration onto its position $\boldsymbol{x}$ at time t.

The mapping $\mathcal{F}$ is therefore defined,

$$\mathcal{F} : \Omega \times [0, T] \longrightarrow \mathbb{R}^3,$$

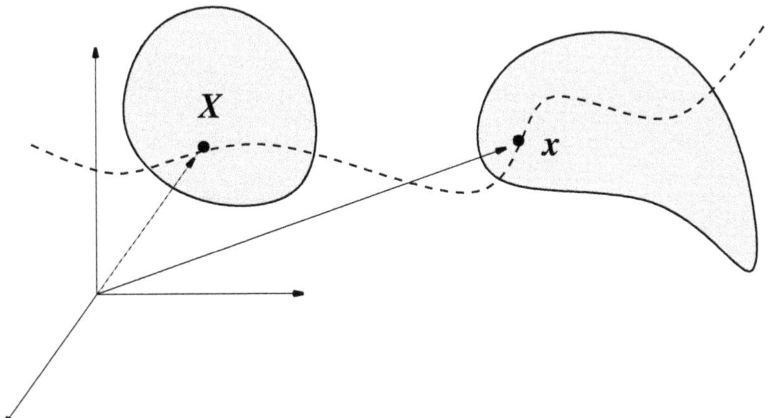

Figure 1.1. The initial and the actual configuration of a body and the position of a particle on its path.

where Ω denotes the initial configuration of the body. The domain $\Omega_t = \mathcal{F}(\Omega, t)$ is referred to as the actual configuration at time t.

This description of motion is referred to as the *Lagrangian* description.

We shall assume that matter is neither created nor removed, that noninterpenetrability of particles is respected, and that the continuity of material orientation is conserved during motion.

These assumptions imply that there exists a one-to-one relation between material particles and points $\boldsymbol{X}$ in the reference configuration, as well as between the initial and actual positions of particles $\boldsymbol{X}$ and $\boldsymbol{x}$, respectively.

Particle path

The trajectory of a given particle in the fixed laboratory frame is the curve that is also referred to as the *particle path* (see Figure 1.1). The particle path is the geometrical locus of the points occupied by the material particle at different times during deformation and can be mathematically expressed as the following set:

$$\mathcal{P}(\boldsymbol{X}) = \{\mathcal{F}(\boldsymbol{X}, t) | t \in [0, T]\}. \tag{1.2}$$

Consider as an example the particle paths of points on a rigid 'railway' wheel that is rolling without slipping along a surface represented by a straight horizontal line.

In order to illustrate particle paths in MATHEMATICA, first define the transformation $\mathcal{F}$ that at time t is given by the superposition of translation of the wheel centre by the distance $\boldsymbol{v}t$ and rotation of the wheel around its centre by the angle ωt. This is done by introducing vector positions of the wheel centre **`cent`**, the particle point a and velocity vector v, and the rotation matrix **`rot`**.

```
a = {a1, a2, a3}; v = {v1, 0, 0};

rot[phi_] :=
```

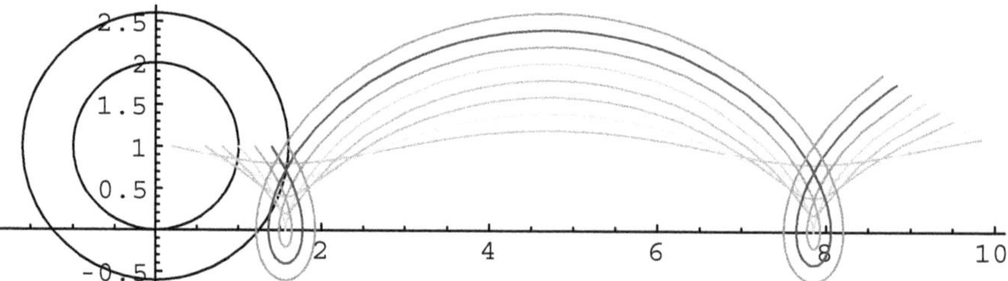

Figure 1.2. Trajectories of points on a rigid 'railway' wheel rolling along a horizontal surface without slipping.

```
{{ Cos[phi], Sin[phi], 0},
 {-Sin[phi], Cos[phi], 0},
 {        0,        0, 1}}

F[a_, t_]:= v t + cent + rot[omega t].(a - center)
F[a, t]
```

The condition of rolling without slipping is ensured by the fact that the total velocity of the point in instantaneous contact with the surface is equal to zero, due to the fact that the contributions to this velocity from the translation and rotation parts of the motion are equal and opposite; that is,

$$v_1 = \omega R.$$

The points selected for particle path tracking are obtained as a double-indexed list using **Table**. **Flatten** transforms the double-indexed list into a single-indexed list.

The **wheel** and **wheel1** represent the 'railway' wheel with an outsized 'tyre' that is allowed to pass below the surface. The **traject** set of particle paths is obtained using the standard **ParametricPlot** command. The form of the command represents the application (**Map**) of the **ParametricPlot** command to all initial **points**. The **Drop** command eliminates the third coordinate for two-dimensional plotting.

All trajectories and the wheel and tyre are displayed in Figure 1.2 using **Show**. The particle paths can be recognised as *cycloids*. Classical implicit equations for these curves can be obtained after some additional manipulations.

```
cent = {0, R, 0};
R = 1; omega = 1; vl = R omega;

points  = Flatten[
 Table[
  center + r {Cos[alpha], Sin[alpha], 0},
    {r, 0.2, 1.6, 0.2}, {alpha, 0, 0}], 1]

wheel = ParametricPlot[
   {R Cos[theta], 1 + R Sin[theta]},
```

```
      {theta,0, 2 \[Pi]},
      DisplayFunction -> Identity ];

wheel1 = ParametricPlot[
   {1.6R Cos[theta], 1 + 1.6R Sin[theta]},
      {theta,0, 2 \[Pi]},
      DisplayFunction -> Identity ];

traject  =
 Map[
  ParametricPlot[
   Evaluate[ Drop[
      F[cent + (0. + 0.2 #){Cos[0], Sin[0], 0}, t],
      -1 ]],
      {t, 0, 10},
       PlotStyle -> {Evaluate[Hue[0.1 #]]} ,
       DisplayFunction -> Identity ]   &#,
     Range[8] ];

Show[wheel, wheel1, Sequence[traject] ,
  DisplayFunction ->  $DisplayFunction,
  AspectRatio -> Automatic ]
```

Eulerian description

Practical experience shows that it is not always possible to track the path of all particles from the initial to the actual configuration. This is generally the case with fluid flows, as one notices when observing the flow of particles in a river from a bridge.

In a situation such as this one can imagine instead that we are able to make two snapshots of the particles at two consecutive time instants. The difference in particle positions in the snapshots depends on the time interval between them. If this interval is sufficiently small (for a particular flow), than the particle displacements can be used to obtain approximate velocities of the particles.

Developing this idea, we shall suppose that the motion at each time instant is described by the velocity field with respect to the *actual configuration:*

$$\boldsymbol{v}(\boldsymbol{x}, t) : \Omega_t \longrightarrow \mathbb{R}^3.$$

In order to recover the particle path defined previously, one has to integrate the velocities of a given particle during time. This leads to a new definition of the particle path for particle $\boldsymbol{X}$ as the solution of the following ordinary differential equation:

$$\frac{d\boldsymbol{x}}{dt} = \boldsymbol{v}(\boldsymbol{x}(t), t) \qquad t \in [0, T] \tag{1.3}$$

$$\boldsymbol{x}(0) = \boldsymbol{X}. \tag{1.4}$$

Streamline

A *streamline* is a curve defined at a particular fixed moment in time so that at each point along the curve the tangent line points in the direction of the instantaneous velocity field. Because a curve can be defined using either a *parametric* or an *implicit description*, the following two descriptions of a streamline arise.

Consider a vector field of velocities $\boldsymbol{v}(\boldsymbol{x}, t)$ and a streamline defined in the parametric form as the curve $\boldsymbol{a}(s)$, with $s \in \mathbb{R}$ the curvilinear coordinate. As the tangent line is in the direction of velocity field, it follows that there exists a variable parameter $\lambda(s)$ that provides the following proportionality:

$$\frac{d\boldsymbol{a}}{ds} = \lambda(s)\,\boldsymbol{v}(\boldsymbol{a}(s), t) \qquad \forall s \in \mathbb{R}.$$

Choosing $\lambda(s) = 1$, one obtains the streamline passing through the point $\boldsymbol{X}$ at time $t = 0$ through integration of the ordinary differential equation

$$\frac{d\boldsymbol{a}}{ds} = \boldsymbol{v}(\boldsymbol{a}(s), t) \qquad t \in [0, T] \tag{1.5}$$

$$\boldsymbol{a}(0) = \boldsymbol{X}. \tag{1.6}$$

An implicit expression for a two-dimensional surface in $\mathbb{R}^3$ can be given by the locus of the solutions of an equation

$$\psi(\boldsymbol{x}) = \text{const}$$

for a scalar-valued function $\varphi : \mathbb{R}^3 \longrightarrow \mathbb{R}$.

A streamline consisting of the points $\boldsymbol{a}$ can also be defined as the intersection of the two surfaces, and therefore corresponds to the solution of the implicit system of equations

$$\psi_1(\boldsymbol{a}) = 0, \qquad \psi_2(\boldsymbol{a}) = 0, \tag{1.7}$$

where ψ_1, ψ_2 are two scalar-valued functions.

For planar flows that take place in the (x_1, x_2) plane, the second equation can be taken to be the equation of a plane, $\psi_2(\boldsymbol{a}) = \boldsymbol{a} \cdot \boldsymbol{e}_3 = 0$, and the analysis can be carried out in terms of only one remaining function, ψ_1.

The gradient

$$\nabla\psi_1(\boldsymbol{a}) = \frac{\partial\psi_1}{\partial x_1}(\boldsymbol{a})\boldsymbol{e}_1 + \frac{\partial\psi_1}{\partial x_2}(\boldsymbol{a})\boldsymbol{e}_2$$

defines a vector field normal to the streamline $(\boldsymbol{a}) = 0$, and therefore also normal to the tangent line of the streamline $\boldsymbol{t}$,

$$\nabla\psi_1(\boldsymbol{a}) \cdot \boldsymbol{t}(\boldsymbol{a}) = 0,$$

with

$$\boldsymbol{t}(\boldsymbol{a}) = -\frac{\partial\psi_1}{\partial x_2}(\boldsymbol{a})\boldsymbol{e}_1 + \frac{\partial\psi_1}{\partial x_1}(\boldsymbol{a})\boldsymbol{e}_2.$$

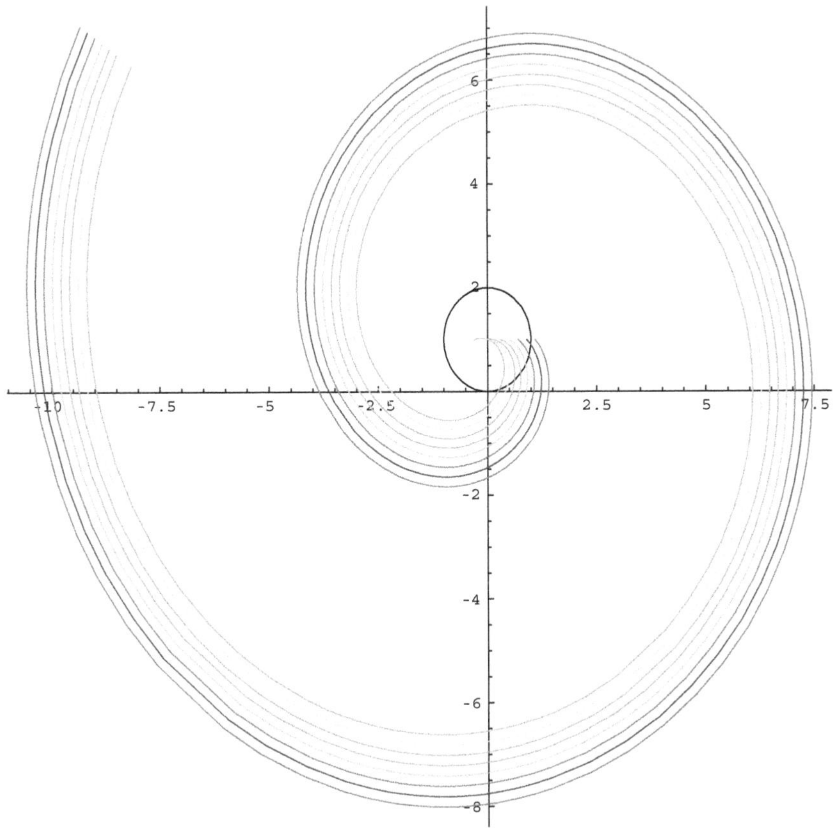

Figure 1.3. Streaklines of the points in a plane rigidly attached to a wheel rolling along a straight line without slipping.

Streakline

A *streakline* associated with point $\boldsymbol{P}$ at the time instant t is the geometrical locus of all particles that passed through point $\boldsymbol{P}$ at an instant $\tau \leq t$. This locus can be written explicitly in mathematical form as the following set:

$$\mathcal{S}(\boldsymbol{P}, t) = \{\mathcal{F}(\mathcal{F}^{-1}(\boldsymbol{P}, \tau), t) | \tau \in (-\infty, t]\}. \tag{1.8}$$

The concept of streaklines stems from their application in fluid mechanics. In flow visualisation experiments one often releases smoke or coloured particles from a certain point. The image of these particles at any time instant is the streakline of the release point at that moment.

The computation of streaklines is illustated next using MATHEMATICA on the basis of the examples already discussed with the turning wheel.

To compute the streakline, we first compute the inverse transformation as a superposition of the inverse translation and rotation. This can also by done using the application of `Solve` to the corresponding equation.

The same command grouping as in the case of particle paths makes it possible to generate the streaklines for a whole series of points, as displayed in Figure 1.3.

```
Finv[alpha_, tau_] :=
  cent + rot[- omega tau]  .   (alpha - v tau - cent)

alpha = {alpha1, alpha2, alpha3}
finv = Simplify[ Solve[ F[alpha, t] == a, alpha] ]

Simplify[
  v t + cent +
    rot[omega t]  .   (Finv[ {alpha1, alpha2, alpha3}, tau] -
          cent) ]

a0 = {0, 3, 0}

F[ Finv[a0 , tau] , 1]  // MatrixForm

ParametricPlot[
Drop[
  F[ Finv[a0 , tau] , 1]
  ,  -1],
 {tau, -10, 0}, AspectRatio -> 1 ]

streaklines  = Map[
  ParametricPlot[
    Evaluate[Drop[
 F[ Finv[ cent + (-0.5 + 0.2 # ) {Cos[0], Sin[0], 0}, t] , 0 ],
    -1 ]] , {t,-10,0},
    PlotStyle -> {Evaluate[Hue[0.1 #]]} ,
    DisplayFunction -> Identity ]  & ,
    Range[8] ];


Show[ wheel, streaklines, DisplayFunction ->  $DisplayFunction ,
  AspectRatio -> 1]
```

The plot of a streakline in the present case appears to be a spiral. Although the wheel has a finite extent within the plane, the streaklines can be understood by imagining the flow pattern due to an infinitely extended plane attached to the wheel and undergoing translation and rotation with it.

Ideal inviscid potential flow

Complete analysis and description of fluid flow requires the introduction of the notion of mass and of conservation laws of mass and momentum, in addition to the description of the kinematics of continuous motion presented in the previous section.

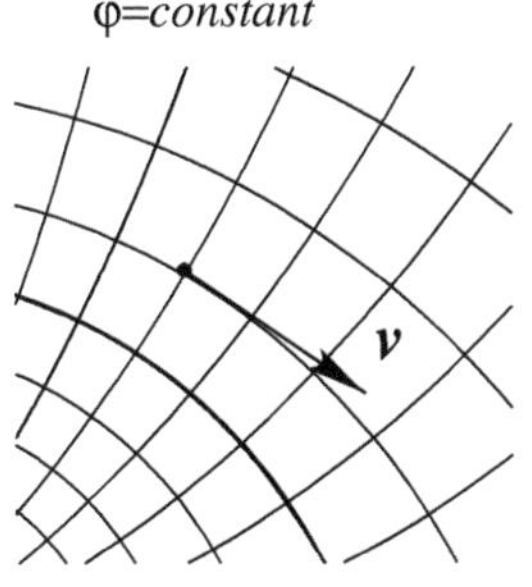

Figure 1.4. Schematic illustration of equipotential lines and streamlines.

For the purposes of the present discussion we omit these theoretical concepts and move directly to the consideration of irrotational flows of an ideal incompressible, homogeneous fluid in the presence of conservative body forces. For fluid flows of this class, the velocity field can be expressed as the gradient of a potential function, and they are therefore referred to as potential flows. The details of these concepts can be found in classical textbooks on continuum and fluid mechanics (Huerre, 2001; Malvern, 1969; Salençon, 2001). Here we shall simply use potential flows to illustrate displacement and velocity fields.

The velocity field of a *two-dimensional potential flow* can be characterized by either the velocity potential φ or the stream function ψ:

$$\boldsymbol{v} = \operatorname{grad}\varphi \qquad \boldsymbol{v} = \operatorname{curl}\psi \tag{1.9}$$

$$\boldsymbol{v} = \frac{\partial\varphi}{\partial x_1}\boldsymbol{e}_1 + \frac{\partial\varphi}{\partial x_2}\boldsymbol{e}_2 \qquad \boldsymbol{v} = -\frac{\partial\psi}{\partial x_2}\boldsymbol{e}_1 + \frac{\partial\psi}{\partial x_1}\boldsymbol{e}_2. \tag{1.10}$$

These equations show that the velocity potential φ and the stream function ψ are conjugate complex functions of the complex variable $z = x + iy$. The continuity equation ensures further that both functions are also harmonic.

The flow can therefore be characterised by the *complex potential*:

$$f(z) = f(x + iy) = \varphi(x, y) + i\psi(x, y).$$

Two important orthogonal families of lines characterising the flow are the

- equipotential lines: $\varphi(x, y) = \text{const}$
- streamlines: $\psi(x, y) = \text{const}$,

(see Figure 1.4).

Some complex potentials charaterizing basic flow patterns are

- *Uniform flow* of uniform velocity U at an angle α with the x axis:

$$f(z) = U\exp(-i\alpha)\,z.$$

- *Point source* of intensity Q at z_0:

$$f(z) = \frac{Q}{2\pi}\log(z - z_0).$$

- *Point vortex* of intensity Γ at z_0:

$$f(z) = \frac{i\Gamma}{2\pi} \log(z - z_0).$$

- *Doublet* of intensity μ at the orientation angle α (which can be obtained by differentiation of the point source):

$$f(z) = -\frac{\mu \exp(i\alpha)}{z - z_0}.$$

- *Flow in a corner* of intensity Γ at z_0:

$$f(z) = \frac{c}{m+1} z^{m+1} \qquad c \in \mathbb{R}, m \in [-1, \infty).$$

Linear combinations or conformal mappings of these potentials lead to a series of classical solutions for potential flows, including the flow around a cylinder and the Kutta–Joukowsky flow around a wing.

We demonstrate here how to use MATHEMATICA to visualise the particular case of flow around a cylinder.

The complex potential `f` for this problem is obtained by the superposition of the fundamental potentials for a uniform flow and those of a doublet and a vortex.

Let the parameters of the flow be as follows: `Uinf` – the velocity at infinity, `Gam` – the inensity of the source-sink doublet and the vortex creating the circulation, and `rc` the radius of the cylinder.

```
rc = 1; Uinf = 1; Gam = - 5.  2 Pi;
Gam / (4  Pi Uinf rc)

f = Uinf (z + rc^2 / z)  - I Gam / (2 Pi) Log[ z / rc]
```

The complex variable z is expressed as a sum of its real and imaginary parts, $z = x + iy$. The velocity potential φ and the stream function ψ are the real and imaginary parts of the complex potential.

The harmonicity of the potential is verified by the Cauchy–Riemann equations:

$$\frac{\partial \varphi}{\partial x} = -\frac{\partial \psi}{\partial y} \qquad \frac{\partial \varphi}{\partial y} = \frac{\partial \psi}{\partial x}.$$

Because MATHEMATICA does not automatically simplify expressions such as $a + i0$ to a when a is a real number, corresponding simplification rules are defined and applied to the expressions.

Finally, the streamlines are plotted as level lines, or contours, of the stream function.

```
f = f /. z ->  x + I y

phi = ComplexExpand[Re[  f ]]
psi = ComplexExpand[Im[  f ]]

phi = Simplify[  phi   /. Complex[a_, 0.`] -> a  /.
```

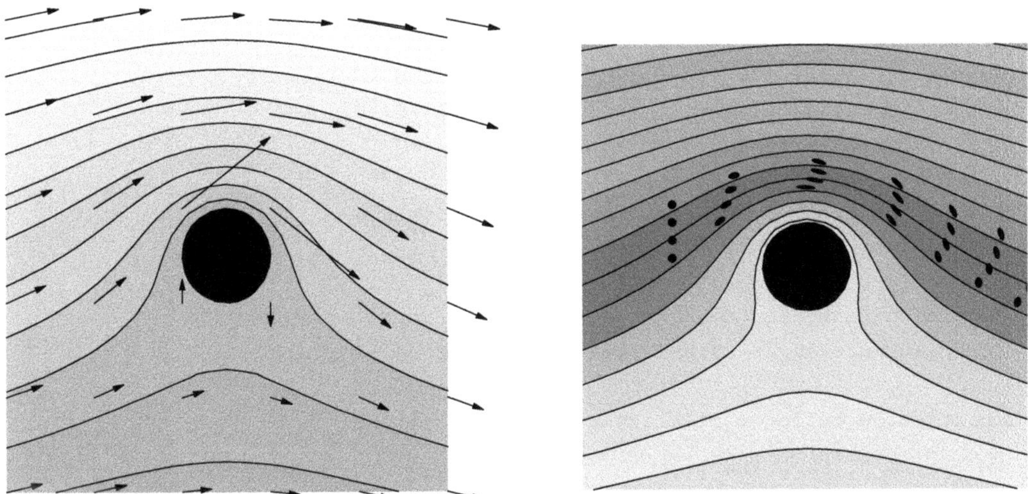

Figure 1.5. Velocity field, streamlines, and deformation of particles during potential flow around a rigid cylinder.

```
      Arg[x +  I y] -> ArcTan[x, y] ]

psi = Simplify[ psi  /. Complex[a_, 0.`] -> a ]

Simplify[  D[phi, x] - D[psi, y] /. 0.` -> 0]
Simplify[  D[phi, y] + D[psi, x] /. 0.` -> 0]


cplot = ContourPlot[ psi , {x, -5, 5}, {y, -5, 5},
    AspectRatio ->  Automatic,
    Contours -> 20, Frame -> False]
```

The velocity field is computed as the gradient of the velocity potential $\boldsymbol{v} = \text{grad}\,\varphi$. The vectors are plotted using the standard package **PlotField**. Options allowing the aspect ratio of the plot to be adjusted are explained in detail in the help notes for that package.

To superimpose the plot of streamlines and the plot of the velocity field, the **Show** command is used. The result of such superposition is displayed in Figure 1.5. The image file was created using the **Export** command.

A useful MATHEMATICA trick is the use of the option **DisplayFunction -> Identity** that permits the creation of a plot in memory without displaying the image. The option **DisplayFunction -> $DisplayFunction** restores the default setting and displays the complete plot.

```
v[x_, y_] = {D[phi, x], D[phi, y]}

<< Graphics`PlotField`
```

```
vplot = Show[
 Graphics[Circle[ {0, 0}, rc], DisplayFunction -> Identity],
 PlotVectorField[ v[x, y] , {x, -5, 5}, {y, -5, 5},
   PlotPoints -> 6, ScaleFactor -> None ,
   DisplayFunction -> Identity],
 DisplayFunction -> $DisplayFunction, AspectRatio -> 1]

shvelocity = Show[ cplot, vplot,
  Graphics[Disk[ {0, 0}, rc], DisplayFunction -> Identity],
  DisplayFunction -> $DisplayFunction, AspectRatio -> 1]

Export["flow_cylinder_velocity.pdf", shvelocity]
```

We now proceed to integrate the velocity field **eqn** at a series of points in order to compute the trajectories of these points. The points are then also grouped into circular particles allowing their deformation to be displayed.

```
eqn = Thread[{x'[t], y'[t]} == v[x[t], y[t]] ]

tmax = 16.0; dt = 1.;

r = 0.1; n = 10;
circlepts =
 Table[N[ r {Sin[2 Pi k/n], Cos[2 Pi k/n]}], {k,n} ];
```

The integration is performed using the **NDSolve** command and the commands applied to a series of points are grouped together in the **flows** function. The position of the deformed particles is extracted using the **flowpts** function.

Using the **Map** command gives the advantage of not having to keep track of the number of points involved in the operations. The programming style used in this part of the example follows the idea of the **circlepts** function described by Bahder (1994).

```
flows[{a_, b_}] :=
 Apply[Flatten[
  NDSolve[ Join[eqn, {x[0] == a + #1, y[0] == b + #2}],
             {x, y}, {t,0,tmax}] ]&,
   circlepts, 1 ]

flowpts[ center:{_,_}] :=
          Map[ ({x[#],y[#]} /. flows[center]) &,
          Range[0,tmax,dt] ]

points = Table[ flowpts[{-3, y0}], {y0, 0.2 , 1.4 , 0.4}];
```

The particles obtained as sets of points are transformed into polygonal dots using the `Graphics` command. All plots are superposed in order to obtain the image shown in Figure 1.5.

```
Show[
   ContourPlot[ Y , {x, -5, 5}, {y, -5, 5},
      DisplayFunction -> Identity , Frame -> False,
      ColorFunction -> Hue,   Contours -> 30],
   Graphics[{GrayLevel[0.] , Disk[ {0, 0}, rc]}],
   Graphics[Map[Polygon, points, {2}]],
   PlotRange -> {{-5, 5}, {-10, 5}},
   AspectRatio -> Automatic,
   DisplayFunction -> $DisplayFunction
   ]
```

1.2 STRAIN

Deformation gradient

The examples in the preceeding section served to show how one can describe the most general motion of material particles leading to complex trajectories, using, for example, the Lagrangian deformation function $\mathcal{F}(\boldsymbol{X}, t)$.

To simplify the analysis of particle motion in the neighbourhood of a material point $\boldsymbol{X}$ it is usual to employ Taylor series expansion of the deformation function with respect to the spatial variable $\boldsymbol{X}$, considering time t to be fixed:

$$\mathcal{F}(\boldsymbol{X} + d\boldsymbol{X}, t) = \mathcal{F}(\boldsymbol{X}, t) + \nabla_{\boldsymbol{X}}\mathcal{F}(\boldsymbol{X}, t) \cdot d\boldsymbol{X} + o(d\boldsymbol{X}^2). \tag{1.11}$$

The widely accepted hypothesis in continuum mechanics is that the first-order term in the above expansion contains sufficient information to explain and predict a large range of phenomena. We define the *deformation gradient* as

$$\boldsymbol{F}(\boldsymbol{X}, t) = \nabla_{\boldsymbol{X}}\mathcal{F}(\boldsymbol{X}, t). \tag{1.12}$$

This is the spatial gradient of a vector-valued function $\boldsymbol{F}$ and is therefore a second-order tensor. This means that in a particular coordinate system $\boldsymbol{F}(\boldsymbol{X}, t)$ can be represented by a time-varying matrix field. A somewhat more detailed discussion of tensor representations in different orthogonal curvilinear coordinate systems in given in Appendix 1.

Let us illustrate some of the properties of the deformation gradient in relation to other physical quantities.

- If an infinitesimal material vector $d\boldsymbol{X}$ originating at point $\boldsymbol{X}$ is considered in the reference configuration, and $d\boldsymbol{x}$ is its image in the actual configuration due to the motion, then in the first-order approximation of the Taylor expansion one obtains

$$d\boldsymbol{x} = \boldsymbol{F} \cdot d\boldsymbol{X}.$$

In other words, the deformation gradient maps infinitesimal material vectors originating at a chosen point $\boldsymbol{X}$ from initial into actual configuration. Note that the translation of the initial point $\boldsymbol{X}$ to its actual position $\boldsymbol{x}$ is not captured by the deformation gradient.

- Because the motion obeys the requirement of *material impenetrability*, the local volume may only be scaled by a finite positive number. This is expressed in terms of the deformation gradient as

$$0 < J < \infty \qquad J = |\det \boldsymbol{F}|. \tag{1.13}$$

More precisely, if one considers an initial material volume Ω which is transformed in the actual configuration into volume ω, then the actual volume is given by the integral

$$\int_{\omega} dv.$$

The volume element defined by three vectors, $d\boldsymbol{x} = \boldsymbol{F}d\boldsymbol{X}$, $d\boldsymbol{y} = \boldsymbol{F}d\boldsymbol{Y}$, $d\boldsymbol{z} = \boldsymbol{F}d\boldsymbol{Z}$, is the volume defined by these vectors and expressed using the determinant of the coordinate matrix $[\cdot,\cdot,\cdot]$,

$$dv = |[d\boldsymbol{x}, d\boldsymbol{y}, d\boldsymbol{z}]| == |[\boldsymbol{F}d\boldsymbol{X}, \boldsymbol{F}d\boldsymbol{Y}, \boldsymbol{F}d\boldsymbol{Z}]| = |\det \boldsymbol{F}| dV = J\,dV,$$

with dV the volume element defined by the three vectors $d\boldsymbol{X}$, $d\boldsymbol{Y}$, $d\boldsymbol{Z}$ in the initial configuration. By the change of variable one obtains

$$\int_{\omega} dv = \int_{\Omega} J\, dV.$$

The rotation and stretch tensors

The *Polar Decomposition Theorem* states that all positive definite tensors can be uniquely decomposed into products of *stretch* and *rotation* tensors. For a complete discussion of the polar decomposition theorem see for example the monograph by Malvern (1969) or the mathematical proofs given in (Halmos, 1959; Soos and Teodosiu, 1983).

The theorem can be applied to the deformation tensor $\boldsymbol{F}$, because $0 < \det \boldsymbol{F} < \infty$. Hence $\boldsymbol{F}$ can be represented as products:

$$\boldsymbol{F} = \boldsymbol{R}\boldsymbol{U} = \boldsymbol{V}\boldsymbol{R}.$$

Here $\boldsymbol{U}$ and $\boldsymbol{V}$ are referred to as the *right* and *left stretch tensors*, respectively, and $\boldsymbol{R}$ denotes the *rotation tensor*, which is orthogonal, that is,

$$\boldsymbol{R}\boldsymbol{R}^T = \boldsymbol{R}^T\boldsymbol{R} = \boldsymbol{I}.$$

Both $\boldsymbol{U}$ and $\boldsymbol{V}$ are symmetric positive definite tensors such that

$$\boldsymbol{U}^2 = \boldsymbol{F}^T\boldsymbol{F} \qquad \boldsymbol{U}^T = \boldsymbol{U} \tag{1.14}$$

$$\boldsymbol{V}^2 = \boldsymbol{F}\boldsymbol{F}^T \qquad \boldsymbol{V}^T = \boldsymbol{V}. \tag{1.15}$$

The squares of the stretch tensors are commonly referred to as the $\boldsymbol{C}$ and $\boldsymbol{B}$ deformation tensors:

$$\boldsymbol{C} = \boldsymbol{U}^2, \qquad \boldsymbol{B} = \boldsymbol{V}^2.$$

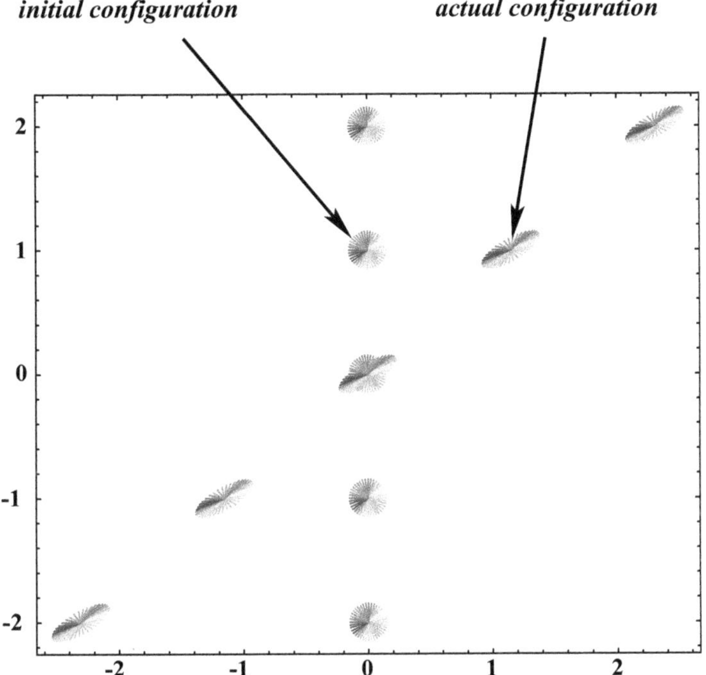

Figure 1.6. Transformation of infinitesimal material vectors from initial into actual configuration in simple shear.

The decomposition of the deformation gradient also produces a decomposition of material vector mapping,

$$d\boldsymbol{x} = \boldsymbol{F}\, d\boldsymbol{X} = \boldsymbol{RU}\, d\boldsymbol{X},$$

into two sequential operations (see Figure 1.6):

- a stretch of the material vector $d\boldsymbol{X}$ into $\boldsymbol{U} \cdot d\boldsymbol{X}$
- a rigid-body rotation of $\boldsymbol{U} \cdot d\boldsymbol{X}$ into $d\boldsymbol{x} = \boldsymbol{RU} \cdot d\boldsymbol{X}$.

This sequence is illustrated in Figure 1.6, where different material vectors are denoted by their colour, which remains unchanged during the process.

A similar illustration can be obtained using the other decomposition $\boldsymbol{F} = \boldsymbol{VR}$. In this case the order of steps in the sequence is reversed: rigid-body rotation due to $\boldsymbol{R}$ is followed by the stretch $\boldsymbol{V}$.

The eigenvalues and eigenvectors of the stretch tensors $\boldsymbol{U}$ and $\boldsymbol{V}$ have particular significance. Recall that the eigenvectors $\boldsymbol{v}$ of tensor $\boldsymbol{T}$ correspond to directions that are preserved during the linear transformation defined by $\boldsymbol{T}$, that is,

$$\boldsymbol{T} \cdot \boldsymbol{v} = \lambda \boldsymbol{v}.$$

Eigenvalue λ represents therefore the relative length change of $\boldsymbol{v}$ through the transformation. A positive definite tensor has three eigenvalues obtained as solutions of the equation

$$\det(\boldsymbol{T} - \lambda \boldsymbol{I}) = 0.$$

Eigenvectors $\boldsymbol{u}_i$, $i = 1, 2, 3$, of the stretch tensor are directions that remain invariant through the stretch and are therefore referred to as the *principal stretches*. Eigenvectors of a positive definite tensor form an orthogonal vector basis. Therefore a positive definite tensor such as $\boldsymbol{U}$ can be written explicitly as

$$\boldsymbol{U} = \sum_{i=1,3} \lambda_i \boldsymbol{u}_i \otimes \boldsymbol{u}_i \tag{1.16}$$

and also as

$$\boldsymbol{U}^2 = \sum_{i=1,3} \lambda_i^2 \boldsymbol{u}_i \otimes \boldsymbol{u}_i. \tag{1.17}$$

The left stretch tensor and the rotation tensor can be computed using the following steps:

- compute $\boldsymbol{U}^2 = \boldsymbol{F}^T\boldsymbol{F}$;
- compute eigenvalues λ_i^2, $i = 1, 2, 3$, by solving

  $$\det\left(\boldsymbol{U}^2 - \lambda^2\boldsymbol{I}\right) = 0$$

  and find the corresponding eigenvectors $\boldsymbol{u}_i$;
- construct $\boldsymbol{U}$ using (1.16);
- compute $\boldsymbol{R} = \boldsymbol{F}\boldsymbol{U}^{-1}$.

The eigenvectors and eigenvalues of the right stretch tensor $\boldsymbol{V}$ are λ_i and $\boldsymbol{R}\boldsymbol{u}_i$ respectively, as is readily deduced from the series of equalities

$$\boldsymbol{V}\boldsymbol{R}\boldsymbol{u}_i = \boldsymbol{F}\,\boldsymbol{u}_i = \boldsymbol{R}\,\boldsymbol{U}\boldsymbol{u}_i = \lambda_i\boldsymbol{R}\,\boldsymbol{u}_i.$$

Geometrical interpretation of the stretch tensors

Let us consider an infinitesimal sphere centred at $\boldsymbol{X}$ and of radius R_ϵ in the reference configuration. Its points denoted by $\boldsymbol{X} + d\boldsymbol{X}$ satisfy the following equation:

$$d\boldsymbol{X} \cdot d\boldsymbol{X} = R_\epsilon^2.$$

The sphere is transformed into an ellipsoid in the actual configuration defined by the material points $\boldsymbol{x} + d\boldsymbol{x}$. Because $d\boldsymbol{X} = \boldsymbol{F}^{-1}d\boldsymbol{x}$, it follows that the equation of this ellipsoid is

$$d\boldsymbol{x}\left(\boldsymbol{F}^{-1^T}\boldsymbol{F}^{-1}\right)d\boldsymbol{x} = d\boldsymbol{x}\,\boldsymbol{V}^{-2}d\boldsymbol{x} = R_\epsilon^2.$$

The principal axes are those of $\boldsymbol{V}$ defined by $\boldsymbol{R}\boldsymbol{u}_i$ and the principal axes have length proportional to λ_i (Figure 1.6).

Linear stretch of infinitesimal material vectors

The lengths of infinitesimal material vectors in the initial and actual configurations are given by

$$dL^2 = d\boldsymbol{X} \cdot d\boldsymbol{X} \qquad dl^2 = d\boldsymbol{x} \cdot d\boldsymbol{x}$$

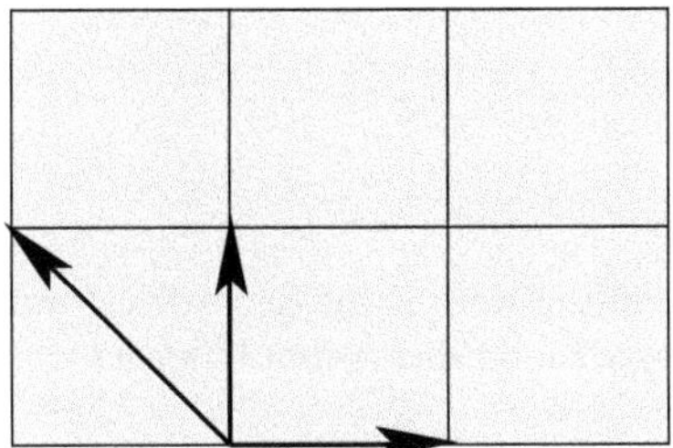
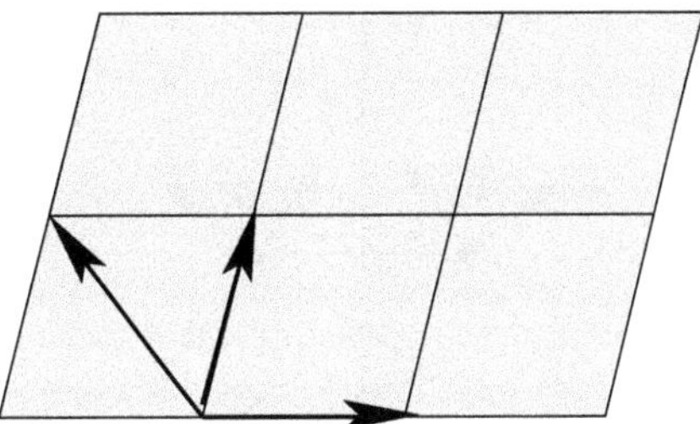

Figure 1.7. The transformation of different directions from initial to actual configuration in simple shear.

(See Figure 1.7). Using the relation between $d\boldsymbol{X}$ and $d\boldsymbol{x}$ given by the deformation gradient, one obtains the following series of equalities for the difference of squared length of the two infinitesimal vectors:

$$dl^2 - dL^2 = d\boldsymbol{x} \cdot d\boldsymbol{x} - d\boldsymbol{X} \cdot d\boldsymbol{X} = (d\boldsymbol{X}\boldsymbol{F}^T) \cdot (\boldsymbol{F} d\boldsymbol{x}) - d\boldsymbol{X} \cdot d\boldsymbol{X}.$$
$$= d\boldsymbol{X}\left(\boldsymbol{F}^T \cdot \boldsymbol{F} - \boldsymbol{I}\right) d\boldsymbol{X} = d\boldsymbol{X}\frac{1}{2}\left(\boldsymbol{U}^2 - \boldsymbol{I}\right) d\boldsymbol{X}.$$

The tensor

$$\boldsymbol{G} = \frac{1}{2}\left(\boldsymbol{F}^T\boldsymbol{F} - \boldsymbol{I}\right) = \frac{1}{2}\left(\boldsymbol{U}^2 - \boldsymbol{I}\right)$$

is called the *finite strain tensor*, also known as the *Green–Lagrange* strain tensor.

A transformation which preserves the distance between any two material points is referred to as *rigid body motion*. It can be shown mathematically that the most general expression for rigid body motion is given by

$$\boldsymbol{x} = \mathcal{F}(\boldsymbol{X}, t) = \boldsymbol{p}(t) + \boldsymbol{Q}(t)\left(\boldsymbol{X} - \boldsymbol{X}_0\right), \tag{1.18}$$

where $\boldsymbol{p}(t)$, $\boldsymbol{Q}(t)$, and $\boldsymbol{X}_0$ denote the time-dependent translation vector, the rotation tensor, and the centre of rotation, respectively. The rotation tensor is an orthogonal tensor at each time instant; that is,

$$\boldsymbol{Q}(t)\boldsymbol{Q}^T(t) = \boldsymbol{I} \qquad \det \boldsymbol{Q}(t) = 1.$$

Simple shear

To develop better insight into these concepts, let us consider in detail the case of simple shear deformation. Deformation is defined by

$$\boldsymbol{x} = \mathcal{F}(\boldsymbol{X}, t) = \boldsymbol{X} + 2a(t)\, X_2 \boldsymbol{e}_1. \tag{1.19}$$

An easy way to visualise this deformation is to imagine the top of a stack of playing cards being pushed to one side, making cards glide on top of each other. (See Figure 1.8.)

Computation of strain, stretch, and rotation tensors

The squares of the right and left stretch tensors can be readily computed using Equations (1.14) and (1.15).

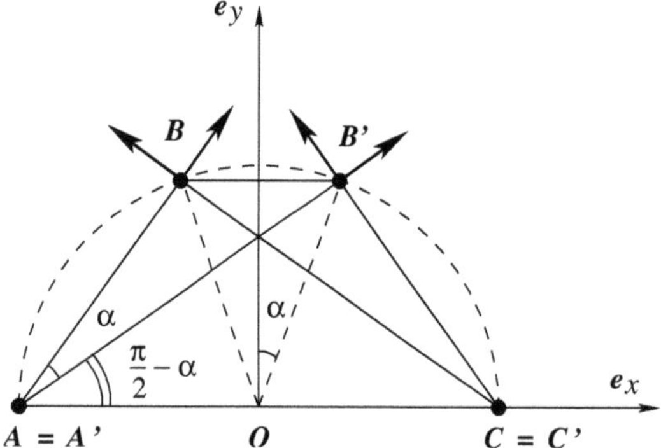

Figure 1.8. Illustration of the geometrical method employed for the discussion of simple shear.

```
T[ x_, y_, z_ ] = {x + 2 y a, y, z}.
(F = Grad[ T[x, y, z]] ) // MatrixForm

( V2 = F . Transpose[F]  ) // MatrixForm
( U2 = Transpose[F] . F  ) // MatrixForm
```

The next step is the determination of $\boldsymbol{U}$ from $\boldsymbol{U}^2$.

```
eigenU2 = Eigensystem[U2]

eigenU2 =
     Simplify[
      eigenU2/.(a^2 + a^4)^(1/2) -> a ( 1 + a^2 )^(1/2)
      ]
```

The construction of $\boldsymbol{U}$ from $\boldsymbol{U}^2$ is performed using representations (1.16) and (1.17) through the function `EigenToTensor` specially defined here.

We recall that `eigen[[1]]` and `eigen[[2]]` represent the eigenvalues and eigenvectors, respectively. At the end, a check is performed to confirm that $\boldsymbol{U}^2 = \boldsymbol{U} \cdot \boldsymbol{U}$.

```
EigenToTensor[ eigen_] :=
  Sum[ eigen[[1,i]] / ( eigen[[2,i]] . eigen[[2,i]] )^2
  Outer[ Times, eigen[[2,i]] , eigen[[2,i]] ], {i,3}]

U = EigenToTensor[eigenU2]

U = Simplify[ U ]

Simplify[ U . U - U2 ]
```

Another way to compute $\boldsymbol{U}$ from $\boldsymbol{U}^2$ is to apply the MATHEMATICA function **MatrixPower**.

The result requires simplification using a series of replacement rules:

$$\sqrt{a^2+a^4} \longrightarrow a\sqrt{1+a^2}$$
$$\sqrt{1+2a^2-2a\sqrt{1+a^2}} \longrightarrow \sqrt{1+a^2}-a$$
$$\sqrt{1+2a^2+2a\sqrt{1+a^2}} \longrightarrow \sqrt{1+a^2}+a.$$

Note that a is assume to be positive and that the positive value of the root has been chosen in the second expression.

A practical trick useful in determining which expression to use in a replacement rule is to inspect the MATHEMATICA internal expression using the **FullForm** command.

```
U = MatrixPower[ U2 , 1/2 ]

U  =  U /. Sqrt[ a^2 + a^4) ]-> a (1 + a^2)^(1/2)

U  =  U  /.
     Sqrt[ (1 + 2 a^2 - 2 a  Sqrt[ 1 + a^2 ]]
                    -> Sqrt[ 1 + a^2 ] - a /.
     Sqrt[ (1 + 2 a^2 + 2 a  Sqrt[ 1 + a^2 ]]
                    -> Sqrt[ 1 + a^2 ] + a

U  =  Simplify[ U ]  /.
      1/Sqrt[ a^2 + a^4 ] :> 1 / ( a  (1  +  a^2)^(1/2) )

Simplify[ U . U - U2 ]
```

The rotation tensor is computed as

$$\boldsymbol{R} = \boldsymbol{F} \cdot \boldsymbol{U}^{-1}$$

and simple inspection of the result shows that it corresponds to rotation by the angle

$$\theta(t) = -\arctan a(t).$$

```
R = F .  Inverse[ U ]

Simplify[ % ]
```

Trigonometric representation of strain, stretch, and rotation tensors

Trigonometric formulae provide a convenient derivation tool, particularly when calculations are performed by hand. In the case of simple shear the classical notation is

$$a(t) = \tan \alpha t$$

so that material deformation is described by the expression

$$x = \mathcal{F}(X, t) = X + 2\tan\alpha(t)\, X_2 e_1. \tag{1.20}$$

Instead of repeating complete strain analysis with MATHEMATICA in trigonometric notation, we only obtain eigenvalues and eigenvectors of U here, to allow easy geometric interpretation of the deformation.

We seek to compute eigenvalues of U^2. To simplify the expressions we use the rule

$$\sqrt{\sec\alpha^2 \tan\alpha^2} = \frac{\sin\alpha}{\cos^2\alpha}.$$

Then we construct substitution formulae for $\tan\left(\frac{\pi}{2} \pm \frac{\alpha}{4}\right)$.

Computations can be continued further and are left as an exercise for the reader.

```
eU2trig = Simplify[ Eigensystem[ U2 /. a -> Tan[ alpha ]] ]

eU2trig = TrigFactor[  eU2trig /.
       (Sec[alpha]^2 Tan[alpha]^2)^(1/2)   ->
           Sin[alpha] Cos[alpha]^2
  ]

tpiplusa = Factor[ Expand[ TrigFactor[
                Tan[ Pi / 4 +  alpha / 2 ]]]]
tpiminusa = Factor[ Expand[ TrigFactor[
                Tan[ Pi / 4 -  alpha / 2 ]]]]

eU2trig =
  eU2trig /. tpiplusa -> Tan[ Pi / 4 +  alpha / 2 ] /.
      tpiplusa^2 -> Tan[ Pi / 4 +  alpha / 2 ]^2 /.
      tpiminusa -> Tan[ Pi / 4 +  alpha / 2 ] /.
      tpiminusa^2 -> Tan[ Pi / 4 +  alpha / 2 ]^2
```

Eigenvalues and eigenvectors of U are given by the following expressions:

$$\lambda_1 = \frac{1+\sin\alpha}{\cos\alpha} = \frac{1+\tan\frac{\alpha}{2}}{1-\tan\frac{\alpha}{2}} \tag{1.21}$$

$$\lambda_2 = \frac{1-\sin\alpha}{\cos\alpha} = \frac{1-\tan\frac{\alpha}{2}}{1+\tan\frac{\alpha}{2}} \tag{1.22}$$

$$\lambda_3 = 1 \tag{1.23}$$

$$u_1 = \cos\left(\frac{\alpha}{2} - \frac{\pi}{4}\right) e_1 + \sin\left(\frac{\alpha}{2} - \frac{\pi}{4}\right) e_2 \tag{1.24}$$

$$u_2 = \cos\left(\frac{\alpha}{2} + \frac{\pi}{4}\right) e_1 + \sin\left(\frac{\alpha}{2} + \frac{\pi}{4}\right) e_2 \tag{1.25}$$

$$u_3 = e_3. \tag{1.26}$$

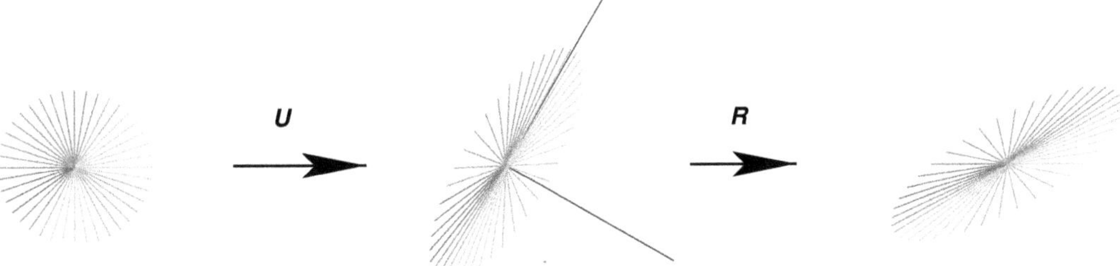

Figure 1.9. An illustration of the polar decomposition of the deformation gradient $\boldsymbol{F}$. Material vectors preserve their colour though the stretch by $\boldsymbol{U}$ and rotation by $\boldsymbol{R}$. Images from left to right represent the decomposed evolution from initial to actual configuration.

Other proofs

In the particular case of simple shear, the analysis of eigenvectors and eigenvalues can be completed using a special technique based Hamilton–Cayley theorem in linear algebra, as described in the monograph by Marsden and Hughes (1994).

Another elegant proof can be obtained by analysing the deformation using the geometric schematic diagram displayed in Figure 1.9. From it one notes that

- Point B is transported to B' and $|BB'| = 2a = 2\tan\alpha$.
- Points on the Ox axis remain stationary: $A = A'$ $C = C'$.
- The angle of shear is $\alpha = \sphericalangle BOy = \sphericalangle B'Oy$.
- Lines AB and BC are transformed into AB' and $B'C$, respectively.
- Simple computations of the circular arc length lead to

$$\sphericalangle BAB' = \alpha, \quad \sphericalangle OAB' = \frac{\pi}{2} - \frac{\alpha}{4}.$$

Plotting stretch and rotation of material vectors

Information about stretch and rotation tensors can now be used to plot the deformation of infinitesimal material vectors under simple shear. For simplicity, two-dimensional motion is considered.

We begin by defining a series of parameters, including the angle, α, the particle radius, r, and the number of vertices of the polygonal particle, n.

colorpts is the set of coordinates of the vertices, where a colour property has been ascribed using the **Hue** command.

Numerical values of the tensors are defined next

$$\texttt{ff} = \boldsymbol{F} \quad \texttt{invrr} = \boldsymbol{R}^{-1} \quad \texttt{uu} = \boldsymbol{U}$$

and eigenvectors are defined as line segments.

```
alpha = Pi / 6;
r = 0.15
n = 40

colorpts =
```

```
  Table[
    {Hue[N[k/n]], N[ r {Sin[2 Pi k/n], Cos[2 Pi k/n]}]},
    {k,n} ];

 ff = N[ {{1, 2 Tan[alpha]}, {0, 1}} ]
 invrr = N[{{Cos[alpha],-Sin[alpha]}, {+Sin[alpha],Cos[alpha]}} ]

 uu = invrr . ff

eigenvector1 =
  Line[{{0,0}, 2.5r {Cos[Pi/4-alpha/2], -Sin[Pi/4-alpha/2]}}];
eigenvector2 =
  Line[{{0,0}, 2.5r {Cos[Pi/4+alpha/2], +Sin[Pi/4+alpha/2]}}];
```

The `vector` map defined here permits one to construct the deformed particle as a bundle of coloured segments.

The set `colorpts` can be considered as representing the initial segments that have been 'deformed' by the action of tensor `tens` and translated by the distance defined by the vector `center`. Note that the point colour expressed here as `#[[1]]` is inherited by the corresponding line.

```
vector[center:{_,_}, tens_] :=
Map[
   {#[[1]], Line[{center, center + tens . #[[2]]}]} & ,
    colorpts ];
```

The initial particle with the segments representing the infinitesimal vectors $d\boldsymbol{X}$, the stretched particle $\boldsymbol{U}\,d\boldsymbol{X}$, and the directions of the eigenvectors and the deformed particle after stretching and rotation $\boldsymbol{F}\,d\boldsymbol{X} = \boldsymbol{R}\boldsymbol{U}\,d\boldsymbol{X}$ can now be plotted as shown in Figure 1.7.

Eigenvector directions do not change through the stretch tranformation and can be readily identified from the plot by this property.

```
Show[ Graphics[{eigenvector1, eigenvector2 ,
          vector[{0,0}, uu],
          vector[{-4 r,0}, IdentityMatrix[2] ],
          vector[{4 r , 0}, ff]}],
     AspectRatio -> Automatic,
     Axes -> False]
```

1.3 SMALL STRAIN TENSOR

Prior to this stage in the analysis of strains, no assumption has been made regarding strain magnitude. Let us now assume that the deformation gradient is close to the identity tensor, that is, that the norm of the displacement gradient is small compared to unity:

$$|\mathrm{grad}\,\boldsymbol{u}| \ll 1.$$

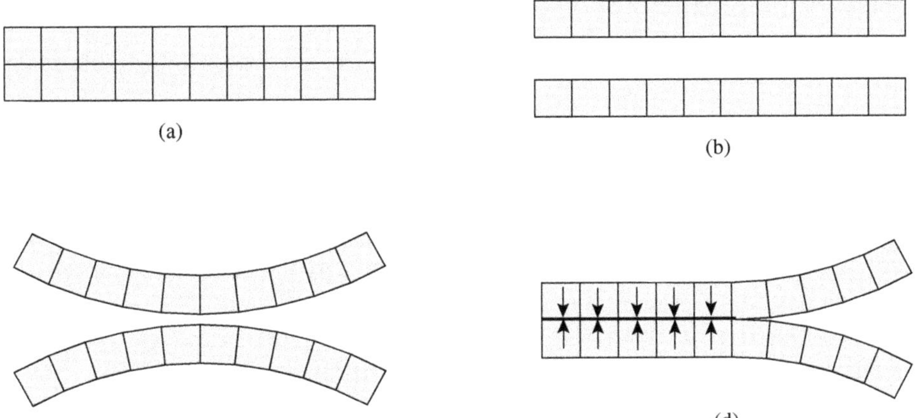

Figure 1.10. An illustration of the compatibility conditions for deformation. Consideration proceeds via the following stages: (a) initial geometry of the beam; (b) the beam being separated into pieces; (c) the **imposition** of local strain; (d) the object cannot be reassembled due to the incompatibility of strains – additional forces have to be applied.

Then definitions of strain, stretch, and rotation tensors can be linearized. A series of algebraic manipulations that are left to the reader as an exercise lead to the equalities

$$\boldsymbol{C} = \boldsymbol{B} = \boldsymbol{I} + \operatorname{grad}\boldsymbol{u} + \operatorname{grad}^T\boldsymbol{u} = \boldsymbol{I} + 2\boldsymbol{\varepsilon} \tag{1.27}$$

$$\boldsymbol{U} = \boldsymbol{V} = \boldsymbol{I} + \boldsymbol{\varepsilon} \tag{1.28}$$

$$\boldsymbol{R} = \boldsymbol{I} + \frac{1}{2}\left(\operatorname{grad}\boldsymbol{u} - \operatorname{grad}^T\boldsymbol{u}\right) = \boldsymbol{I} + \boldsymbol{\omega}, \tag{1.29}$$

where

$$\boldsymbol{\varepsilon} = \frac{1}{2}\left(\operatorname{grad}\boldsymbol{u} + \operatorname{grad}^T\boldsymbol{u}\right) \qquad \text{and} \qquad \boldsymbol{\omega} = \frac{1}{2}\left(\operatorname{grad}\boldsymbol{u} - \operatorname{grad}^{\mathrm{T}}\boldsymbol{u}\right)$$

denote the *small strain tensor* $\boldsymbol{\varepsilon}$ and *small rotation tensor* $\boldsymbol{\omega}$, respectively.

1.4 COMPATIBILITY EQUATIONS AND INTEGRATION OF SMALL STRAINS

In the preceding sections we discussed how strains can be computed from a given displacement function. However, in many practical situations the inverse problem is often encountered, that is, if and how one can recover the displacement field from the knowledge of the strain tensor field. The answer to this question is only positive if the strain tensor satisfies certain conditions. An illustration of what these conditions must be can be obtained if one considers putting together a broken jar. If all pieces remain undistorted then the jar can be glued back together. However, if each piece is deformed on its own in an independent way, then it is clear that reconstruction is only possible provided certain *compatibility* conditions between the deformed pieces are satisfied. (See Figure 1.10.) The purpose of this section is to introduce the rigorous mathematical form of compatibility conditions and to present the algorithm for displacement field reconstruction from strains within MATHEMATICA.

Cauchy–Schwarz integrability conditions

As the first step, examine the following classical result. Consider a differentiable scalar function

$$f : \Omega \subset \mathbb{R}^n \longrightarrow \mathbb{R}.$$

One can readily compute the vector gradient field $\boldsymbol{g} = \operatorname{grad} f$.

Let us now analyse the inverse operation. Suppose that a vector field $\boldsymbol{g}$ is given and the task is to identify its potential f. We would like to identify the condition that ensures the existence of a function f such that $\boldsymbol{g} = \operatorname{grad} f$.

Let the vector field $\boldsymbol{g}$ be described in cartesian coordinates as $\boldsymbol{g} = g_i \boldsymbol{e}_i$. It is well known (Spivak, 1965), that integrability conditions are expressed as

$$\frac{\partial g_i}{\partial x_j} = \frac{\partial g_j}{\partial x_i} \tag{1.30}$$

provided that the domain of integration Ω is star-shaped (Figure 1.11). An invariant vector form of this condition is

$$\operatorname{curl} \boldsymbol{g} = 0.$$

Using the Stokes theorem (Spivak, 1965),

$$\int_\gamma \boldsymbol{g} \cdot d\boldsymbol{x} = \int_\Gamma \operatorname{curl} \boldsymbol{g} \cdot \boldsymbol{n}\, ds,$$

where Γ is the area enclosed by the path γ and $\boldsymbol{n}$ is the normal to this area, this result also guarantees that the integral of $\boldsymbol{g}$ on any closed path $\gamma \subset \Omega$ vanishes:

$$\int_\gamma \boldsymbol{g} \cdot d\boldsymbol{x} = \int_\gamma g_i\, dx_i = 0.$$

Compatibility conditions for small strains

The problem addressed in this section is the reconstruction of the displacement field $\boldsymbol{u}$ from a given tensor field of small strains $\boldsymbol{\varepsilon}$. We seek to formulate the conditions in invariant form similarly to the Cauchy–Schwarz conditions for vector functions given in the previous section.

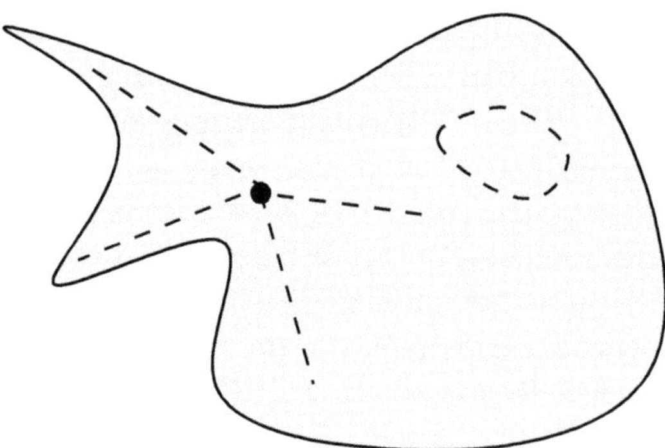

Figure 1.11. A star-shaped domain and a closed integration path (closed dashed curve).

We first observe that the small displacement gradient can be split into symmetric and antisymmetric parts, $\boldsymbol{\varepsilon}$ and $\boldsymbol{\omega}$, respectively,

$$\operatorname{grad}\boldsymbol{u} = \boldsymbol{\varepsilon} + \boldsymbol{\omega},$$

where

$$\boldsymbol{\varepsilon} = \frac{1}{2}\left(\operatorname{grad}\boldsymbol{u} + (\operatorname{grad}\boldsymbol{u})^T\right) \qquad \boldsymbol{\omega} = \frac{1}{2}\left(\operatorname{grad}\boldsymbol{u} - (\operatorname{grad}\boldsymbol{u})^T\right).$$

We first discuss some general properties of the strain and rotation tensors. The antisymmetric part of the displacement gradient $\boldsymbol{\omega}$, like all antisymmetric tensors, can be represented by a vector that will be referred to here as the *infinitesimal rotation vector* $\boldsymbol{\vartheta}$. In cartesian coordinates the relationship is given by

$$\omega_{ij} = \epsilon_{ijk}\vartheta_k \qquad i,j = 1,2,3,$$

where ϵ_{ijk} is the Levi-Cività fully antisymmetric third rank symbol and summation of repeated indexes is assumed (here k). In MATHEMATICA terminology corresponds to the signature of the permuation ijkcomputed by the command `Signature[i,j,k]`. Knowing the infinitesimal rotation vector $\boldsymbol{\vartheta}$ is equivalent to knowing $\boldsymbol{\omega}$.

The relationship between $\boldsymbol{\vartheta}$ and the displacement field is given by

$$\boldsymbol{\vartheta} = -\frac{1}{2}\operatorname{curl}\boldsymbol{u} \qquad \vartheta_k = \frac{1}{2}\epsilon_{klm}u_{l,m} \quad k = 1,3. \tag{1.31}$$

Computing the gradient of the infinitesimal rotation vector one observes that

$$\operatorname{grad}\boldsymbol{\vartheta} = -\operatorname{curl}\boldsymbol{\varepsilon}. \tag{1.32}$$

This relation can be readily verified with MATHEMATICA.

A general form of displacement field is specified in a given system of coordinates, here `Cartesian`. Second-order strain and rotation tensors are computed in different ways.

```
<<Tensor2Analysis.m
SetCoordinates[Cartesian[x,y,z]]

u = {u1[x, y, z], u2[x, y, z], u3[x, y, z]}
gu = Grad[u]

eps = ( Grad[u] + Transpose[Grad[ u]] ) / 2
ome = ( Grad[u] - Transpose[Grad[ u]] ) / 2

theta = - Curl[ u ] / 2

omega =
  Table[
    Sum[ Signature[ {i, j, k}] theta[[k]] ,
         {k, 3}], {i, 3}, {j, 3}
    ]
```

Formulas (1.31) and (1.32) are thus verified.

```
Simplify[ omega - ome]

Simplify[ Curl[eps] +  Grad[ theta ]]
```

We now proceed to present a constructive proof of the main result of this section. The proof consists of the following two steps:

- First, show that given the displacement field $\boldsymbol{\varepsilon}$, one can reconstruct the antisymmetric part of the displacement gradient $\boldsymbol{\omega}$.
- Second, show that given a tensor displacement gradient $\boldsymbol{g} = \boldsymbol{\varepsilon} + \boldsymbol{\omega}$ that satisfies certain conditions, one can determine the displacement vector field $\boldsymbol{u}$ such that $\boldsymbol{g} = \operatorname{grad} \boldsymbol{u}$.

Given the tensor strain field $\boldsymbol{\varepsilon}$, one can readily compute $\boldsymbol{h} = -\operatorname{curl} \boldsymbol{\varepsilon}$. Generalisation of the theorem referred to in the previous section provides the result that the tensor field $\boldsymbol{h}$ is indeed the gradient of a vector field $\boldsymbol{\vartheta}$ if and only if $-(\operatorname{curl} \boldsymbol{h}^T)^T = 0$, implying that

$$(\operatorname{curl} (\operatorname{curl} \boldsymbol{\varepsilon})^T)^T = 0. \tag{1.33}$$

This is the invariant form of the strain compatibility condition.

The value of the infinitesimal rotation vector $\boldsymbol{\vartheta}$ at point $\boldsymbol{p}$ can be recovered up to a constant-vector term using the integral formula

$$\boldsymbol{\vartheta}(\boldsymbol{p}) = -\int_{\gamma_{p_o p}} (\operatorname{curl} \boldsymbol{\varepsilon}) \cdot d\boldsymbol{x},$$

where $\gamma_{p_o p}$ is a path connecting points $\boldsymbol{p}_o$ and $\boldsymbol{p}$. Having determined the infinitesimal rotation vector ϑ, one can readily compute the rotation tensor ω and therefore construct the small displacement gradient

$$\boldsymbol{g} = \boldsymbol{\varepsilon} + \boldsymbol{\omega}.$$

It now remains to show that there exists $\boldsymbol{u}$ such that $\boldsymbol{g} = \operatorname{grad} \boldsymbol{u}$. Once again applying the theorem of the previous section, one concludes that $\boldsymbol{u}$ exists provided $-(\operatorname{curl} \boldsymbol{g}^T)^T = 0$, implying that

$$(\operatorname{curl} \boldsymbol{g}^T)^T = (\operatorname{curl} \boldsymbol{\varepsilon}^T)^T + (\operatorname{curl} \boldsymbol{\omega}^T)^T = 0.$$

This is verified by construction of $\boldsymbol{\omega}$ from $\boldsymbol{\varepsilon}$ via $\boldsymbol{\vartheta}$ together with the following equalities:

$$(\operatorname{curl} \boldsymbol{\varepsilon}^T)^T + (\operatorname{curl} \boldsymbol{\omega}^T)^T = (\operatorname{curl} \boldsymbol{\varepsilon})^T + (\operatorname{grad} \boldsymbol{\vartheta})^T - \boldsymbol{I} \operatorname{div} \boldsymbol{\vartheta} = 0.$$

We also used the fact that

$$\operatorname{div} \boldsymbol{\vartheta} = \frac{1}{2} \epsilon_{klm} u_{m,lk} = 0.$$

The displacement vector $\boldsymbol{u}$ is now obtained by integration:

$$\boldsymbol{u}(\boldsymbol{p}) = \int_{\gamma_{p_O p}} (\boldsymbol{\varepsilon} + \omega) \cdot d\boldsymbol{x}.$$

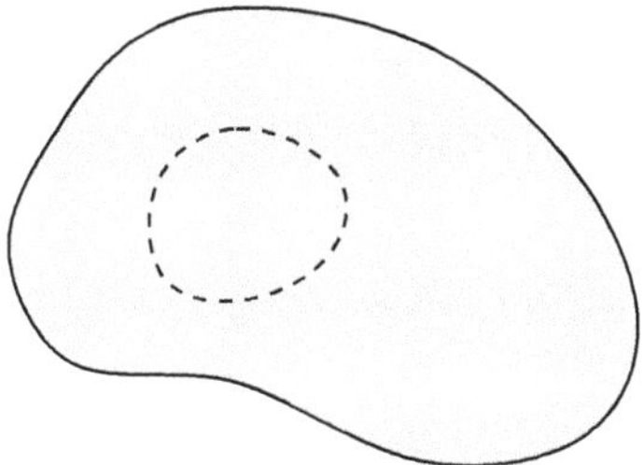
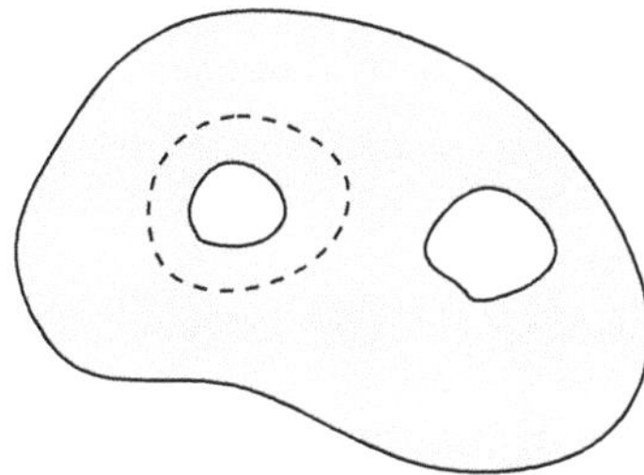

Figure 1.12. Simply connected (left) and multiply connected (right) domains.

It is convenient to define the operator applied to the strain tensor in equation (1.33) as the *incompatibility operator*, that is,

$$\mathrm{inc}\,\boldsymbol{\varepsilon} = -(\mathrm{curl}\,(\mathrm{curl}\,\boldsymbol{\varepsilon})^T)^T.$$

If incompatibility of strain is zero then the strain field is compatible, and the displacement field can be reconstructed. If incompatibility is nonzero, unique displacement field reconstruction is impossible. See Figure 1.12.

We now code the described steps within MATHEMATICA and check the validity of the proposed integration procedure for a general displacement field. The main tool in the procedure is the **`IntegrateGrad`** operator, which integrates the gradient of a given scalar or vector field and which is provided as part of the **`Tensor2Analysis`** package and explained in Appendix 1.

Consider a given strain tensor $\boldsymbol{\varepsilon}$ that for convenience can be computed in terms of the arbitrary displacement field $\boldsymbol{u}$. The validation of the proposed procedure would then consist of the reconstruction of $\boldsymbol{u}$.

First, compatibility of strain is verified using the **`Inc`** operator from the package.

Displacement reconstruction begins with the computation of the gradient of the infinitesimal rotation vector and then of the vector itself by integration.

A check is performed for the correctness of integration.

```
Inc[eps]
```

```
gradtheta = - Curl[eps]
```

```
theta = IntegrateGrad[gradtheta]
```

```
Simplify[Grad[theta] - gradtheta]
```

From the computed infinitesimal rotation vector $\boldsymbol{\vartheta}$ the rotation tensor $\boldsymbol{\omega}$ is computed. The small displacement gradient $\boldsymbol{\varepsilon} + \boldsymbol{\omega}$ is then integrated to obtain the displacement vector $\boldsymbol{u}$.

The result is finally verified.

```
omega = Table[Sum[Signature[{i,j,k}] theta[[k]],
                                    {k,3}],{i,3},{j,3}]
```

```
Simplify[omega + (Grad[u] - Transpose[Grad[u]])/2]
```

```
uint = IntegrateNabla[eps + omega]

Simplify[eps - (Grad[uint] - Transpose[Grad[uint]])/2 ]
Simplify[u - uint]
```

For convenience of use, the sequence of commands for reconstructing the displacement field from a known strain tensor can be grouped into a single command.

Commands are wrapped in a `Module` and can be called as a function.

Note that displacement reconstruction is only attempted if the strains are found to be compatible.

```
IntegrateStrain[strain_] :=
  Module[ {theta, omega},
    (  theta = IntegrateGrad[-Curl[strain]];
       omega =
         Table[Sum[Signature[{i, j, k}] theta[[k]],
                              {k,3}],{i,3},{j,3}];
       IntegrateGrad[strain + omega]
    ) /; ( Inc[strain] === {{0,0,0},{0,0,0},{0,0,0}})]
```

Strain integration and compatibility operators

`IntegrateGrad[gf]`	Integrates gradient of scalar or vector field `gf` to obtain scalar or vector potential `f`
`IntegrateStrain[eps]`	Integrates small strain field `eps` to obtain the displacement vector field `u`
`Inc[f]`	Incompatibility operator for tensor field `f`

Equivalent compatibility conditions for small strains

The compatibility equation for the strain tensor $\boldsymbol{\varepsilon}$

$$\operatorname{inc}\boldsymbol{\varepsilon} = -(\operatorname{curl}(\operatorname{curl}\boldsymbol{\varepsilon})^T)^T = 0$$

is enforced *if and only if* the following equation is satisfied:

$$\Delta\,\boldsymbol{\varepsilon} + \operatorname{grad}\operatorname{grad}(\operatorname{tr}\boldsymbol{\varepsilon}) - (\operatorname{grad} + \operatorname{grad}^T)\boldsymbol{\varepsilon} = \mathbf{0}. \tag{1.34}$$

A general proof of this statement for all cases is given by Gurtin (1982).

The MATHEMATICA package supplied with this book provides the reader with a practical means of verifying the equivalence of the two above statements for arbitrary tensor fields $\boldsymbol{\varepsilon}$ in any of the orthogonal coordinate systems defined within MATHEMATICA.

Rigid body motion

If the small strain tensor is identically equal to zero,

$$\boldsymbol{\varepsilon} = 0,$$

then the strain integration procedure produces the following displacement field:

$$\boldsymbol{u}(\boldsymbol{p}) = \boldsymbol{u}(\boldsymbol{p}_0) + \boldsymbol{\vartheta}(\boldsymbol{p}_0) \wedge (\boldsymbol{p} - \boldsymbol{p}_0).$$

It is important to remark that the preceeding field is *not* a rigid body motion, that is, a combination of translation and rotation. One can easily check that the corresponding *finite* strain tensor is not equal to zero (as would be expected for rigid body motion):

$$\boldsymbol{e} = \frac{1}{2}\left(\operatorname{grad}\boldsymbol{u} + \operatorname{grad}\boldsymbol{u}^T + \operatorname{grad}\boldsymbol{u}^T \operatorname{grad}\boldsymbol{u}\right)$$

$$\boldsymbol{e} = \frac{1}{2}\left(\boldsymbol{\varepsilon} + \boldsymbol{\omega}^2\right)$$

$$= \frac{1}{2}\boldsymbol{\omega}^2 \neq 0.$$

Only under the assumption of *small strains* does the relation $\boldsymbol{e} \approx \boldsymbol{\epsilon}$ hold.

SUMMARY

This chapter introduced the basic quantities used to describe the kinematics of deformation. The polar decomposition of the deformation gradient tensor into stretch and rotation tensors was presented. The Green–Lagrange finite strain tensor was introduced, followed by the small strain tensor particularly widely used in linear elasticity. Particular attention was devoted to the strain compatibility conditions and the procedure for integrating strain fields to obtain displacements.

EXERCISES

1. Ocean waves – Trochoidal waves in the Lagrangian representation (Coirier, 1997; Germain, 1983)

Consider in cartesian coordinates (x, y, z) the half-space $y \leq 0$ and the following plane particle motion defined in this domain:

$$x = X + A\exp(kY/H)\cos 2\pi\left(\frac{X}{L} - \frac{t}{T}\right)$$

$$y = Y + A\exp(kY/H)\sin 2\pi\left(\frac{X}{L} - \frac{t}{T}\right)$$

$$z = Z.$$

Parameters A, k, H, L, T are real numbers and their physical significance should become clear from consideration of motion. We shall assume only that $A \ll L$ and leave more specific choices to the reader. See Figure 1.13.

(a) Show that the trajectories of particles lying on the plane $X = \text{const}$ are circles and plot the trajectories for a point lying on a vertical line.

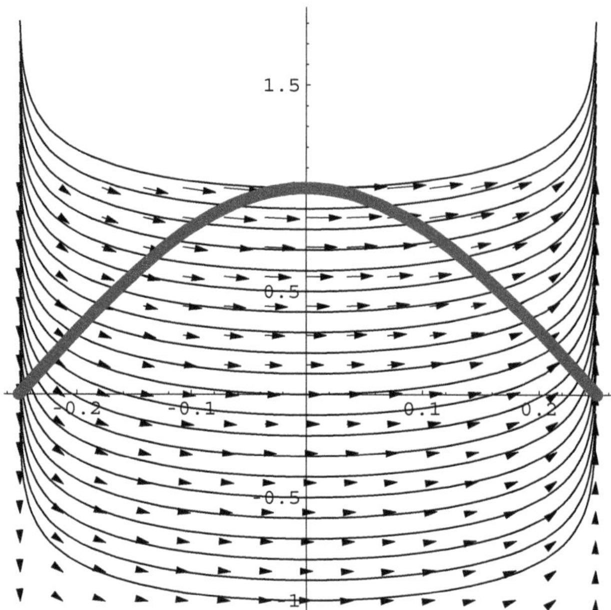

Figure 1.13. Trochoidal waves in the Lagrangian representation: the free surface (solid line), the velocity field, and the streamlines at time $t = 0$, plotted for parameter values $a = 1$, $w = 1$, $k = 1$. Note that velocity vectors and streamlines only have physical significance for points lying below the free surface.

(b) Determine the evolution of the planes $Z = \text{const}$ with time.
(c) Compute the velocity field and plot this field for a given time instant.
(d) Obtain in closed form the equation of streamlines by

- integrating the parametric form

$$\frac{d\boldsymbol{a}}{ds} = \lambda(s)\boldsymbol{v}$$

  using the **DSolve** operator.
- finding a complete integral (see package **Calculus'DSolveIntegrals'** of the equation

$$\operatorname{grad}\varphi \cdot \boldsymbol{v} = 0 \qquad \frac{\partial\varphi}{\partial X}(X, Y)v_X(X, Y) + \frac{\partial\varphi}{\partial Y}(X, Y)v_Y(X, Y) == 0.$$

2. Ocean waves – Trochoidal waves in the Eulerian representation (Coirier, 1997; Germain, 1983)

Consider in cartesian coordinates (x, y, z) the half-space $y \leq 0$ and the following plane particle motion defined by the velocity field:

$$\boldsymbol{v}(x, y, z; t) = v_x(x, y, z; t)\boldsymbol{e}_x + v_y(x, y, z; t)\boldsymbol{e}_y$$
$$v_x = -A\exp(ky/H)\sin 2\pi\left(\frac{x}{L} - \frac{t}{T}\right)$$
$$v_y = A\exp(ky/H)\cos 2\pi\left(\frac{x}{L} - \frac{t}{T}\right).$$

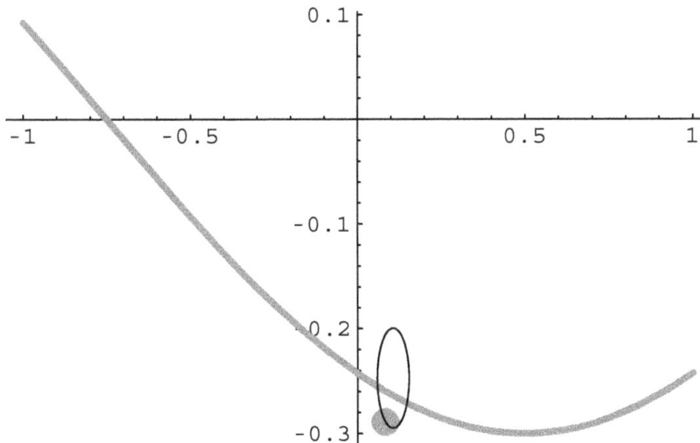

Figure 1.14. Trochoidal waves in the Eulerian representation: free surface (thick solid line), approximate trajectory of a particle (thin elliptical solid line), and position of the particle (filled circle) at time $t = 0$ plotted for parameter values $a = 1$, $w = 1$, $k = 1$.

Parameters A, k, H, L, T are real numbers and their physical significance will become clear from consideration of motion. We again only assume that $A \ll L$, leaving specific choices to the reader. See Figure 1.14.

(a) Show that the trajectories of particles situated on the plane $X =$ const are circles and plot the trajectories for a point lying on a vertical line.
(b) Determine the evolution of the planes $Z =$ const with time.
(c) Compute the velocity field and plot the field for a given time instant.

3. Potential flows

Plot the streamlines, streaklines, and velocity fields for the potential flows defined in the text: uniform flow, point source, doublet, flow in a corner.

4. Constraints: Inextensibility and incompressibility

Propose a function φ applied to the stretch tensor $\boldsymbol{C}$ such that the equation

$$\varphi(\boldsymbol{C}) = 0$$

expresses

(a) the condition of material inextensibility in the direction $\boldsymbol{m}$.
(b) the condition of material incompressibility, that is, that the material is not allowed to undergo a volume change.

Express the condition of incompressibility in terms of the principal stretches.

5. Small strains in the shear experiment

Discuss the shear experiment under the assumption of small strains:

(a) define a necessary and sufficient condition to impose on $a(t) = \tan\,\alpha(t)$ to ensure small strains.
(b) compute the polar decomposition, define the eigenvalues and the eigenvectors of the strain ellipsoid.

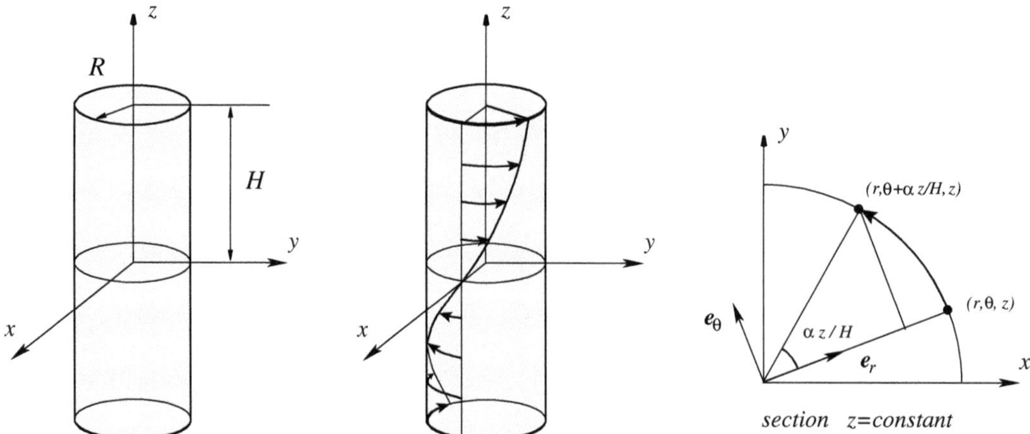

Figure 1.15. The torsion displacement.

(c) define the small strains and the small rotation tensors.
Hint: See end of notebook `C01_glide_shear_strain.nb`.

6. Particular strain states

Compute the strain tensors and the principal stretches and their directions assuming large and infinitesimal strains for the following transformations:
(a) uniform dilatation: $\boldsymbol{x} = \lambda \boldsymbol{X}$
(b) uniaxial extension: $\boldsymbol{x} = \lambda X \boldsymbol{e}_x + Y \boldsymbol{e}_y + Z \boldsymbol{e}_z$
(c) uniaxial extension with lateral contraction: $\boldsymbol{x} = \lambda X \boldsymbol{e}_x + \lambda^{-\frac{1}{2}} Y \boldsymbol{e}_y + \lambda^{-\frac{1}{2}} Z \boldsymbol{e}_z$.

7. Isochoric spherically symmetric deformation

Let us consider a spherically symmetric deformation in a body defined by

$$\boldsymbol{x} = f(R)\boldsymbol{X},$$

where (R, Θ, Φ) are the coordinates of the vector $\boldsymbol{X} = R\boldsymbol{e}_R$ in the spherical system.
(a) Find $f(R)$ if the deformation is isochoric, that is, volume is conserved.
(b) Compute the strain tensors and the principal stretches in this case.

8. Torsion of a cylinder

Consider a right circular cylinder of radius R and height $2H$, illustrated in Figure 1.15.

Torsional deformation is defined in the cylindrical polar coordinate system of the figure in the following way,

$$(r, \theta, z) \longrightarrow \left(r, \theta + \alpha(t)\frac{z}{L}, z\right),$$

where $\alpha(t)$ is angle of torsion, generally a time-dependent function.
(a) Define the position vectors $\boldsymbol{X}, \boldsymbol{x}(\boldsymbol{X}, t)$ and the vector field of displacements $\boldsymbol{u}(\boldsymbol{X}, t)$ in the cylindrical polar coordinate system.
(b) Compute the gradient of the transformation $\boldsymbol{F}$.

(c) Verify that it can be written in the following form:

$$F = R\left(\frac{\alpha r}{L}\right) F_1.$$

(d) Determine the polar decomposition of F, using the similarity between F_1 and the solution obtained in the case of pure shear.
(e) Define the conditions for small strains
(f) Compute the polar decomposition in the case of small strains:
  - using the formulas of small strains and rotations
  - by linearizing $F = R \cdot U$.

Note that the rotation operator is commutative under the condition of small rotations.
Hint: See notebook `C01_torsion.nb`.

9. Extension of a cylinder

Consider a right circular cylinder of radius R and height $2H$.

Extensional deformation is defined in the cylindrical polar coordinate system in Figure 1.15 in the following way;

$$(r, \theta, z) \longrightarrow (\lambda^{-\frac{1}{2}} r, \theta, \lambda z),$$

where λ characterises the amount of extension.
(a) Define position vectors $X, x(X, t)$ and the vector field of displacements $u(X, t)$ in the cylindrical polar coordinate system.
(b) Compute the gradient of transformation F.
(c) Verify that it can be written in the following form:

$$F = R\left(\frac{\alpha r}{L}\right) F_1.$$

(d) Determine the polar decomposition of F, using the similarity between F_1 and the solution obtained in the case of pure shear.
(e) Define the conditions for small strains.

10. Compatibility and integration of strains (Polytechnique Collective, 1990–2005)

Reproduce the reasoning for the integration of small strains in different coordinates systems (spherical, paraboloidal, cylindrical, etc.) and check the validity of the results.

11. General form of compatible small-strain tensors

Find the general expression of the components of a compatible strain tensor under the following conditions:
(a) cartesian coordinate system:

$$\boldsymbol{\varepsilon}(x, y, z) = \varepsilon_{xy}(x, y, z)\,(\boldsymbol{e}_x \otimes \boldsymbol{e}_y + \boldsymbol{e}_y \otimes \boldsymbol{e}_x)\,.$$

(b) cylindrical polar coordinate system:

$$\boldsymbol{\varepsilon}(r, \theta, z) = \varepsilon_{r\theta}(\theta, z)\,(\boldsymbol{e}_z \otimes \boldsymbol{e}_\theta + \boldsymbol{e}_\theta \otimes \boldsymbol{e}_z)\,.$$

(c) cylindrical polar coordinate system:

$$\boldsymbol{\varepsilon}(r, \theta, z) = \varepsilon_{r\theta}(r, \theta, z)\,(\boldsymbol{e}_r \otimes \boldsymbol{e}_\theta + \boldsymbol{e}_\theta \otimes \boldsymbol{e}_r)\,.$$

(d) cylindrical polar coordinate system:

$$\boldsymbol{\varepsilon}(r, \theta, z) = \varepsilon_{rz}(r, \theta, z)\,(\boldsymbol{e}_r \otimes \boldsymbol{e}_z + \boldsymbol{e}_z \otimes \boldsymbol{e}_r)\,.$$

Hint: See notebook `C01_ex_compatibility.nb`.

12. Compatibility of small strains generated by a temperature field (Polytechnique Collective 1990–2005)

Let us assume that a body is heated to a temperature field $\theta(x, y, z)$. Assuming that thermal dilatation is isotropic, that is, material stretch in all directions is the same due to temperature change, the small strain tensor due to the thermal dilatation is

$$\boldsymbol{\varepsilon}(x, y, z) = \alpha\theta(x, y, z)\boldsymbol{I},$$

where α is the coefficient of linear thermal expansion.

(a) Find the general expression of the temperature field such that the dilatation strains are compatible.
(b) Integrate the preceding strain and determine the general form of the displacement field.
(c) Suppose that the body is a cylinder of height $2H$ and radius R subject to a temperature gradient along its axis of $2\Delta\theta$.
  - Under which condition is the small strain condition still valid?
  - Compute the radius of curvature of the medium section of the cylinder and define an order of magnitude for thin sheets of usual materials.

Hint: See notebook `C01_temperature_compatibility.nb`.

13. Rigid body translation and rotation

Using MATHEMATICA packages `Tensor2Analysis.m`, `Displacement.m` and `IntegrateStrain.m`, obtain explicit expressions for displacements due to rigid body translation and rotation in the following coordinate systems:

- in the cartesian coordinate system
- in the cylindrical polar coordinate system
- in the spherical coordinate system.

Show that neither strains nor stresses arise due to these displacements, that is, check by differentiation (using the `Grad` operator) that the corresponding strains (and hence stresses) are zero. Also, obtain expressions for the small rotation tensor. Consider its structure and establish its relationship with the rotation vector.

Hint: See notebook `C01_rigid_displacements.nb`.

2 Dynamics and statics: stresses and equilibrium

OUTLINE

In this chapter we consider the force equilibrium in a continuous body under the assumption that the underlying deformation is adequately described by the small strain hypothesis. The principle of virtual power occupies a central place in this treatment, since it offers a rational basis for developing equations that apply to a continuum in a state of equilibrium. Furthermore, the concept of stress arises naturally from this analysis as dual to small strain for a solid continuum. Stress equilibrium and traction boundary conditions also appear in the most convenient invariant form. For illustration, virtual power expressions are given for systems obeying different kinematics, such as inviscid fluid flow or beams in simple bending, and the resulting stresslike quantities and their equilibrium equations are readily derived. Cauchy's stress principle and the Cauchy–Poisson theorem are also given.

Once it is established that equlibrium stress states in continuum solids in the absence of body forces are given by divergence-free tensors, the representation of such tensor fields is addressed. Beltrami potential representation of divergence-free tensors is considered, and Donati's theorem is introduced to illustrate a certain duality that exists between stress equilibrium and strain compatibility conditions.

2.1 FORCES AND MOMENTA

A body may be subject to a system of *exterior forces* of the following types:

- *Body force* $\boldsymbol{f}$ is described by a vector field distributed over the entire volume of body Ω. Denoting body force by the vector $\boldsymbol{f}(\boldsymbol{x}, t)$ represents the fact that the force $\boldsymbol{f}(\boldsymbol{x}, t)dv$ acts on an infinitesimal volume dv at point $\boldsymbol{x}$ at time t. An example of the body force is the force of gravity.
- *Surface traction* $\boldsymbol{t}$ is described by a vector field distributed over the surface boundary $\partial\Omega$ of body Ω. Denoting the surface traction by the vector $\boldsymbol{t}(\boldsymbol{x}, t)$ represents the fact that the force $\boldsymbol{t}(\boldsymbol{x}, t)ds$ acts on an infinitesimal element of surface area ds at point $\boldsymbol{x}$ at time t.

A particular type of body force arises when inertial effects during motion are considered.

Let mass density be denoted by ρ, which is thought to be a given function over the entire body Ω, such that for every part of the body $\Pi \subset \Omega$ the mass of Π is given by

$$\int_{\Pi} \rho\, dv.$$

The *linear* and *angular momentum* of the subset Π are expressed as

$$\int_\Pi \rho \dot{\mathbf{u}}\, dV \quad \text{and} \quad \int_\Pi \boldsymbol{x} \times \rho \dot{\mathbf{u}}\, dv,$$

respectively, where $\dot{u}$ denotes velocity.

Linear and *angular inertial forces* are body forces related to the rates of change of linear and angular momentum and expressed as $\rho \ddot{\boldsymbol{u}}$ and $\boldsymbol{x} \times \rho \ddot{\boldsymbol{u}}$, respectively, where $\ddot{\boldsymbol{u}}$ denotes acceleration.

Laws of momentum balance and change (Newton's second law) in the case of continuous bodies apply for each subset of the body. Moreover, we require that, for a body in motion, exterior and inertial forces are balanced by the internal forces arising due to the material response to deformation. To understand how an appropiate model can be constructed for the description of internal forces, we shall follow two independent routes presented in the next two sections.

In the first method the concept of stress arises from considerations of balance of virtual power. In the second method the concept of stress arises from geometrical arguments and the balance of momentum. For a historical overview of the development of the concept of stress we recommend the essay by Truesdell (1968).

2.2 VIRTUAL POWER AND THE CONCEPT OF STRESS

Virtual power p_f of a body force $\boldsymbol{f}$ is defined as the scalar product of the force vector $\boldsymbol{f}$ and a virtual velocity field $\boldsymbol{v}$:

$$p_f = \boldsymbol{f} \cdot \boldsymbol{v}.$$

Virtual power is thus a linear map over the space of virtual velocities. The equivalence that exists between such linear map and the vector field $\boldsymbol{f}$ can be established by mathematical arguments such as the Riesz theorem (Kestelman, 1960) which will not, however, be further discussed here. On this basis, the balance of forces and moments can be established in terms of the balance of virtual powers generated by the system of forces under consideration.

For a continous body Ω we shall assume that *virtual velocity* fields $\boldsymbol{v}$ are continuous differentiable vector fields forming a vector space $\mathbb{V}$. Actual body motion, including velocity fields corresponding to rigid body motion, are included in the virtual velocity fields. We recall that the general form of the velocity field for rigid body motion is

$$\boldsymbol{v}(\boldsymbol{x}, t) = \boldsymbol{a}(t) + \boldsymbol{b}(t) \times \boldsymbol{x},$$

where $\boldsymbol{a}$ and $\boldsymbol{b}$ are two time-dependent vector fields corresponding to translation and rotation, respectively. A straightforward strain rate computation shows that the motion corresponding to this velocity field is rigid only in the *infinitesimal* sense. However, because the theory developed here is the infinitesimal theory, this does not lead to any contradiction.

Rigid body motion velocity fields form a vector subspace in $\mathbb{V}$, denoted by $\mathbb{V}_0$.

For systems of forces discussed in Section 2.1 one can therefore define the virtual power of *internal and external forces* $\mathcal{P}_i$ and $\mathcal{P}_e$ and of *inertial forces* $\mathcal{P}_a$, respectively, as linear integral maps over virtual velocity fields $\boldsymbol{v}$ defined as

$$\mathcal{P}_e(\boldsymbol{v}) = \int_\Omega \boldsymbol{f} \cdot \boldsymbol{v}\, dv + \int_{\partial\Omega} \boldsymbol{t} \cdot \boldsymbol{v}\, ds$$

$$\mathcal{P}_a(\boldsymbol{v}) = \int_\Omega \rho \ddot{\boldsymbol{u}} \cdot \boldsymbol{v}\, dv.$$

Momentum balance (Newton's second law) is equivalent to the *principle of virtual power.*

Principle of virtual power
For every virtual velocity field $\mathbf{v}$ *the following equality holds:*

$$\mathcal{P}_e(\boldsymbol{v}) + \mathcal{P}_i(\boldsymbol{v}) = \mathcal{P}_a(\boldsymbol{v}). \tag{2.1}$$

for a body in static equilibrium *inertial forces vanish, because* $\ddot{\mathbf{u}} = 0$, *and the principle of virtual power is expressed as:*

$$\mathcal{P}_e(\boldsymbol{v}) + \mathcal{P}_i(\boldsymbol{v}) = 0 \tag{2.2}$$

The virtual power of the internal forces $\mathcal{P}_i(\boldsymbol{v})$ remains to be defined by specifying explicitly the nature of the linear map over the space of virtual velocity fields. A rational condition should be imposed that this power (and hence the underlying internal forces) should be independent of the choice of Galilean frame of reference used by the observer. Hence virtual power of internal forces should vanish when applied to virtual velocity fields corresponding to rigid body motion:

$$\mathcal{P}_i(\boldsymbol{v}) = 0 \qquad \forall \boldsymbol{v} \in \mathbb{V}_0. \tag{2.3}$$

In this way the form of the vector subspace of virtual velocity fields $\boldsymbol{V}$ derived from the kinematic analysis of the previous chapter results in certain conditions on the system of internal forces $\mathcal{P}_i$. This approach can be used to develop internal force models for ideal fluids, solid bodies, columns, beams, plates, etc. A detailed discussion of this approach can be found in Salençon (2001).

Ideal fluid

Select the space $\mathbb{V}$ of virtual velocity fields to be continuous differentiable vector fields over Ω. Define the virtual power as the following linear map:

$$\mathcal{P}_i(\boldsymbol{v}) = -\int_\Omega p \operatorname{div} \boldsymbol{v}\, dv. \tag{2.4}$$

Here p is a scalar field that characterises internal forces. The scalar p is dual to the rate of volume change defined by $\operatorname{div} \boldsymbol{v}$ and represents therefore the internal pressure field within the body. It is easy to check that this form of internal virtual power vanishes for virtual velocities corresponding to rigid body motion.

The virtual power expression (2.4) can also be rewritten by the application of the Stokes theorem in the equivalent form

$$\mathcal{P}_i(\boldsymbol{v}) = \int_\Omega \operatorname{grad} p \cdot \boldsymbol{v}\, dv + \int_{\partial\Omega} p\,\boldsymbol{n} \cdot \boldsymbol{v}\, ds, \tag{2.5}$$

where $\boldsymbol{n}$ is the outward unit normal to the boundary $\partial\Omega$. The principle of virtual power becomes

$$\int_{\Omega}(-\operatorname{grad} p + \boldsymbol{f})\cdot\boldsymbol{v}\,dv + \int_{\partial\Omega}(\boldsymbol{t} + p\,\boldsymbol{n})\cdot\boldsymbol{v}\,ds = \int_{\Omega}\rho\ddot{\boldsymbol{u}}\cdot\boldsymbol{v}\,dv. \tag{2.6}$$

Because the above equality holds for an arbitrary virtual velocity field $\boldsymbol{v}$, this leads to the following equilibrium equations in the local form:

- for each interior point $\boldsymbol{x}\in\Omega$: $-\operatorname{grad} p + \boldsymbol{f} = \rho\ddot{\boldsymbol{u}}$
- for each boundary point $\boldsymbol{x}\in\partial\Omega$: $\boldsymbol{t} = +p\,\boldsymbol{n}$.

The condition on the boundary shows that the ideal fluid model only allows boundary loading in the form of normal surface tractions. This anomaly was first observed by d'Alembert and has since been known as the d'Alembert paradox (Truesdell, 1968).

Continuum solid

Once again we start our consideration with the space $\mathbb{V}$ of virtual velocity fields given by continuous differentiable vector fields over Ω. The linear map representing virtual power is in this case given by

$$\mathcal{P}_i(\boldsymbol{v}) = -\int_{\Omega}\boldsymbol{\sigma} : \operatorname{grad}\boldsymbol{v}\,dv, \tag{2.7}$$

where $\boldsymbol{\sigma}$ is a symmetric tensor field, referred to as the *stress field*.

In granular materials such as sand, the internal work is done not only due to particle deformation, but also due to the mutual rotation of particles. The stress tensor in this case may not be symmetric. The model that arises if this is taken into consideration is generally known as a Cosserat material. One must consider the possibility of the existence of an additional external force producing a distribution of body moments. This, for example, is the case if particles forming the body are magnetic and the body is subject to an externally applied magnetic field.

In classical continuum mechanics, the symmetry of stresses is assumed, $\boldsymbol{\sigma} = \boldsymbol{\sigma}^T$, implying that $\boldsymbol{\sigma} : \operatorname{grad}\boldsymbol{v} = \boldsymbol{\sigma} : \epsilon[\boldsymbol{v}]$. Thus the symmetric stress tensor is dual to the small strain tensor $\boldsymbol{\varepsilon}$ and does not act on rotations. As for an ideal fluid, one can readily check that the internal virtual power defined in (2.7) vanishes for virtual velocities corresponding to rigid body motion.

Applying the Stokes theorem to (2.7) as before, one obtains

$$\mathcal{P}_i(\boldsymbol{v}) = \int_{\Omega}\operatorname{div}\boldsymbol{\sigma} : \varepsilon[\boldsymbol{v}]\,dv - \int_{\partial\Omega}\boldsymbol{v}\cdot\boldsymbol{\sigma}\cdot\boldsymbol{n}\,ds, \tag{2.8}$$

where $\boldsymbol{n}$ is the outward unit normal to the boundary $\partial\Omega$. The principle of virtual power becomes

$$\int_{\Omega}(\operatorname{div}\boldsymbol{\sigma} + \boldsymbol{f})\cdot\boldsymbol{v}\,dv + \int_{\partial\Omega}(\boldsymbol{t} - \boldsymbol{\sigma}\cdot\boldsymbol{n})\cdot\boldsymbol{v}\,ds = \int_{\Omega}\rho\ddot{\boldsymbol{u}}\cdot\boldsymbol{v}\,dv. \tag{2.9}$$

Because the above equality holds for an arbitrary virtual velocity field $\boldsymbol{v}$, it implies that the following local balance equations are satisfied:

- for each interior point $\boldsymbol{x}\in\Omega$: $\operatorname{div}\boldsymbol{\sigma} + \boldsymbol{f} = \rho\ddot{\boldsymbol{u}}$
- for each boundary point $\boldsymbol{x}\in\partial\Omega$: $\boldsymbol{t} = \boldsymbol{\sigma}\cdot\boldsymbol{n}$.

The boundary condition is no longer restricted to a particular type of surface traction: a solid body can carry surface tractions in the form of both pressures and shear forces.

Balance of linear and angular momentum for a continuum solid

The balance of linear and angular momentum is given by the requirement that for any part of the body, $\Pi \subset \Omega$, the resultant of internal and external forces is equal to the rate of change of linear momentum:

$$\int_{\partial\Pi} \boldsymbol{s}\, ds + \int_{\Pi} \boldsymbol{f}\, dv = \frac{d}{dt}\int_{\Pi} \rho \dot{\boldsymbol{u}}\, dv \tag{2.10}$$

$$\int_{\partial\Pi} \boldsymbol{x} \times \boldsymbol{s}\, ds + \int_{\Pi} \boldsymbol{x} \times \boldsymbol{f}\, dv = \frac{d}{dt}\int_{\Pi} \boldsymbol{x} \times \rho \dot{\boldsymbol{u}}\, dv. \tag{2.11}$$

Here we denote the surface boundary traction on part Π by $\boldsymbol{s}$ and use the form $\boldsymbol{s} = \boldsymbol{\sigma} \cdot \boldsymbol{n}$ on $\partial\Pi$.

The above equations are obtained by integrating the local form of the balance equations and applying the formula for the time derivative of a moving domain.

As a particular case of the above expressions, the principle of virtual work can be applied to the virtual velocity field corresponding to general rigid body movement, giving the balance of linear and angular momentum for the entire body Ω.

Bending of thin plates

A plate can be defined as a solid occupying a domain of the form $\Omega = \omega \times [-h/2, h/2]$. Here $\omega \subset \mathbb{R}^2$ is the domain occupied by the body in the planar median surface, and h is the thickness. The Love–Kirchhoff kinematic assumption restricts the virtual velocity fields to the form

$$\boldsymbol{v}(\hat{\boldsymbol{x}}, x_3) = v(\hat{\boldsymbol{x}})\boldsymbol{e}_3 - x_3\, \widehat{\text{grad}}\, u,$$

where $\hat{x} \in \mathbb{R}^2$, and $\hat{grad}$ denotes the two-dimensional gradient with respect to $\hat{\boldsymbol{x}}$.

The choice of the form of internal virtual power is made as

$$\mathcal{P}_i(\boldsymbol{v}) = -\int_{\Omega} \boldsymbol{M} : \widehat{\text{grad}}\, \widehat{\text{grad}}\, \boldsymbol{v}\, dv, \tag{2.12}$$

where $\boldsymbol{M}$ is a second-order tensor field. $\boldsymbol{M}$ acts on the tensor field of bending rates $\widehat{\text{grad}}\, \widehat{\text{grad}}\, \boldsymbol{v}$ that is a measure of the rate of change of the geometric curvature of a bent plate.

Applying the Stokes theorem twice to the preceding form of $\mathcal{P}_i$ leads, after a series of operations similar to those discussed previously, to the following equations in local form:

- for each interior point $\hat{\boldsymbol{x}} \in \omega$: $\widehat{\text{div}}\, \widehat{\text{div}} \boldsymbol{M} + b = \rho \ddot{u}$
- for each boundary point $\hat{\boldsymbol{x}} \in \partial\omega$: $m = \boldsymbol{n} \cdot \boldsymbol{M} \cdot \boldsymbol{n} \quad q = \boldsymbol{n} \cdot \left(2\widehat{\text{div}} \boldsymbol{M} - \widehat{\text{grad}} \boldsymbol{M} : (\boldsymbol{n} \otimes \boldsymbol{n})\right)$.

The resultant body force f per unit area, resultant bending moment m per unit length, and resultant shear force q per unit length are related to the body force $\boldsymbol{f}$ and surface tractions $\boldsymbol{t}$ by the formulas

$$f(\hat{\boldsymbol{x}}) = \int_{-h/2}^{h/2} \boldsymbol{e}_3 \cdot \boldsymbol{f}(\hat{\boldsymbol{x}}, x_3)\, dx_3 \qquad m(\hat{\boldsymbol{x}}) = \int_{-h/2}^{h/2} x_3 \boldsymbol{t}(\hat{\boldsymbol{x}}, x_3)\, dx_3 \qquad q(\hat{\boldsymbol{x}}) = \int_{-h/2}^{h/2} \boldsymbol{e}_3 \cdot \boldsymbol{t}(\hat{\boldsymbol{x}}, x_3)\, dx_3.$$

The above expressions demonstrate that force fields acting within thin plates are given by integral moments of different orders of the three-dimensional body force and surface tractions.

A similar expression relates the second-rank tensor of bending moments to the stress tensor:

$$\boldsymbol{M}(\hat{\boldsymbol{x}}) = \int_{-h/2}^{h/2} x_3 \boldsymbol{\sigma}(\hat{\boldsymbol{x}}, x_3)\, dx_3. \tag{2.13}$$

2.3 THE STRESS TENSOR ACCORDING TO CAUCHY

A different approach to the concept of stress was first announced in 1822 by Cauchy as his famous *stress principle*. It leads to the same concept of stress as a second-order tensor field introduced above for the description of internal forces within a continuum solid body.

Consider a small arbitrary part $\Pi \subset \Omega$ of a solid body and introduce the *stress vector* $\boldsymbol{s}_n(\boldsymbol{x}, t)$ at point $\boldsymbol{x} \in \partial\Pi$ at time t. For an outward unit normal $\boldsymbol{n}$ on the boundary $\partial\Pi$, this is defined as the force per unit area exerted by the portion of Ω outside Π on the portion of Ω inside Π. If the point lies on the boundary of the body, $\boldsymbol{x} \in \partial\Omega$, then the stress vector coincides with the vector of surface traction, $\boldsymbol{s}_n = \boldsymbol{t}$.

The balance of linear and angular momentum on part Π requires that

$$\int_{\partial\Pi} \boldsymbol{s}_n\, ds + \int_{\Pi} \boldsymbol{f} = \frac{d}{dt}\int_{\Pi} \rho \dot{\boldsymbol{u}}\, dv \tag{2.14}$$

$$\int_{\partial\Pi} \boldsymbol{x} \times \boldsymbol{s}_n\, ds + \int_{\Pi} \boldsymbol{x} \times \boldsymbol{f} = \frac{d}{dt}\int_{\Pi} \boldsymbol{x} \times \rho \dot{\boldsymbol{u}}\, dv. \tag{2.15}$$

From these equations one can derive the following theorem.

Theorem: Cauchy–Poisson Theorem: The balance of linear and angular momentum in part Π in equations (2.14) and (2.15) is satisifed if and only if

- The stress vector is a linear function of the outward normal and can therefore be expressed as

$$\boldsymbol{s}_n(\boldsymbol{x}, t) = \boldsymbol{\sigma}(\boldsymbol{x}, t) \cdot \boldsymbol{n}(\boldsymbol{x}),$$

  where $\boldsymbol{\sigma}$ is a symmetric tensor field.
- Fields $\boldsymbol{u}, \boldsymbol{\sigma}, \boldsymbol{f}$ satisfy the following equation:

$$\operatorname{div} \boldsymbol{\sigma} + \boldsymbol{f} = \rho \ddot{\boldsymbol{u}}.$$

The direct argument in this statement was proved by Cauchy in 1823, whereas the correctness of the inverse argument was established by Poisson in 1829.

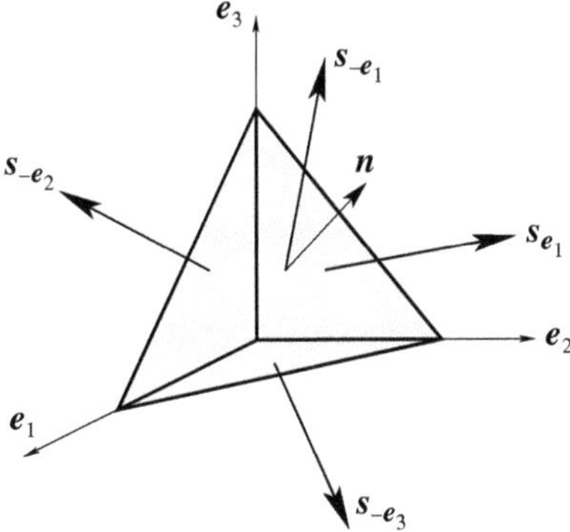

Figure 2.1. The Cauchy tetrahedron.

Without providing a complete proof of this theorem, we wish to mention that the principal steps taken in the proof are to observe the following facts:

- $\boldsymbol{s}_n = -\boldsymbol{s}_{-n}$ follows from the fact that mutual forces between the bodies must always balance each other.
- $\boldsymbol{s}_n = n_1\boldsymbol{s}_{e_1} + n_2\boldsymbol{s}_{e_2} + n_3\boldsymbol{s}_{e_3}$, which follows from the balance of forces on an infinitesimal tetrahedron (Figure 2.1), where $\boldsymbol{n} = n_1\boldsymbol{e}_1 + n_2\boldsymbol{e}_2 + n_3\boldsymbol{e}_3$. As a consequence, one can define

$$\boldsymbol{\sigma} = \boldsymbol{s}_{e_1} \otimes \boldsymbol{e}_1 + \boldsymbol{s}_{e_2} \otimes \boldsymbol{e}_2 + \boldsymbol{s}_{e_3} \otimes \boldsymbol{e}_3.$$

- Balance of linear momentum over an infinitesimal volume element leads to $\text{div}\boldsymbol{\sigma} + \boldsymbol{f} = \rho\ddot{\boldsymbol{u}}$.
- Symmetry of $\boldsymbol{\sigma}$, that is, $\boldsymbol{\sigma} = \boldsymbol{\sigma}^T$, is established by the consideration of balance of angular momentum and the absence of a volume distribution of moments.

The symmetry of the stress tensor is not observed in the case of granular materials, as mentioned in the preceding section. The balance of angular momentum is enriched in this particular case by additional terms corresponding to distribution of moments within the body.

Example stress states

Suppose that body forces are negligible and that a homogeneous stress tensor $\boldsymbol{\sigma}$ is in equilibrium with external surface tractions. Some classical examples of fundamental stress states are discussed here.

- *Uniaxial tension* in direction $\boldsymbol{k}$:

$$\boldsymbol{\sigma} = \sigma\boldsymbol{k} \otimes \boldsymbol{k} \qquad \boldsymbol{\sigma} \cdot \boldsymbol{n} = \sigma(\boldsymbol{k} \cdot \boldsymbol{n})\boldsymbol{k}.$$

Suppose the coordinate system is selected so that $\boldsymbol{k} = \boldsymbol{e}_3$. Then the matrix form of the stress tensor corresponding to the state of uniaxial tension is

$$\boldsymbol{\sigma} = \begin{bmatrix} 0 & 0 & 0 \\ 0 & 0 & 0 \\ 0 & 0 & \sigma \end{bmatrix}.$$

- *Hydrostatic compression*:

$$\boldsymbol{\sigma} = -p\,\mathbf{1} \qquad \boldsymbol{\sigma}\cdot\boldsymbol{n} = -p\boldsymbol{n}.$$

The matrix form of the stress tensor for material under hydrostatic compression is

$$\boldsymbol{\sigma} = -p\begin{bmatrix} 1 & 0 & 0 \\ 0 & 1 & 0 \\ 0 & 0 & 1 \end{bmatrix}.$$

For this stress state the virtual power of internal forces reduces to the form defined for perfect fluids, that is, $-\boldsymbol{\sigma} : \operatorname{grad}\boldsymbol{v} = p\,\mathbf{1} : \operatorname{grad}\boldsymbol{v} = p\operatorname{div}\boldsymbol{v}$.

- *Pure shear* τ in a plane is defined by a pair of orthogonal vectors $\boldsymbol{k}, \boldsymbol{m}$:

$$\boldsymbol{\sigma} = \tau\,(\boldsymbol{k}\otimes\boldsymbol{m} + \boldsymbol{m}\otimes\boldsymbol{k}) \qquad \boldsymbol{\sigma}\cdot\boldsymbol{n} = \tau\,((\boldsymbol{m}\cdot\boldsymbol{n})\boldsymbol{k} + (\boldsymbol{k}\cdot\boldsymbol{n})\boldsymbol{m}).$$

If $\boldsymbol{k} = \boldsymbol{e}_1$ and $\boldsymbol{m} = \boldsymbol{e}_2$, the matrix form of the stress state of pure shear is

$$\boldsymbol{\sigma} = \begin{bmatrix} 0 & \tau & 0 \\ \tau & 0 & 0 \\ 0 & 0 & 0 \end{bmatrix}.$$

2.4 POTENTIAL REPRESENTATIONS OF SELF-EQUILIBRATED STRESS TENSORS

We shall call $\boldsymbol{\sigma}$ a *self-equilibrated stress tensor* if the linear and angular moments due to its tractions vanish on an arbitrary closed surface $\Sigma \subset \Omega$:

$$\int_\Sigma \boldsymbol{\sigma}\cdot\boldsymbol{n}\,ds = 0 \qquad \int_\Sigma \boldsymbol{x}\times\boldsymbol{\sigma}\cdot\boldsymbol{n}\,ds = 0. \tag{2.16}$$

Application of the Stokes theorem readily shows that a self-equilibrated stress field also satisfies

$$\operatorname{div}\boldsymbol{\sigma} = 0. \tag{2.17}$$

We shall call the stress field that satisfies this equation a *divergence-free stress tensor*.

We have seen above that a *self-equilibrated stress tensor* is always *divergence-free*. The reverse assertion is only true if Σ is a simple closed regular surface, that is, it is not the boundary surface of a domain with holes or cavities. Gurtin (1982) provides an example for a spherical shell $\Omega = (r, \theta, \varphi) \in [a, b] \times [-\pi/2, \pi/2] \times [0, 2\pi]$.

We now consider possible potential representations of self-equilibrated and divergence-free stress tensors by merely stating results without proof. A complete treatment may be found in the monograph of Gurtin (1982).

The Beltrami stress potential

If Ω is bounded by a simple closed surface, then the following statements are equivalent:

(i) symmetric stress tensor field $\boldsymbol{\sigma}$ is divergence-free,

$$\operatorname{div}\boldsymbol{\sigma} = 0$$

(ii) symmetric stress tensor field $\boldsymbol{\sigma}$ is self-equilibrated,

$$\int_\Sigma \boldsymbol{\sigma}\cdot\boldsymbol{n}\,ds = 0 \qquad \int_\Sigma \boldsymbol{x}\times\boldsymbol{\sigma}\cdot\boldsymbol{n}\,ds = 0$$

(iii) there exists a second-order symmetric tensor field $\boldsymbol{B}$ such that

$$\boldsymbol{\sigma} = \operatorname{inc}\boldsymbol{B}.$$

The equivalence of the first two statements for a simply bounded body has already been established. The equivalence of statements (i) and (iii) is illustrated by the following sequence of equalities (see also Gurtin (1982)):

$$\operatorname{div}\boldsymbol{\sigma} = \operatorname{div}\operatorname{inc}\boldsymbol{B} = \operatorname{div}\left(\operatorname{curl}(\operatorname{curl}\boldsymbol{B})^T\right)^T = 0.$$

Particular cases of the Beltrami stress potential are conventionally referred to by different names. These are most simply expressed in cartesian coordinates.

The Airy stress potential

$$\boldsymbol{B} = \phi\,\boldsymbol{e}_3\otimes\boldsymbol{e}_3.$$

The Maxwell stress potential

$$\boldsymbol{B} = \phi_1\,\boldsymbol{e}_1\otimes\boldsymbol{e}_1 + \phi_2\,\boldsymbol{e}_2\otimes\boldsymbol{e}_2 + \phi_3\,\boldsymbol{e}_3\otimes\boldsymbol{e}_3.$$

The Morera stress potential

$$\boldsymbol{B} = \omega_1(\boldsymbol{e}_2\otimes\boldsymbol{e}_3 + \boldsymbol{e}_3\otimes\boldsymbol{e}_2) + \omega_2(\boldsymbol{e}_3\otimes\boldsymbol{e}_1 + \boldsymbol{e}_1\otimes\boldsymbol{e}_3) + \omega_3(\boldsymbol{e}_1\otimes\boldsymbol{e}_2 + \boldsymbol{e}_2\otimes\boldsymbol{e}_1).$$

The form of stress components due to these potentials in arbitrary orthogonal coordinate systems can be readily obtained using MATHEMATICA.

The Beltrami–Schaeffer stress potential

For a body occupying the domain Ω bounded by a regular surface,

the symmetric stress tensor $\boldsymbol{\sigma}$ is divergence-free,

$$\operatorname{div}\boldsymbol{\sigma} = 0,$$

if and only if

there exist a second-order symetric tensor field $\boldsymbol{B}$ and a harmonic vector field $\boldsymbol{h}$ such that

$$\boldsymbol{\sigma} = \operatorname{inc}\boldsymbol{B} + \operatorname{grad}\boldsymbol{h} + \operatorname{grad}^T\boldsymbol{h} - (\operatorname{div}\boldsymbol{h})\mathbf{1}.$$

On the role of the incompatibility operator in elasticity

From the discussions in the preceding sections and the previous chapter it becomes apparent that a fundamental role in the analysis of strains and stresses is played by the incompatility operator, inc. Among various continuous symmetric second-rank tensor fields, only the divergence-free tensors provide candidates for stress fields in a continuum solid. The Beltrami stress potential provides a constructive formalism for obtaining divergence-free stress tensors $\boldsymbol{\sigma}$ from *arbitrary* continuous tensor fields $\boldsymbol{B}$ by the application of the incompatibility operator, inc.

The incompatibility operator, inc, was first introduced in Section 1.4 in connection with the criterion of integrability of a given strain field for the reconstruction of a unique displacement field.

Using potential representations of symmetric stress tensors, the following proposition can be established (Germain, 1983; Gurtin, 1982):

Donati's theorem Let $\boldsymbol{\varepsilon}$ be a symmetric tensor field on Ω such that the statement

$$\int_{\Omega} \boldsymbol{\sigma} : \boldsymbol{\varepsilon} \, dv = 0$$

holds true for every symmetric tensor field $\boldsymbol{\sigma}$ that vanishes on $\partial\Omega$ and satisfies

$$\operatorname{div} \boldsymbol{\sigma} = 0.$$

Then $\boldsymbol{\varepsilon}$ satisfies the compatibility equation

$$\operatorname{inc} \boldsymbol{\varepsilon} = 0.$$

Donati's theorem establishes a certain duality between strain compatibility and stress equilibrium conditions in the mechanics of continuum solids.

SUMMARY

In this chapter key concepts of virtual power and hence of stress are introduced. Equilibrium and boundary conditions arise naturally from the principle of virtual power. The definition is found to be equivalent to the concept of stress introduced using the Cauchy tetrahedron. The Beltrami potential representation of self-equilibrating stress fields is introduced. The role of the incompatibility operator in elastostatics is discussed.

EXERCISES

1. Balance of linear and angular momentum

Show that the balance of the linear and angular momentum for an Ω can be obtained as a direct consequence of the principle of virtual work using an infinitesimal rigid velocity field:

$$\boldsymbol{v} = \boldsymbol{a} + \boldsymbol{b} \times \boldsymbol{x}.$$

2. Mean stress

Let us consider a body of arbitrary shape Ω supporting a system of surface tractions defined by

$$\boldsymbol{t} = \boldsymbol{A} \cdot \boldsymbol{n},$$

where $\boldsymbol{A}$ is a given symmetric second-order tensor.

(a) Define the conditions for $\boldsymbol{A}$ in order to ensure equilibrium of the body.

(b) Show by applying the principle of virtual work with the appropiate virtual velocity field that mean stress equals $\mathbf{A}$,

$$\frac{1}{|\Omega|} \int_{\Omega} \boldsymbol{\sigma} \, dv = \boldsymbol{A},$$

where $|\Omega|$ denotes the volume of Ω.

3. Stress transformation between coordinate systems

The transformation of coordinates, as well as scalar, vector, and tensor fields, between different coordinate systems is implemented in MATHEMATICA using the commands `CoordinatesFromCartesian`, `FieldFromCartesian`, and so forth.

(a) Use the command `SetCoordinates[Cylindrical[r,t,z]]` to define a cylindrical polar coordinate system (r, θ, z). Determine the transformation between cylindrical polar coordinates and cartesian coordinates using `CoordinatesFromCartesian[x, y, z]`.

(b) Specify a stress state with respect to a cylindrical polar coordinate system and transform it into the corresponding cartesian coordinate system (x, y, z) using the command `FieldToCartesian`.

(c) Now define a spherical coordinate system using the command `SetCoordinates[Spherical[r,p,t]]`, and use `FieldFromCartesian` to obtain the stress tensor in the spherical coordinate system. Assuming a simple axisymmetric stress state, verify that the result obtained in this way is correct.

Hint: See notebook `C02_cyl_to_sph.nb.`

4. Stress balance on infinitesimal volume elements (I)

The purpose of this exercise is to construct the stress balance equations for an infinitesimal volume element $\Omega = [x - dx, x + dx] \times [y - dy, y + dy] \times [z - dz, z + dz]$ in cartesian coordinates (x, y, z) following classical reasoning using a series expansion of the stress tensor and the balance of tractions acting on the surface of Ω.

(a) Compute the stress tensor at middle points on the faces of an infinitesimal cubic volume Ω using first-order series expansion (MATHEMATICA operator `Series`) .

(b) Using the constant midface values for each face, compute traction vectors acting on each elementary face and the corresponding linear momentum values.

(c) Compute the total linear momentum of all traction vectors.

(d) Compute $8 \times (\operatorname{div} \boldsymbol{\sigma}) dx \, dy \, dz$ and compare with the preceeding result. Comment on the observation.

Hint: See notebook `C02_balance_cartesian.nb.`

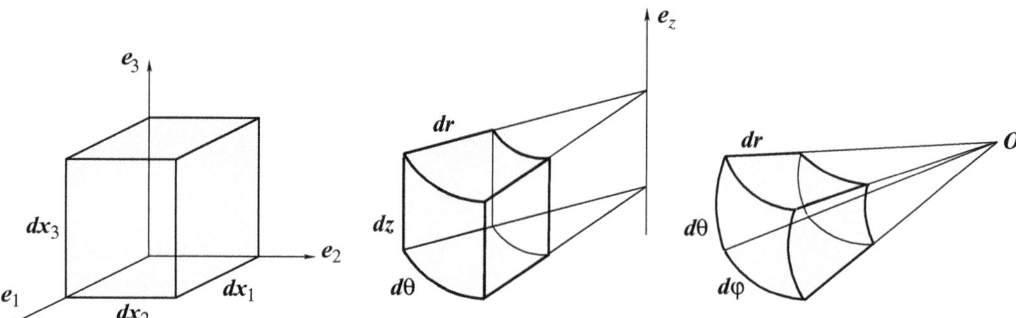

Figure 2.2. Volume elements in cartesian, cylindrical, and spherical coordinate systems.

5. Stress balance on infinitesimal volume elements (II)

Perform the same exercise as before for the following cases:

(a) in cylindrical coordinates (r, θ, z) by considering the infinitesimal volume $\Omega = [r - dr, r + dr] \times [\theta - d\theta, \theta + d\theta] \times [z - dz, z + dz]$.

(b) in spherical coordinates (r, θ, ϕ) by considering the infinitesimal volume $\Omega = [r - dr, r + dr] \times [\theta - d\theta, \theta + d\theta] \times [\phi - d\phi, \phi + d\phi]$.

See, Figure 2.2.

Hint: See notebook `C02_balance_cylindrical.nb`.

6. Stress tensors in a cylindrical pipe

Consider a tubular cylinder $\Omega = [R_i, R_e] \times [0, 2\pi] \times [0, H]$ in the cylindrical coordinate system (r, θ, z).

The tube is subject to a series of surface tractions: inner and outer pressures p_i and p_e, respectively, on the cylindrical surfaces, and pressure f on both end sections.

(a) Justify why the tube is in equilibrium state.

(b) Write down boundary conditions and balance equations for the stress tensor $\boldsymbol{\sigma}$.

(c) Simplify the preceding system of partial differential equations under the assumptions of cylindrical symmetry. Show that this system admits an infinite number of solutions.

(d) What can be concluded about the uniqueness of the stress field?

(e) Use the principle of virtual work with an appropiate virtual velocity field to compute the mean value of $\sigma_{\theta\theta}$.

Hint: Show that for each function σ_{rr} satisfying boundary conditions, a complete solution of the equations can be obtained.

7. Stress tensors in a spherical shell

Consider a spherical shell $\Omega = [R_i, R_e] \times [0, \pi] \times [0, 2\pi]$ in the spherical coordinate system (r, θ, φ).

The tube is subject to pressures p_i and p_e on the inner and outer surfaces, respectively.

(a) Justify why the shell is in equilibrium state.

(b) Write down boundary conditions and balance equations for the stress tensor $\boldsymbol{\sigma}$.

(c) Simplify the preceding system of partial differential equations under the assumptions of spherical symetry. Show that this system admits an infinite number of solutions.

(d) What can be concluded about the uniqueness of the stress field?

Hint: See notebook `C02_stress_spherical.nb`.

8. Polynomial Airy stress function Timoshenko and Goodier (1951)

Consider a rectangular domain $\Omega = [-a, a] \times [-b, b] \times -c, c]$. Using MATHEMATICA allows one to explore body forces and surface traction in equilibrium with the stress field defined by Airy stress potentials in the form of polynomials of order m:

$$\phi_m = \sum_{i=0}^{m} \alpha_i \, x^i y^{m-i}.$$

Carry out the calculations for the cases $m = 2, 3, 4$, and provide a physical interpretation of the individual boundary loading patterns.

9. Local stress representation using Mohr's circles (Salencon, 2001; Timoshenko, 1951)

Assume that the stress tensor in the neighbourhood of point $\boldsymbol{x} \in \Omega$ considered with respect to a cartesian coordinate system and expressed in the matrix form appears to be a diagonal matrix; that is,

$$\boldsymbol{\sigma} = \sigma_1 \, \boldsymbol{e}_1 \otimes \boldsymbol{e}_1 + \sigma_2 \, \boldsymbol{e}_2 \otimes \boldsymbol{e}_2 + \sigma_3 \, \boldsymbol{e}_3 \otimes \boldsymbol{e}_3,$$

such that $\sigma_i \in \mathbb{R}, i = 1, 2, 3$, are distinct from each other.

(a) Consider an infinitesimal surface element defined by the unit normal

$$\boldsymbol{n} = n_1 \, \boldsymbol{e}_1 + n_2 \, \boldsymbol{e}_2 + n_3 \, \boldsymbol{e}_3, \qquad n_1^2 + n_2^2 + n_3^2 = 1,$$

passing through point $\boldsymbol{x}$. Compute the traction vector

$$\boldsymbol{t} = \boldsymbol{\sigma} \cdot \boldsymbol{n}$$

as well as the normal stress

$$\sigma = \boldsymbol{n} \cdot \boldsymbol{\sigma} \cdot \boldsymbol{n}$$

and the tangential shear stress

$$\tau^2 + \sigma^2 = |\boldsymbol{t}|^2.$$

The normal and the tangential shear stress are the lengths of the normal and tangential components of the traction vector with respect to the plane defined by the normal $\boldsymbol{n}$ (Figure 2.3).

(b) Show that the preceeding equations can be assembled into the following system of equations:

$$n_1^2 = \frac{\tau^2 + (\sigma - \sigma_2)(\sigma - \sigma_3)}{(\sigma_1 - \sigma_2)(\sigma_1 - \sigma_3)}$$

$$n_2^2 = \frac{\tau^2 + (\sigma - \sigma_3)(\sigma - \sigma_1)}{(\sigma_2 - \sigma_3)(\sigma_2 - \sigma_1)}$$

$$n_3^2 = \frac{\tau^2 + (\sigma - \sigma_1)(\sigma - \sigma_2)}{(\sigma_3 - \sigma_1)(\sigma_3 - \sigma_2)}.$$

(c) Using the `ParametricMesh.m` package provided with this book and described in Appendix 3, plot the domain spanned by the normal and shear stresses (σ, τ) on the

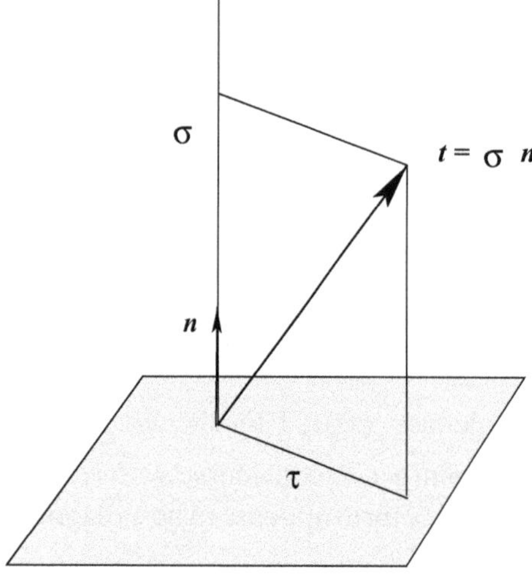

Figure 2.3. Traction vector $\boldsymbol{t} = \boldsymbol{\sigma} \cdot \boldsymbol{n}$ on an infinitesimal suface element with unit normal $\boldsymbol{n}$, and its normal and shear components.

(σ, τ) plane, also referred to as *Mohr's plane* when $\boldsymbol{n}$ covers all possible orientations in $\mathbb{R}^3$, that is, its tip moves across the entire surface of the unit sphere (Figure 2.4).

(d) Commands used in **`ParametricMesh.m`** use colour function that is defined with respect to the coordinates of the "deformed" mesh. Modify the commands presented in Appendix 3 so that the colour corresponding to the position of a normal vector is determined as a function of the spherical coordinate angles (θ, φ):

$$\boldsymbol{n}(\theta, \varphi) = \cos\varphi \sin\theta \boldsymbol{e}_1 + \sin\varphi \sin\theta \boldsymbol{e}_2 + \cos\theta \boldsymbol{e}_3.$$

(e) Assuming that $\sigma_1 < \sigma_2 < \sigma_3$ and using $n_i^2 \geq 0, i = 1, 2, 3$, conclude that

$$\tau^2 + (\sigma - \sigma_2)(\sigma - \sigma_3) \geq 0$$
$$\tau^2 + (\sigma - \sigma_3)(\sigma - \sigma_1) \leq 0$$
$$\tau^2 + (\sigma - \sigma_1)(\sigma - \sigma_2) \geq 0.$$

Consider and interpret the plot obtained with Mathematica (Figure 2.4) and deduce a formal proof.

Hint: See notebook: **`C02_mohr_circles.nb`**.

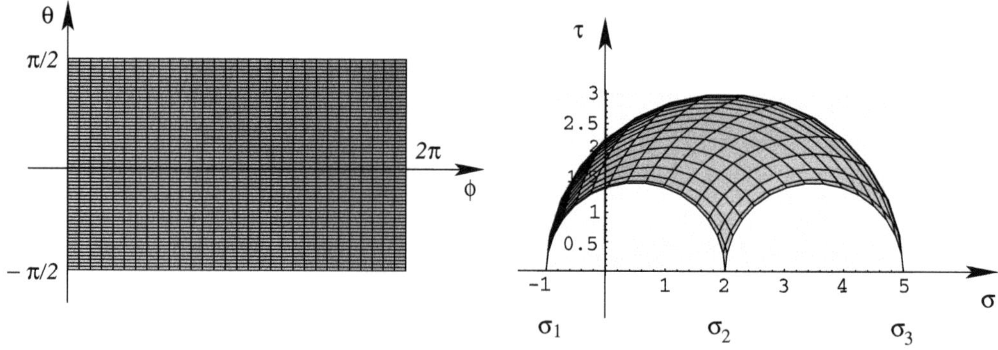

Figure 2.4. The domain of the spherical coordinate angles defining unit normals $\boldsymbol{n}(\theta, \varphi)$ (left panel) and the domain covered in the Mohr plane by the normal and shear stress (σ, τ) (right panel) .

10. Stress balance and the Beltrami potential

Verify using MATHEMATICA that a stress tensor derived from a Beltrami potential,

$$\boldsymbol{\sigma} = \operatorname{inc} \boldsymbol{B},$$

is divergence-free, that is,

$$\operatorname{div} \boldsymbol{\sigma} = 0.$$

Is it necessary to check this relation for different coordinate systems?

3 Linear elasticity

OUTLINE

In the preceding chapters we have discussed, on the one hand, the *kinematics* of deformation of continuous media, where the principal unknowns are the displacement vector field $\boldsymbol{u}$ and the strain tensor field $\boldsymbol{\epsilon}$. On the other hand, we have introduced the *dynamics* of deformation, representing the balance of forces in terms of the stress tensor field $\boldsymbol{\sigma}$ as the principal unknown.

Until now we have made no attempt to relate the strain and stress fields to each other. Before we begin the discussion of the detailed nature of this relationship, we can make the following general remarks:

- Description so far is clearly incomplete, because we have at our disposal only 6 kinematic relations and 3 force balance equations for the determination of the $3 + 6 + 6$ unknown functions, that is, the components of displacement $\boldsymbol{u}$, strain $\boldsymbol{\epsilon}$, and stress $\boldsymbol{\sigma}$.
- We are so far unable to distinguish between different materials which might assume different deformed configurations under the same external loading. Clearly structures produced out of wood, steel, or ceramic may deform in different ways, so that the complete solutions are different.

The purpose of this chapter is to establish a class of relationships between strains and stresses known as the linear elastic constitutive law and to discuss a series of basic properties of these relationships.

3.1 LINEAR ELASTICITY

We shall simply assume for now that there exists a linear relationship between the components of stress and strain tensors at each point in the body. This observation is well supported by experimental evidence, provided that the strains remain sufficiently small. We shall refer to this theoretical framework as *linear elasticity* and refer to the relationship between stresses and strains as *Hooke's law*.

In a given orthonormal basis the linear elastic stress–strain relastionship is expressed as

$$\sigma_{ij} = C_{ijkl}\epsilon_{kl}, \tag{3.1}$$

where C_{ijkl} can be referred to as elastic stiffnesses, elastic constants, or elastic moduli.

We recall that all lower case Latin subscripts assume the values 1,2,3 and that summation over a repeated index is implied, which is known as the Einstein summation convention. In the above equation this applies to the indices k and l on the right hand side, so that, for example, for component σ_{23} the expanded form is

$$\sigma_{23} = C_{2311}\varepsilon_{11} + C_{2312}\varepsilon_{12} + C_{2313}\varepsilon_{13} + C_{2321}\varepsilon_{21} + \dots. \tag{3.2}$$

Equation (3.1) can also be expressed in tensorial form as

$$\boldsymbol{\sigma} = \boldsymbol{C} : \boldsymbol{\varepsilon}, \tag{3.3}$$

where $\boldsymbol{C}$ is the fourth rank order of the *elastic moduli*, also known as *Hooke's tensor*. The colon between the two tensors represents the 'double dot' operator, which involves convolution (product and summation) over two pairs of indices. The corresponding MATHEMATICA operator `DDot` can be constructed out of the operators `Dot` and `Tr` (or their generalised counterparts `GDot` and `GTr`) (Section 3.5).

Hooke's tensor can be written with respect to a cartesian coordinate system in tensorial form as

$$\boldsymbol{C} = C_{ijkl}\, \boldsymbol{e}_i \otimes \boldsymbol{e}_j \otimes \boldsymbol{e}_k \otimes \boldsymbol{e}_l. \tag{3.4}$$

We remark that, as $\boldsymbol{C}$ is a fourth rank tensor in the three-dimensional space, it will possess $3 \times 3 \times 3 \times 3 = 81$ components. However, due to the symmetry of the strain tensor $\boldsymbol{\varepsilon}$, $\varepsilon_{kl} = \varepsilon_{lk}$, we may assume for simplicity that

$$C_{ijkl} = C_{ijlk}.$$

Similarly, due to the symmetry of the stress tensor $\boldsymbol{\sigma}$, $\sigma_{kl} = \sigma_{lk}$, we can assume that

$$C_{ijkl} = C_{jikl}.$$

These two symmetry relations reduce the number of independent components of $\boldsymbol{C}$ from the initial 81 to $6 \times 6 = 36$.

Moreover, we shall impose a third symmetry relation:

$$C_{ijkl} = C_{klij}.$$

This condition expresses the statement that for every pair of strain tensors we have

$$\boldsymbol{\varepsilon}_1 : \boldsymbol{C} : \boldsymbol{\varepsilon}_2 = \boldsymbol{\varepsilon}_2 : \boldsymbol{C} : \boldsymbol{\varepsilon}_1.$$

The physical meaning of this statement is that no net mechanical work is done in a closed loading cycle and it will be further discussed in Chapter 4. This symmetry requirement ensures the existence of an elastic energy potential. As a consequence of this final symmetry statement we remark that in the most general case $\boldsymbol{C}$ may have the maximum number of 21 independent components (see next section).

Moreover, the existence of the elastic energy potential, together with the fact that the body cannot instantenously release energy to the environment, imposes the hypothesis of material stability, which translates into the positive definiteness of $\boldsymbol{C}$:

$$\boldsymbol{\varepsilon} : \boldsymbol{C} : \boldsymbol{\varepsilon} \geq 0 \qquad \forall \boldsymbol{\varepsilon}.$$

Instead of representing the stresses as linear combinations of strains, we could have formulated their linear relationship the opposite way around, by putting

$$\varepsilon_{ij} = S_{ijkl}\sigma_{kl} \qquad \boldsymbol{\varepsilon} = \boldsymbol{S} : \boldsymbol{\sigma} \tag{3.5}$$

$$\boldsymbol{S} = S_{ijkl}\,\boldsymbol{e}_i \otimes \boldsymbol{e}_j \otimes \boldsymbol{e}_k \otimes \boldsymbol{e}_l, \tag{3.6}$$

where $\boldsymbol{S}$ denotes the fourth-order tensor of *elastic compliances*.

It is accepted within the linear elastic framework that the strains and stresses are connected through a one-to-one relation. The operation of re-expressing the stresses as a linear combination of strains is equivalent to the inversion of the linear system (3.1). The elastic compliance tensor $\boldsymbol{S}$ is therefore the inverse tensor of $\boldsymbol{C}$; that is, $\boldsymbol{S} = \boldsymbol{C}^{-1}$.

By repeating the arguments of symmetry, similar requirements can be imposed on the compliance tensor,

$$S_{ijkl} = S_{klij} = S_{jikl}, \tag{3.7}$$

thus establishing that $\boldsymbol{S}$ also has the maximum of 21 independent components; and its positive definiteness:

$$\boldsymbol{\sigma} : \boldsymbol{S} : \boldsymbol{\sigma} \geq 0 \qquad \forall\boldsymbol{\sigma}.$$

3.2 MATRIX REPRESENTATION OF ELASTIC COEFFICIENTS

The above representation of the elastic coefficients as fourth rank tensors is the most natural way of introducing the linear dependence between the second rank stress and strain tensors.* However, it is not particularly convenient for visualising the relationships imposed by Hooke's law or the effect of special material symmetries on its form.

Instead, the relationships between the components of stress and strain can be expressed by collecting the terms and eliminating the redundant equations due to symmetry, leading to the following system:

$$\begin{pmatrix} \sigma_{11} \\ \sigma_{22} \\ \sigma_{33} \\ \sigma_{23} \\ \sigma_{31} \\ \sigma_{12} \end{pmatrix} = \begin{pmatrix} C_{1111} & C_{1122} & C_{1133} & C_{1123} & C_{1131} & C_{1112} \\ C_{1122} & C_{2222} & C_{2233} & C_{2223} & C_{2231} & C_{2212} \\ C_{1133} & C_{2233} & C_{3333} & C_{3323} & C_{3331} & C_{3312} \\ C_{1123} & C_{2233} & C_{3323} & C_{2323} & C_{2331} & C_{2312} \\ C_{1131} & C_{2231} & C_{3323} & C_{2323} & C_{3131} & C_{3112} \\ C_{1112} & C_{2212} & C_{3323} & C_{2312} & C_{3112} & C_{1212} \end{pmatrix} \cdot \begin{pmatrix} \varepsilon_{11} \\ \varepsilon_{22} \\ \varepsilon_{33} \\ 2\,\varepsilon_{23} \\ 2\,\varepsilon_{31} \\ 2\,\varepsilon_{12} \end{pmatrix}. \tag{3.8}$$

This method establishes the link between the 6-vector of stress and the 6-vector of strain via a 6×6 symmetric matrix. The above expression demonstrates explicitly how the 21 independent components of the tensor $\boldsymbol{C}$ populate this matrix.

It is now convenient to replace with a single index each of the first and second pairs of indices of C_{ijkl} that identify the stress and strain components. We introduce the following set of rules:

$$11 \leftrightarrow 1 \quad 22 \leftrightarrow 2 \quad 33 \leftrightarrow 3 \quad 23 \leftrightarrow 4 \quad 31 \leftrightarrow 5 \quad 12 \leftrightarrow 6.$$

* A complete discussion of matrix representation, as well as the issues of material symmetry, is given in Cowin and Mehrabadi (1990, 1992).

Table 3.1. *Relationship between strains and stresses in the fourth rank, Voigt, and second rank notation*

Fourth rank tensor		Voigt		Second rank tensor	Fourth rank tensor		Voigt		Second rank tensor
ε_{11}	=	ε_1	=	$\hat{\varepsilon}_1$	σ_{11}	=	σ_1	=	$\hat{\sigma}_1$
ε_{22}	=	ε_2	=	$\hat{\varepsilon}_2$	σ_{22}	=	σ_2	=	$\hat{\sigma}_2$
ε_{33}	=	ε_3	=	$\hat{\varepsilon}_3$	σ_{33}	=	σ_3	=	$\hat{\sigma}_3$
ε_{23}	=	$\frac{1}{2}\varepsilon_4$	=	$\frac{1}{\sqrt{2}}\hat{\varepsilon}_4$	σ_{23}	=	σ_4	=	$\frac{1}{\sqrt{2}}\hat{\sigma}_4$
ε_{31}	=	$\frac{1}{2}\varepsilon_5$	=	$\frac{1}{\sqrt{2}}\hat{\varepsilon}_5$	σ_{31}	=	σ_5	=	$\frac{1}{\sqrt{2}}\hat{\sigma}_5$
ε_{12}	=	$\frac{1}{2}\varepsilon_6$	=	$\frac{1}{\sqrt{2}}\hat{\varepsilon}_6$	σ_{12}	=	σ_6	=	$\frac{1}{\sqrt{2}}\hat{\sigma}_6$

Thus for example

$$C_{1231} \leftrightarrow c_{31} \quad C_{2233} \leftrightarrow c_{23},$$

so that the system is rewritten as

$$\begin{pmatrix} \sigma_{11} \\ \sigma_{22} \\ \sigma_{33} \\ \sigma_{23} \\ \sigma_{31} \\ \sigma_{12} \end{pmatrix} = \begin{pmatrix} c_{11} & c_{12} & c_{13} & c_{14} & c_{15} & c_{16} \\ c_{21} & c_{22} & c_{23} & c_{24} & c_{25} & c_{26} \\ c_{31} & c_{32} & c_{33} & c_{34} & c_{35} & c_{36} \\ c_{41} & c_{42} & c_{43} & c_{44} & c_{45} & c_{46} \\ c_{51} & c_{52} & c_{53} & c_{54} & c_{55} & c_{56} \\ c_{61} & c_{62} & c_{63} & c_{64} & c_{65} & c_{66} \end{pmatrix} \cdot \begin{pmatrix} \varepsilon_{11} \\ \varepsilon_{22} \\ \varepsilon_{33} \\ 2\varepsilon_{23} \\ 2\varepsilon_{31} \\ 2\varepsilon_{12} \end{pmatrix}. \tag{3.9}$$

According to this convention, known as the *Voigt notation*, single index notation is also introduced for strains and stresses. The last relation becomes

$$\begin{pmatrix} \sigma_1 \\ \sigma_2 \\ \sigma_3 \\ \sigma_4 \\ \sigma_5 \\ \sigma_6 \end{pmatrix} = \begin{pmatrix} c_{11} & c_{12} & c_{13} & c_{14} & c_{15} & c_{16} \\ c_{21} & c_{22} & c_{23} & c_{24} & c_{25} & c_{26} \\ c_{31} & c_{32} & c_{33} & c_{34} & c_{35} & c_{36} \\ c_{41} & c_{42} & c_{43} & c_{44} & c_{45} & c_{46} \\ c_{51} & c_{52} & c_{53} & c_{54} & c_{55} & c_{56} \\ c_{61} & c_{62} & c_{63} & c_{64} & c_{65} & c_{66} \end{pmatrix} \cdot \begin{pmatrix} \varepsilon_1 \\ \varepsilon_2 \\ \varepsilon_3 \\ 2\varepsilon_4 \\ 2\varepsilon_5 \\ 2\varepsilon_6 \end{pmatrix}. \tag{3.10}$$

There are two major drawbacks of the Voigt notation. First, it is apparent that single-index stresses and strains enter the 6-vector object differently (note the factor of 2 that appears in front of the shear strain components). Second, the 6-vector and matrix objects created in this way are no longer tensors, and therefore the transformation of coordinate systems and differential equations cannot be accomplished in a natural way in this notation.

The situation can be corrected if, instead of an arbitrary choice of transition from second rank tensors to 6-vectors, one chooses an *orthonormal* vectorial base in the 6-dimensional vector space. This will lead to associated symmetric matrices and give rise to the so-called *second-rank tensor* notation. See Table 3.1 for a comparison of these notations.

The following vectors are naturally associated with the set of basis tensors in the linear space of symmetric second-order tensors:

$$\boldsymbol{v}_1 = \boldsymbol{e}_1 \otimes \boldsymbol{e}_1 \tag{3.11}$$

$$\boldsymbol{v}_2 = \boldsymbol{e}_2 \otimes \boldsymbol{e}_2 \tag{3.12}$$

$$\boldsymbol{v}_3 = \boldsymbol{e}_3 \otimes \boldsymbol{e}_3 \tag{3.13}$$

$$\boldsymbol{v}_4 = \boldsymbol{e}_2 \otimes \boldsymbol{e}_3 + \boldsymbol{e}_3 \otimes \boldsymbol{e}_2 \tag{3.14}$$

$$\boldsymbol{v}_5 = \boldsymbol{e}_3 \otimes \boldsymbol{e}_1 + \boldsymbol{e}_1 \otimes \boldsymbol{e}_3 \tag{3.15}$$

$$\boldsymbol{v}_6 = \boldsymbol{e}_1 \otimes \boldsymbol{e}_2 + \boldsymbol{e}_2 \otimes \boldsymbol{e}_1. \tag{3.16}$$

The vectors in this basis span the complete space of second rank tensors and are linearly independent. However, the natural norm for some of them is different from unity:

$$|\,\boldsymbol{v}_4\,| = |\,\boldsymbol{v}_5\,| = |\,\boldsymbol{v}_6\,| = \sqrt{2}. \tag{3.17}$$

Renormalising these vectors, we can now develop a consistent second rank representation using the orthonormal basis for both strains and stresses:

$$\boldsymbol{v}_1 = \boldsymbol{e}_1 \otimes \boldsymbol{e}_1 \tag{3.18}$$

$$\boldsymbol{v}_2 = \boldsymbol{e}_2 \otimes \boldsymbol{e}_2 \tag{3.19}$$

$$\boldsymbol{v}_3 = \boldsymbol{e}_3 \otimes \boldsymbol{e}_3 \tag{3.20}$$

$$\boldsymbol{v}_4 = \frac{1}{\sqrt{2}}(\boldsymbol{e}_2 \otimes \boldsymbol{e}_3 + \boldsymbol{e}_3 \otimes \boldsymbol{e}_2) \tag{3.21}$$

$$\boldsymbol{v}_5 = \frac{1}{\sqrt{2}}(\boldsymbol{e}_3 \otimes \boldsymbol{e}_1 + \boldsymbol{e}_1 \otimes \boldsymbol{e}_3) \tag{3.22}$$

$$\boldsymbol{v}_6 = \frac{1}{\sqrt{2}}(\boldsymbol{e}_1 \otimes \boldsymbol{e}_2 + \boldsymbol{e}_2 \otimes \boldsymbol{e}_1). \tag{3.23}$$

Hooke's law in this basis is written in the form

$$\begin{pmatrix} \sigma_{11} \\ \sigma_{22} \\ \sigma_{33} \\ \sqrt{2}\sigma_{23} \\ \sqrt{2}\sigma_{31} \\ \sqrt{2}\sigma_{12} \end{pmatrix} = \begin{pmatrix} C_{1111} & C_{1122} & C_{1133} & \sqrt{2}C_{1123} & \sqrt{2}C_{1131} & \sqrt{2}C_{1112} \\ C_{1122} & C_{2222} & C_{2233} & \sqrt{2}C_{2223} & \sqrt{2}C_{2231} & \sqrt{2}C_{2212} \\ C_{1133} & C_{2233} & C_{3333} & \sqrt{2}C_{3323} & \sqrt{2}C_{3331} & \sqrt{2}C_{3312} \\ \sqrt{2}C_{1123} & \sqrt{2}C_{2233} & \sqrt{2}C_{3323} & 2C_{2323} & 2C_{2331} & 2C_{2312} \\ \sqrt{2}C_{1131} & \sqrt{2}C_{2231} & \sqrt{2}C_{3323} & 2C_{2323} & 2C_{3131} & 2C_{3112} \\ \sqrt{2}C_{1112} & \sqrt{2}C_{2212} & \sqrt{2}C_{3323} & 2C_{2312} & 2C_{3112} & 2C_{1212} \end{pmatrix} \cdot \begin{pmatrix} \varepsilon_{11} \\ \varepsilon_{22} \\ \varepsilon_{33} \\ \sqrt{2}\varepsilon_{23} \\ \sqrt{2}\varepsilon_{31} \\ \sqrt{2}\varepsilon_{12} \end{pmatrix}, \tag{3.24}$$

which is equivalent to the *second rank tensor* notation

$$\begin{pmatrix} \hat{\sigma}_1 \\ \hat{\sigma}_2 \\ \hat{\sigma}_3 \\ \hat{\sigma}_4 \\ \hat{\sigma}_5 \\ \hat{\sigma}_6 \end{pmatrix} = \begin{pmatrix} \hat{c}_{11} & \hat{c}_{12} & \hat{c}_{13} & \hat{c}_{14} & \hat{c}_{15} & \hat{c}_{16} \\ \hat{c}_{21} & \hat{c}_{22} & \hat{c}_{23} & \hat{c}_{24} & \hat{c}_{25} & \hat{c}_{26} \\ \hat{c}_{31} & \hat{c}_{32} & \hat{c}_{33} & \hat{c}_{34} & \hat{c}_{35} & \hat{c}_{36} \\ \hat{c}_{41} & \hat{c}_{42} & \hat{c}_{43} & \hat{c}_{44} & \hat{c}_{45} & \hat{c}_{46} \\ \hat{c}_{51} & \hat{c}_{52} & \hat{c}_{53} & \hat{c}_{54} & \hat{c}_{55} & \hat{c}_{56} \\ \hat{c}_{61} & \hat{c}_{62} & \hat{c}_{63} & \hat{c}_{64} & \hat{c}_{65} & \hat{c}_{66} \end{pmatrix} \cdot \begin{pmatrix} \hat{\varepsilon}_1 \\ \hat{\varepsilon}_2 \\ \hat{\varepsilon}_3 \\ \hat{\varepsilon}_4 \\ \hat{\varepsilon}_5 \\ \hat{\varepsilon}_6 \end{pmatrix}. \tag{3.25}$$

In a similar way, the two different notations introduced above can similarly be used for the compliance tensor $\boldsymbol{S}$, leading to the expressions for compliance in the *Voigt* notation, $\boldsymbol{s}$, and the *second rank tensor* notation, $\hat{\boldsymbol{s}}$.

It is important to remark that in all representations the following relations hold between the matrices:

$$\boldsymbol{C} = \boldsymbol{S}^{-1}, \quad \boldsymbol{c} = \boldsymbol{s}^{-1}, \quad \hat{\boldsymbol{c}} = \hat{\boldsymbol{s}}^{-1}. \tag{3.26}$$

Mathematica Programmaing

Creation of tensors

`SymIndex[2,number], SymIndex[4]`	creates a list of symmetric indices for second or fourth rank list
`MakeName[myexp]`	joins strings over elements of lists
`MakeTensor[mystring, 2, dim]`	creates a second rank tensor with head `mystring`
`MakeTensor[mystring, 4]`	creates a fourth rank tensor with head `mystring`

Let us now explore how the foregoing definitions can be efficiently implemented using MATHEMATICA.

As a start, we create tensorial symbolic expressions which will automatically assign names to the components. Our interest lies in creating symmetric second- and fourth-order tensors in the forms presented in the previous section and defining procedures for the passage between different tensor notations, that is, the fourth rank tensor, Voigt, and second rank tensor notations (denoted by `4,V`, and `2`, respectively), first for the Hooke stiffness tensor, and then for its inverse, the compliance tensor.

`SymIndex` creates a list of indices of a second-order tensor in a space of dimension `dim`. An array of indices is created using `Array[List, Array[dim &, 2]]` and then rearranged, taking symmetry into account by applying the `Sort` command. Here `##` stands for `SlotSequence` and makes it possible to accept a sequence of objects as input in a `Pure Function`.

The `MakeName` command concatenates a list of strings and/or numbers to an expression.

The `MakeTensor` command combines the previous commands in order to create a tensorial expression having components which respect the symmetry conditions. Some examples illustrate the operation of this command.

```
SymIndex[2, dim_] :=
  Apply[(Sort[List[##]]&),Array[List,Array[dim &,2]],{2}]

SymIndex[2, 3] // MatrixForm
SymIndex[2, 6] // MatrixForm

SymIndex[4] =
  Apply[Flatten[Sort[Map[Sort, Partition[List[##], 2]]]] &,
```

```
   Array[ List, Array[3 &, 4]], 4]

SymIndex[4] // MatrixForm

MakeName[myexp__] := ToExpression //@ (StringJoin
              Map[ToString[#]  &, List[myexp]] )

MakeTensor[mystring_, 2,  dim_]   :=
      Apply[(MakeName[mystring, ##] &) , SymIndex[2, dim], {2}]

MakeTensor[mystring_, 4 ] :=
     Apply[(MakeName[mystring,  ##]  &) , SymIndex[4], {4}]]


( stress = MakeTensor["sig", 2, 3]) // MatrixForm
( strain = MakeTensor["eps", 2, 3]) // MatrixForm
( C4 = MakeTensor["C", 4] ) // MatrixForm
```

For programming the fourth-rank tensor to second-rank tensor transformation, we have chosen to define sets of rules for the forward and reverse substitution of indices between the second rank tensors of dimension 3 and vectors of dimension 6.

```
indexrule2to1 =  {{1, 1} -> 1, {2, 2} -> 2, {3, 3} -> 3,
     {2, 3} -> 4, {3, 1} -> 5, {1, 2} -> 6, {3, 2} -> 4,
     {1, 3} -> 5, {2, 1} -> 6}

indexrule1to2 = Map[Rule[#[[2]], #[[1]] ] & , indexrule2to1]

Index6[1] = Range[6] /. indexrule1to2
```

The passage between the fourth-rank tensors and the second-rank tensors is now easily accomplished by picking any desired form of tensor using the predefined `indexrule`.

The commands perform the following transformations:

- `HookeVto4` and `Hooke4toV` transform the Voigt notation into the fourth-order tensor notation and back
- `HookeVto2` and `Hooke2toV` transform the Voigt notation into the second-order tensor notation and back
- `Hooke4to2` and `Hooke2to4` transform the fourth-order into the second-order tensor notation, and back.

Note the use of factors 2 and $\sqrt{2}$ in the code that ensure that the tensorial form of the result is maintained whenever appropriate.

```
HookeVto4[ myC_] :=
  Array[  myC[[ {#1, #2}
     /. indexrule2to1, {#3, #4} /. indexrule2to1]] &,
    Array[3 &, 4]]

Hooke4toV[ myC_] :=
  Apply[ Part[ myC,##] &,
    Array[    Join[ #1
       /. indexrule1to2 , #2 /. indexrule1to2] &,
      Array[6 &, 2]] , 2]

Hooke2toV[myc2_] := Table[
      Which[
        i <= 3 && j <= 3 , myc2[[i, j]],
        4 <= i && j <= 3 , myc2[[i, j]]/ 2^(1/2),
        i <= 3 && 4 <= j , myc2[[i, j]]/ 2^(1/2),
        4 <= i && 4 <= j , myc2[[i, j]]/ 2
             ], i, 6, j, 6]

HookeVto2[mycV_] := Table[
      Which[
        i <= 3 && j <= 3 , mycV[[i, j]],
        4 <= i && j <= 3 , mycV[[i, j]] * 2^(1/2),
        i <= 3 && 4 <= j , mycV[[i, j]] * 2^(1/2),
        4 <= i && 4 <= j , mycV[[i, j]] * 2
             ], i, 6, j, 6]

Hooke4to2[myC_] := HookeVto2[Hooke4toV[ myC ]]
Hooke2to4[myC_] := HookeVto4[Hooke2toV[ myC ]]
```

The correctness of **Hooke_to_** definitions can now be checked by creating a tensor and exploring its appearance in different notations.

MatrixForm makes it possible to check the result in a convenient way, even when applied to fourth-order tensors.

```
(C2 = MakeTensor["C", 2, 6]) // MatrixForm
 C2to4 = HookeVto4[C2]
 C2back = Hooke4toV[C2to4]

 C4to2 = Hooke4toV[C4]
```

```
 C4back = HookeVto4[C4to2]
 CV = Hooke2toV[C2]

C2back = HookeVto2[CV]
CV = Hooke4to2[C4]
C4back = Hooke2to4[CV]
```

A similar set of transformations can easily be defined for the compliance tensor.

```
Compliance2toV[mys2_] := Hooke2toV[mys2]
ComplianceVto2[mys2_] := HookeVto2[mys2]

Compliance4toV[myS4_] :=
    ComplianceVto2[ ComplianceVto2[ Hooke4toV[ myS4 ]]]
ComplianceVto4[mysV_] :=
    HookeVto4[ Compliance2toV[ Compliance2toV[ mysV]] ]

( S4 = MakeTensor["S", 4] ) // MatrixForm

( SV = MakeTensor["SV", 2, 6] ) // MatrixForm

Hooke4toV[ ComplianceVto4[ SV]] // MatrixForm

ComplianceVto2[ SV] // MatrixForm
```

To explore the linear elastic law freely we additionally define the left and right double dot product between a fourth-rank and a second-rank tensor.

Manipulation procedures can be used that make use of the generalised dot product command `GDot` described in Appendix 1.

```
CEDot[Ctensor_, straintensor_ ] :=
    GTr[ GDot[Ctensor, straintensor, 4, 1], 3, 4]
ECDot = CEDot

DDot[T4_, t2_] := GTr[GDot[T4, t2, 1, 1], 1, 4]
```

Another, more straightforward method employs the standard MATHEMATICA `Sum` command.

```
CEDot[Ctensor_, straintensor_ ] :=
 Table[
      Sum[ Ctensor[[i,j,k,l]] straintensor[[k,l]],
      {k,3},{l,3}],
```

```
  {i,3},{j,3}]

ECDot[Ctenso_, straintensor_ ] :=
  Table[ Sum[
    straintensor[[k,l]] Ctensor[[k,l,i,j]],
        {k,3},{l,3}],
      {i,3},{j,3}]
```

Manipulations shown here illustrate the definitions introduced in this chapter, namely, the expression of stresses as a function of strains and vice versa, and the symmetries of the elasticity tensor and of the stress tensor.

```
sigma = CEDot[C4, strain]
strain = CEDot[S4, stress]

Simplify[ CEDot[C4, strain] - ECDot[C4, strain] ]
Simplify[ sigma - Transpose[sigma]]
```

Second rank, Voigt and fourth rank tensor notations

`indexrule2to1` `indexrule1to2`	changes indices from 6-vector to 3-matrix
`Hooke2to4[C]`	changes elasticity tensor from second-rank to fourth-rank tensor notation
`HookeVto4[C]`	changes elasticity tensor from Voigt to fourth rank tensor notation
...	
`Compliance2to4[C]`	changes compliance tensor from second-rank to fourth-rank tensor notation
`ComplianceVto4[C]`	changes compliance tensor from Voigt to fourth-rank tensor notation
...	

3.3 MATERIAL SYMMETRY

Elastic moduli C_{ijkl} relating the cartesian components of strains and stresses depend of the orientation of the coordinate system with respect to the orientation of the body.

If the values of all elastic moduli are equal for any two different cartesian coordinate system orientations we shall speak of *material elastic symmetry*.

More precisely, let us suppose that $(\boldsymbol{e}_1, \boldsymbol{e}_2, \boldsymbol{e}_3)$ and $(\boldsymbol{g}_1, \boldsymbol{g}_2, \boldsymbol{g}_3)$ define two systems of cartesian coordinates related through

$$\boldsymbol{e}_i = R_{ij}\boldsymbol{g}_j, \tag{3.27}$$

where the transformation (rotation or reflection) matrix R is

$$R_{ij} = \boldsymbol{e}_i \cdot \boldsymbol{e}_j \quad \text{with} \quad \det[R_{ij}] = \pm 1. \tag{3.28}$$

Then the components of $\boldsymbol{C}$ are written as

$$\boldsymbol{C} = C^g_{ijkl}\boldsymbol{g}_i \otimes \boldsymbol{g}_j \otimes \boldsymbol{g}_k \otimes \boldsymbol{g}_l \tag{3.29}$$

$$\boldsymbol{C} = C^e_{pqrs}\boldsymbol{e}_p \otimes \boldsymbol{e}_q \otimes \boldsymbol{e}_r \otimes \boldsymbol{e}_s \tag{3.30}$$

$$= C^e_{pqrs} R_{pi} R_{qj} R_{rk} R_{sl}\boldsymbol{g}_i \otimes \boldsymbol{g}_j \otimes \boldsymbol{g}_k \otimes \boldsymbol{g}_l. \tag{3.31}$$

The material symmetry demands therefore that

$$C_{ijkl} = C_{pqrs} R_{pi} R_{qj} R_{rk} R_{sl}. \tag{3.32}$$

These relations impose additional restrictions on the elastic moduli and further reduce the number of independent components that describe the elastic properties of materials that conform to the additionally imposed symmetry requirements.

The set of transformations $\boldsymbol{R}$ defining the material symmetry of the elastic body correspond to an algebraic group generally denoted by $\mathcal{G}(\boldsymbol{R})$. Group $\mathcal{G}$ is a subgroup of the group of orthogonal transformations.

It can be shown (Chadwick et al., [2001]) using rigorous mathematical reasoning that there are precisely eight groups of material symmetry. Without going into the details of the mathematical proof, we present these classes below (see Figure 3.1 for illustration).

Triclinic elastic materials

For the vast majority of materials we can assume that in terms of linear elastic properties, no difference can be detected between the positive and negative senses of a certain direction, say $+x$ and $-x$. We therefore assume that all materials respect the material symmetry relation (3.32) if the group of symmetry transformations is defined as

$$\mathcal{G} = \{\boldsymbol{I}, -\boldsymbol{I}\}.$$

In this case no further restrictions can be imposed on the elastic moduli. They can be displayed in the form of a symmetric 6×6 matrix (dots denote symmetry)

$$\begin{pmatrix} C_{11} & C_{12} & C_{13} & C_{14} & C_{15} & C_{16} \\ \cdot & C_{22} & C_{23} & C_{24} & C_{25} & C_{26} \\ \cdot & \cdot & C_{33} & C_{34} & C_{35} & C_{36} \\ \cdot & \cdot & \cdot & C_{44} & C_{45} & C_{46} \\ \cdot & \cdot & \cdot & \cdot & C_{55} & C_{56} \\ \cdot & \cdot & \cdot & \cdot & \cdot & C_{66} \end{pmatrix}. \tag{3.33}$$

Monoclinic elastic materials

Materials with only one plane of reflexion (material symmetry) are called *monoclinic materials*.

Generally we shall denote by

$$\mathcal{N} = \{\boldsymbol{n}_1, \ldots, \boldsymbol{n}_i, \ldots\}$$

the set of normals to the planes of reflexion.

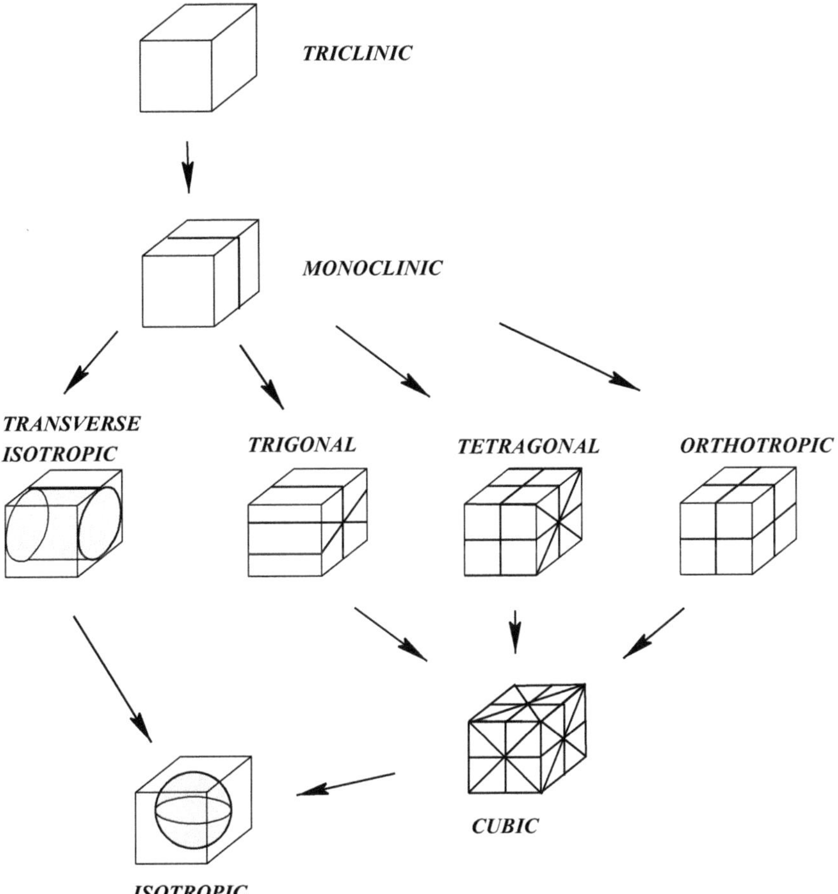

Figure 3.1. The classes of material symmetry represented as symmetry planes for an elementary volume, and the relations between these classes.

In the present case of monoclinic materials only one plane of symmetry is present. The number of independent stiffnesses is 13 in a basis oriented so that the mirror plane normal $\boldsymbol{n}_1$ coincides with one of the basis vectors ($\boldsymbol{e}_2$ or $\boldsymbol{e}_3$). For the stiffness matrix components with respect to this vector basis, we obtain

$$\text{if} \quad \mathcal{N} = \{\boldsymbol{e}_2\} \quad \text{then} \quad C_{14} = C_{16} = C_{24} = C_{26} = C_{34} = C_{36} = C_{45} = C_{56} = 0 \tag{3.34}$$

$$\text{if} \quad \mathcal{N} = \{\boldsymbol{e}_3\} \quad \text{then} \quad C_{14} = C_{15} = C_{24} = C_{25} = C_{34} = C_{35} = C_{46} = C_{56} = 0. \tag{3.35}$$

Tetragonal elastic materials

This type of material possesses five mirror planes defined by five vector normals. Of these normals four are coplanar and at angles of $\pi/4$ to each other, whereas the fifth normal vector is perpendicular to this plane. This type of material is called tetragonal. We select the set of normals

$$\mathcal{N} = \left\{\boldsymbol{e}_2, \frac{1}{\sqrt{2}}(\boldsymbol{e}_2 + \boldsymbol{e}_3), \boldsymbol{e}_3, \frac{1}{\sqrt{2}}(-\boldsymbol{e}_2 + \boldsymbol{e}_3), \boldsymbol{e}_1\right\} \tag{3.36}$$

and obtain the set of elastic stiffnesses

$$\begin{pmatrix} C_{11} & C_{12} & C_{13} & 0 & 0 & 0 \\ \cdot & C_{11} & C_{13} & 0 & 0 & 0 \\ \cdot & \cdot & C_{33} & 0 & 0 & 0 \\ \cdot & \cdot & \cdot & C_{44} & 0 & 0 \\ \cdot & \cdot & \cdot & \cdot & C_{44} & 0 \\ \cdot & \cdot & \cdot & \cdot & \cdot & C_{66} \end{pmatrix}. \tag{3.37}$$

Trigonal elastic materials

An elastic material with three coplanar mirror plane normals separated by an angle of $\frac{\pi}{3}$ is called trigonal.

Let the set of normals be given by

$$\mathcal{N} = \left\{ \boldsymbol{e}_2, \frac{1}{2}\left(\boldsymbol{e}_2 + \sqrt{3}\boldsymbol{e}_3\right), \frac{1}{2}\left(-\boldsymbol{e}_2 + \sqrt{3}\boldsymbol{e}_3\right) \right\}. \tag{3.38}$$

Then the symmetric matrix of the elastic stiffnesses is expressed as

$$\begin{pmatrix} C_{11} & C_{12} & C_{13} & 0 & C_{15} & 0 \\ \cdot & C_{22} & C_{23} & 0 & -C_{15} & 0 \\ \cdot & \cdot & C_{33} & 0 & 0 & 0 \\ \cdot & \cdot & \cdot & C_{44} & 0 & -C_{15} \\ \cdot & \cdot & \cdot & \cdot & C_{55} & 0 \\ \cdot & \cdot & \cdot & \cdot & \cdot & \frac{1}{2}(C_{11} - C_{12}) \end{pmatrix}. \tag{3.39}$$

Orthotropic elastic materials

We shall now consider *orthotropic* materials, for which the three mirror plane normals form an orthogonal basis. If these three normals coincide with the basis vectors of the cartesian system,

$$\mathcal{N} = \{\boldsymbol{e}_1, \boldsymbol{e}_2, \boldsymbol{e}_3\},$$

then the matrix of elastic stiffnesses has the form

$$\begin{pmatrix} C_{11} & C_{12} & C_{13} & 0 & 0 & 0 \\ \cdot & C_{22} & C_{23} & 0 & 0 & 0 \\ \cdot & \cdot & C_{33} & 0 & 0 & 0 \\ \cdot & \cdot & \cdot & C_{44} & 0 & 0 \\ \cdot & \cdot & \cdot & \cdot & C_{55} & 0 \\ \cdot & \cdot & \cdot & \cdot & \cdot & C_{66} \end{pmatrix}. \tag{3.40}$$

Transversely isotropic elastic materials

An elastic material is said to be *transversly isotropic* if a vector $\boldsymbol{e}$ and all vectors orthogonal to $\boldsymbol{e}$ are mirror plane normals. If $\boldsymbol{e} = \boldsymbol{e}_1$, then the stiffness matrix has the form

$$\begin{pmatrix} C_{11} & C_{12} & C_{13} & 0 & 0 & 0 \\ \cdot & C_{11} & C_{13} & 0 & 0 & 0 \\ \cdot & \cdot & C_{33} & 0 & 0 & 0 \\ \cdot & \cdot & \cdot & C_{44} & 0 & 0 \\ \cdot & \cdot & \cdot & \cdot & C_{44} & 0 \\ \cdot & \cdot & \cdot & \cdot & \cdot & \frac{1}{2}(C_{11} - C_{12}) \end{pmatrix}. \tag{3.41}$$

Rules for defining symmetry classes

`triclinic`	triclinic symmetry
`monoclinic2`	monoclinic symmetry along $\boldsymbol{e}_2$
`monoclinic3`	monoclinic symmetry along $\boldsymbol{e}_3$
`tetragonal`	tetragonal symmetry
`trigonal`	trigonal symmetry
`orthotropic`	orthotropic symmetry
`transverse`	transversely isotropic symmetry
`isotropy`	isotropic symmetry

Cubic elastic symmetry

An elastic material for which there are nine mirror plane normals coinciding with the edges and face diagonals of a cube is said to be *cubic*. The set of normals is given by

$$\mathcal{N} = \left\{\boldsymbol{e}_1, \boldsymbol{e}_2, \boldsymbol{e}_3, \frac{(\boldsymbol{e}_2 + \boldsymbol{e}_3)}{\sqrt{2}}, \frac{(-\boldsymbol{e}_2 + \boldsymbol{e}_3)}{\sqrt{2}}, \frac{(\boldsymbol{e}_3 + \boldsymbol{e}_1)}{\sqrt{2}}, \frac{(-\boldsymbol{e}_3 + \boldsymbol{e}_1)}{\sqrt{2}}, \frac{(\boldsymbol{e}_1 + \boldsymbol{e}_2)}{\sqrt{2}}, \frac{(-\boldsymbol{e}_1 + \boldsymbol{e}_2)}{\sqrt{2}}, \right\}. \tag{3.42}$$

The corresponding stiffness matrix is

$$\begin{pmatrix} C_{11} & C_{12} & C_{12} & 0 & 0 & 0 \\ \cdot & C_{11} & C_{12} & 0 & 0 & 0 \\ \cdot & \cdot & C_{11} & 0 & 0 & 0 \\ \cdot & \cdot & \cdot & C_{44} & 0 & 0 \\ \cdot & \cdot & \cdot & \cdot & C_{44} & 0 \\ \cdot & \cdot & \cdot & \cdot & \cdot & C_{44} \end{pmatrix}. \tag{3.43}$$

Isotropic elastic materials

An elastic material for which every direction can be identified with a mirror plane normal is called *isotropic*. In this case the material has no preferred direction in terms of deformation: all directions are entirely equal. Such a material has only two independent elastic moduli, and the stiffness matrix has the form

$$\begin{pmatrix} C_{11} & C_{12} & C_{12} & 0 & 0 & 0 \\ \cdot & C_{11} & C_{12} & 0 & 0 & 0 \\ \cdot & \cdot & C_{11} & 0 & 0 & 0 \\ \cdot & \cdot & \cdot & \frac{1}{2}(C_{11}-C_{12}) & 0 & 0 \\ \cdot & \cdot & \cdot & \cdot & \frac{1}{2}(C_{11}-C_{12}) & 0 \\ \cdot & \cdot & \cdot & \cdot & \cdot & \frac{1}{2}(C_{11}-C_{12}) \end{pmatrix}. \tag{3.44}$$

Mathematica programming

The different symmetry groups can be defined in MATHEMATICA as sets of rules defining the relationships between the elastic moduli. Some examples are given in the form of MATHEMATICA code here.

```
monoclinic2 = Thread[ {C14,C16,C24,C26,C34,C36,C45,C56}->0]

(C2 /. monoclinic2) // MatrixForm

tetragonal = Join[
 Thread[{C14,C15,C16,C24,C25,C26,C34,C35,C36,C45,C46,C56}->0],
    {C22 -> C11, C23 -> C13, C55 -> C44}]

(C2 /. tetragonal) // MatrixForm

isotropic = Join[
 Thread[{C14,C15,C16,C24,C25,C26,C34,C35,C36,C45,C46,C56}->0],
    {C13 -> C12, C22 -> C11, C23 -> C12, C33 -> C11,
    C44 -> (C11 - C12)/2, C55 -> (C11 - C12)/2,
    C66 -> (C11 - C12)/2}]

(C2 /. isotropic) // MatrixForm
```

The compliance matrix $\boldsymbol{s}$ will display the same distribution of zeros. This can be readily established with the help of the theory of symmetric block matrices and even more easily verified using MATHEMATICA. For example:

```
Inverse[(C2 /. isotropic) ] // MatrixForm
```

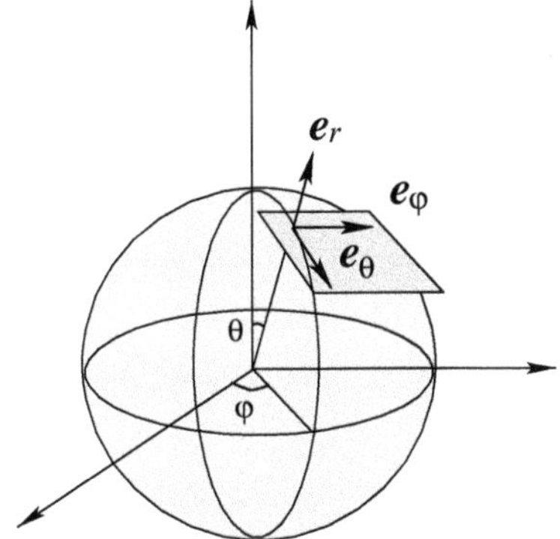

Figure 3.2. Plane tangent to the unit sphere at the point defined by the spherical coordinate angles (θ, φ). The plane normal is $\boldsymbol{e}_r$, and the plane is spanned by the vectors $\boldsymbol{e}_\theta$ and $\boldsymbol{e}_\varphi$.

Displaying the symmetry planes

Visualisation of the planes of symmetry that characterise different classes of material symmetry can be readily achieved in MATHEMATICA using some simple commands.

We note that the basis vector $\boldsymbol{e}_r$ in the spherical coordinate system assumes values associated with all possible directions in the Euclidean space. For a specified value of $\boldsymbol{e}_r$, the other two basis vectors, $\boldsymbol{e}_\theta$ and $\boldsymbol{e}_\varphi$, span a plane with the normal $\boldsymbol{e}_r$. In our plots the planes will be represented by $[-1, 1] \times [-1, 1]$ rectangles in the $(\boldsymbol{e}_\theta, \boldsymbol{e}_\varphi)$ plane.

In MATHEMATICA we define the normal vector of a mirror plane by the spherical coordinate angles (θ, φ). We then define a linear function corresponding to the plane and plot the rectangle using the `ParametricPlot3D` command. See Figure 3.2.

```
SetCoordinates[Spherical[r, t, p]]
CoordinatesToCartesian[{r, t, p}]
base = Transpose[ JacobianMatrix[]] / ScaleFactors[]

plane[t_, p_] = u base[[2]] + v base[[3]]

PlanePlot[ t_, p_, opt___] :=
  ParametricPlot3D[plane[t,p], u,-1,1, v,-1,1, opt]
```

To display an entire set of symmetry planes, we map the preceding commands over the complete set and display the result.

Note the use of the `$DisplayFunction` option to show only the final plot.

Symmetry planes in the case of trigonal and cubic material symmetries are displayed in Figure 3.3.

```
ntrigonal = {{0,Pi/2}, {2Pi/3,Pi/2}, {4Pi/3,Pi/2}}
nortotropic = {{Pi/2, 0}, {Pi/2, Pi/2}, {0, 0}}
ncubic = {{Pi/2,0}, {Pi/2,Pi/2}, 0,0,{Pi/4,Pi/2},
          {3Pi/4,Pi/2}, {Pi/4,Pi}, {3Pi/4,Pi},
          {Pi/4,3Pi/2}, {3Pi/4,3Pi/2}}
```

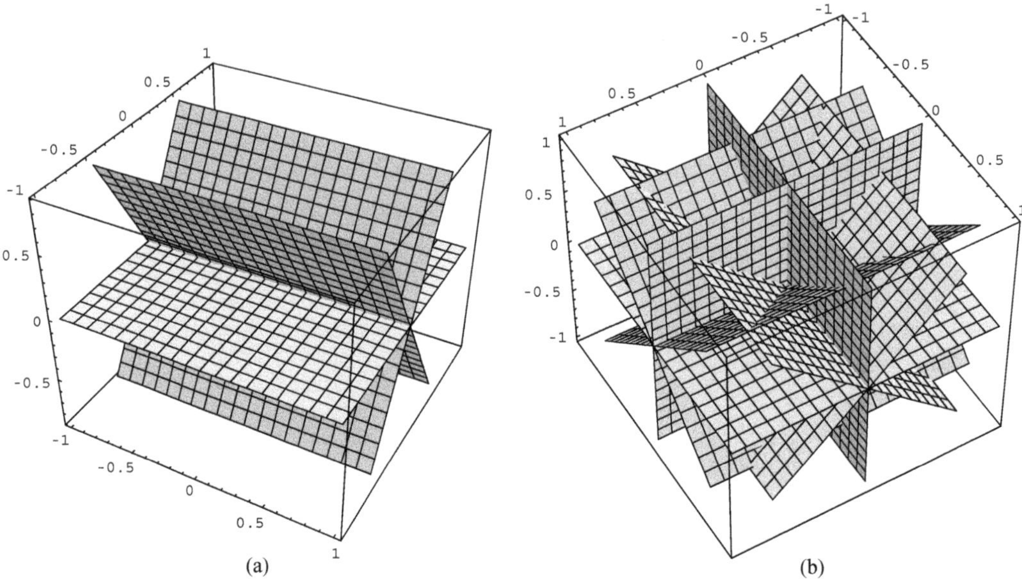

Figure 3.3. Symmetry planes for (a) trigonal and (b) cubic material symmetry.

```
ShowPlanes[ ncubic, ViewPoint -> {3, 5, 7}]

planeset =
  Map[ PlanePlot[#[[1]], #[[2]],
      DisplayFunction -> Identity] &, ntrigonal]

(Show[planeset,
  DisplayFunction -> $DisplayFunction ] &) @@ planeset
```

3.4 THE EXTENSION EXPERIMENT

In order to gain better insight into the meaning of different symmetry groups, let us perform a series of thought experiments involving specimens of different orientations.

From a homogeneous material possessing given material symmetry we shall cut imaginary cylindrical specimens in various directions and perform tensile experiments on each of them with a unit stress applied along the specimen axis. We shall then record the axial extension and hence will be able to determine the apparent elastic compliance and apparent modulus in various directions.

Let us first discuss the corresponding formulae and then present MATHEMATICA programming.

For a specimen cut in the direction $\boldsymbol{n}$ the stress field is defined by

$$\sigma = \boldsymbol{n} \otimes \boldsymbol{n}. \tag{3.45}$$

The corresponding strain field is computed using the elastic constitutive law with the compliance tensor $\boldsymbol{S}$:

$$\boldsymbol{\varepsilon} = \boldsymbol{S} : \boldsymbol{\sigma}. \tag{3.46}$$

The longitudinal strain is proportional to the directional compliance and is computed as

$$e(\boldsymbol{n}) = \boldsymbol{n} \cdot \boldsymbol{\varepsilon} \cdot \boldsymbol{n} = \boldsymbol{n} \otimes \boldsymbol{n} : \boldsymbol{S} : \boldsymbol{n} \otimes \boldsymbol{n}. \tag{3.47}$$

The apparent modulus in the chosen direction is computed as

$$E(\boldsymbol{n}) = \frac{\sigma}{e(\boldsymbol{n})} = \frac{1}{\boldsymbol{n} \otimes \boldsymbol{n} : \boldsymbol{S} : \boldsymbol{n} \otimes \boldsymbol{n}}. \tag{3.48}$$

The preceding formulas are now written in terms of the MATHEMATICA commands. Note the use of the command `Outer` to construct the outer product, ⊗, between two vectors.

```
normal = Cos[p] Sin[t], Sin[p] Sin[t], Cos[t]

(stress = Outer[ Times , normal, normal]) // MatrixForm
```

The illustration of spatial variation of directional compliance is obtained as follows. We begin by recomputing the stiffness and compliance tensor for the case of cubic material symmetry, `C2cubic` and `S2cubic`, respectively.

We can use this occasion to verify the distribution of zeros of the stiffness and the compliance tensors and to check the inverse relationship in this specific case.

```
cubic =  Join[
 Thread[{c14,c15,c16,c24,c25,c26,c34,c35,c36,c45,c46,c56}
         ->0],
  {c13->c12, c22->c11, c23->c12,
   c33->c11, c55->c44, c66->c44}]

(C2cubic = MakeTensor["c", 2,6]/.cubic) //MatrixForm

(S2cubic = Simplify[Inverse[C2cubic]]) //MatrixForm

sol = Solve[ {s11 == S2cubic[[1, 1]],
             s12 == S2cubic[[1, 2]],
             s44 == S2cubic[[4, 4]]},
                  {c11, c12, c44}][[1]]

(S2cubic = Simplify[ S2cubic /. sol ]) // MatrixForm
```

We use the following numerical values of compliance, keeping in mind that these are usually quoted in the Voigt notation:

```
aluminium = {s11->.6 10^-2, s12->-0.58 10^-2, s44->3.5 10^-2}
copper = {s11->1.49 10^-2, s12->-0.63 10^-2, s44->1.33 10^-2}
austenite = {s11->1.49 10^-2, s12->-0.63 10^-2, s44->1.33 10^-2}
TiN = {s11->2.17 10^-3, s12->-3.8 10^-4, s44->5.95 10^-3}

(S4cubic = ComplianceVto4[S2cubic]) // MatrixForm

(strain = CEDot[ S4cubic, stress]) // MatrixForm
```

The axial extension

$$\varepsilon_{nn} = \boldsymbol{n} \cdot \boldsymbol{\varepsilon} \cdot \boldsymbol{n}$$

can be computed using different combinations of MATHEMATICA commands.

Particular cases of material behaviour can then be obtained by using the replacement rule defined previously.

```
epsnn = Simplify[ Nest[ Dot[ # , normal] &, S4cubic , 4] ]
epsnn = Simplify[ normal . strain . normal ]


Simplify[ epsnn /. s44 -> (s11 - s12)/2]
```

Plots shown in Figures 3.4 and 3.5 are obtained using the `ParametricPlot` command and varying the directions $(\theta, \varphi) \in [0, 2\pi) \times (-\frac{\pi}{2}, \frac{\pi}{2})$.

```
(epsnn normal) /. aluminium;
```

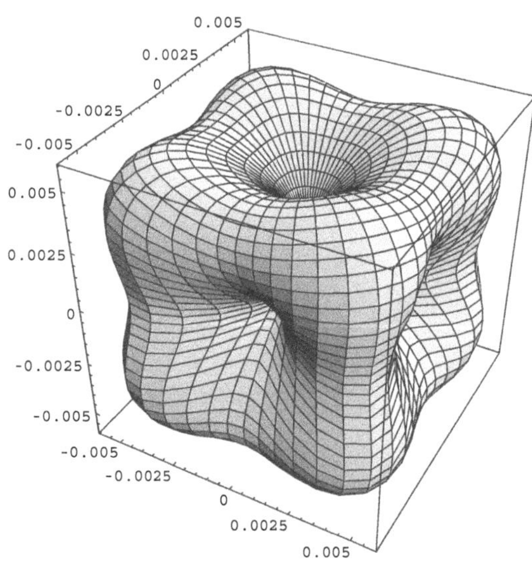

Figure 3.4. The orientational variation of axial extension for TiN.

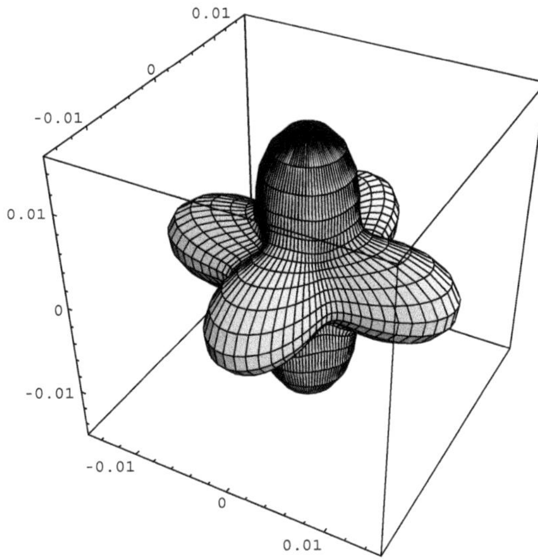

Figure 3.5. The orientational variation of axial extension for fcc iron (austenite).

```
ParametricPlot3D[ Evaluate[%], {t,0,2Pi}, {p,-Pi2,Pi2},
  PlotPoints -> 50]
```

3.5 FURTHER PROPERTIES OF ISOTROPIC ELASTICITY

In the isotropic case the compliance tensor $\boldsymbol{S}$ depends on only two parameters.

The isotropic stiffness tensor is usually expressed in terms of the Lamé moduli (λ, μ), whereas the isotropic compliance tensor is more coveniently defined using Young's modulus E and Poisson's ratio ν.

Please note the use of the variable **EE** for Young's modulus E in MATHEMATICA to avoid confusion with the base of natural logarithms, **E = 2.71828....**

The isotropic stiffness and compliance 4-tensors, $\boldsymbol{C}$ and $\boldsymbol{S}$, respectively, are defined and constructed as follows:

$$C_{ijkl} = \lambda\delta_{ij}\delta_{kl} + \mu(\delta_{ik}\delta_{jl} + \delta_{il}\delta_{jk})$$
$$\boldsymbol{\sigma} = \boldsymbol{C} : \boldsymbol{\varepsilon}$$
$$S_{ijkl} = \frac{-\nu}{E}\delta_{ij}\delta_{kl} + \frac{(1+\nu)}{2Y}(\delta_{ik}\delta_{jl} + \delta_{il}\delta_{jk})$$
$$\boldsymbol{\varepsilon} = \boldsymbol{S} : \boldsymbol{\sigma}.$$

The calculations can be implemented in the form of MATHEMATICA functions **DDot** (double dot product) and **IsotropicCompliance**. It is useful to provide an additional definition of the isotropic compliance tensor as a function of a single parameter **nu**, as it is sometimes convenient to set Young's modulus **EE** to unity.

```
IsotropicStiffness[lambda_, mu_] := Array[
    lambda KroneckerDelta[#1,#2] KroneckerDelta[#3,#4]+
    mu (KroneckerDelta[#1,#3] KroneckerDelta[#2,#4]+
    KroneckerDelta[#1,#4]KroneckerDelta[#2,#3]) &,
  {3, 3, 3, 3}]
```

```
IsotropicCompliance[EE_, nu_] := Array[
 -nu/EE KroneckerDelta[#1, #2] KroneckerDelta[#3, #4]+
 (1+nu)/2/EE( KroneckerDelta[#1,#3] KroneckerDelta[#2,#4]+
   KroneckerDelta[#1, #4]KroneckerDelta[#2, #3]) &,
 {3, 3, 3, 3}]

IsotropicCompliance[nu_] := IsotropicCompliance[1, nu]

SS = IsotropicCompliance[nu]
```

The isotropic elastic constitutive law can be expressed using the particular form of the stiffness and compliance tensors $\boldsymbol{C}$ and $\boldsymbol{S}$ in one of the two equivalent forms

$$\boldsymbol{\sigma} = \lambda \operatorname{tr} \boldsymbol{\varepsilon} + 2\mu\, \boldsymbol{\varepsilon} \qquad \boldsymbol{\varepsilon} = \frac{1+\nu}{E}\boldsymbol{\sigma} - \frac{\nu}{E} \operatorname{tr} \boldsymbol{\sigma}. \tag{3.49}$$

The relationship between Young's modulus E and Poisson's ratio ν, on the one hand, and the Lamé coefficients λ and μ, on the other, is established by

$$E = \frac{\mu(3\lambda + 2\mu)}{\lambda + \mu}, \qquad \nu = \frac{\lambda}{2(\lambda + \mu)}, \tag{3.50}$$

$$\lambda = \frac{\nu E}{(1+\nu)(1-2\nu)}, \qquad \mu = \frac{E}{2(1+\nu)}. \tag{3.51}$$

The extension experiment performed on isotropic material will of course lead to uniform elongation in all directions. As a consequence, the elongation diagram will be a sphere.

Thermal expansion

The foregoing discussion neglected entirely the effect of temperature on material deformation. It is well known that most solid bodies expand upon heating, but the nature of this expansion is different from elastic deformation, in that it is not caused by the application of external or body forces. A generalisation of the elastic constitutive law is therefore required to take into account the presence of thermal strains and their effects on elastic strains and ultimately stresses.

Strictly speaking, to consider the *mutual* influence of temperature and stress, one must note that the application of stress leads to small changes in the temperature of the object. In an adiabatic experiment, an elastic body will generally cool down under tension and heat up in compression, in a manner similar to that for an ideal gas (thermoelastic effect). However, the magnitude of this temperature change is of the order of fractions of Kelvin and can therefore be neglected in most practical engineering situations.

Prominent examples of a heated body expanding are seen in mercury thermometers, or in the extension of rails in the heat, which may lead to buckling unless appropriate measures are taken. The magnitude of thermal strains that arise in such practical situations

Table 3.2. *Thermoelastic properties of some isotropic materials.*

	E [10^9 Pa]	ν [adim]	ρ [kg/m^3]	α [10^{-6}/K]
Aluminium	71	0.34	2.6	23
Steel	210	0.285	7.8	13
Zinc	78	0.21	7.15	30
Copper	100	0.33	7.15	17
Beryllium	300	0.05	1.85	12
Titanium	105	0.34	4.5	9
Granite	60	0.27	2.3-3	20
Marble	26	0.3	2.8	8
Glass	60	0.2–0.3	2.5–2.9	3.4–5.9
PMMA	2.9	0.4	1.8	80–90

is on the order of several percent for temperature changes up to 100 K, and is of significant importance for many engineering applications.

To obtain the *thermoelastic* constitutive equations, on additional corrective term must be introduced into the elasticity equations that is linear in terms of the temperature change θ. The fundamental assumption made at this point is that small thermal and elastic strains are additive, but only elastic strains cause stresses. With $\boldsymbol{\varepsilon}$ denoting the *total* strain, the thermoelastic equations then take the form

$$\boldsymbol{\sigma} = \boldsymbol{C} : (\boldsymbol{\varepsilon} - \boldsymbol{A}\theta) \qquad \boldsymbol{\varepsilon} = \boldsymbol{S} : \boldsymbol{\sigma} + \boldsymbol{A}\theta, \tag{3.52}$$

where $\boldsymbol{A}$ is the symmetric second-order tensor of thermal linear expansion coefficients.

In the case of a body possessing isotropic thermal linear expansion properties, $\boldsymbol{A}$ has the form

$$\boldsymbol{A} = \alpha \boldsymbol{I}.$$

Thermomechanical characteristics for a series of bodies which can be thought to be isotropic are given in Table 3.2.

Residual stresses

The discussion in the preceeding sections was based on the implicit assumption that the body was stress-free in its initial configuration:

$$\boldsymbol{u} = 0 \longrightarrow \boldsymbol{\varepsilon} = 0 \longrightarrow \boldsymbol{\sigma} = \boldsymbol{C} : \boldsymbol{\varepsilon} = 0.$$

Although this assumption holds for a significant fraction of the practical cases of interest to the engineering community, it excludes situations of some considerable importance. Generally, when structures under discussion have been manufactured by processes involving large strains and/or high temperatures, the structures are already stressed in the initial configuration in which the service loading will be applied. This prestress is referred to as *residual stress*, or sometimes also as initial stress. As a simple practical example, one can think of bimaterial assemblies of substances with different thermal linear expansion coefficients that are joined together at high temperatures (for example, by diffusion bonding) and then cooled down to room temperature, or any other lower service temperature. Residual stress states that arise in this case would be relatively simple for simple geometries.

However, in most practical cases, manufacturing is a complex processes involving various operations that create misfit and accommodation strains within the object, leading to complex residual stress distributions. It is therefore necessary to include in our consideration the possible presence of a self-equilibrated residual stress state, $\boldsymbol{\sigma}_0$, in the initial configuration. An alternative way of thinking about residual stresses is in terms of residual strains, or eigenstrains $\boldsymbol{\varepsilon}^*$, which will not be discussed here.

Denoting by $\boldsymbol{\sigma}_0$ the residual stress field, the constitutive law can be written in the general form

$$\boldsymbol{\sigma} = \boldsymbol{\sigma}_0 + \boldsymbol{C} : \boldsymbol{\varepsilon} \tag{3.53}$$

or

$$\boldsymbol{\sigma} - \boldsymbol{\sigma}_0 = \boldsymbol{C} : (\boldsymbol{\varepsilon} - \boldsymbol{A}\theta),$$

if thermal expansion is also taken into consideration.

Because force balance must be satisfied in the body Ω in the initial configuration (when the displacements and strains are identically zero), the residual stress field must be self-equilibrated,

$$\operatorname{div} \boldsymbol{\sigma}_0 = 0 \quad \text{in } \Omega, \qquad \boldsymbol{\sigma}_0 \cdot \boldsymbol{n} = 0 \quad \text{on } \partial\Omega$$

provided no external forces are applied and that no displacements are prescribed in the initial configuration.

A remark must be made about the limitations of this approach to the description of residual stress effects. The approach described above cannot explain or predict the following phenomena:

- *Stress-induced buckling*, that is, the bifurcation of elastic solutions that may occur in a prestressed body, even in the absence of external loading.
- *Change of frequency of a vibrating string due to the presence of prestress*, as observed and used in all stringed musical instruments.

The correct approach in this case consists of adding further terms to equation (3.52) that are obtained in the general case by linearization (first-order series expansion) of the deformation energy under a kinematical description for a nonlinear elastic material around the initial configuration using the hypothesis of small strains. For sources providing more complete coverage of this topic see, for example Ogden (1997) and Truesdell (1968).

3.6 LIMITS OF LINEAR ELASTICITY

The foregoing sections were devoted to a discussion of the linear elastic constitutive model. It is now also appropriate to point out its limitations, that is, the conditions under which the basic assumptions made in the analysis fail, mainly due to the fact that the implied linearity is violated:

- Although the strains remain *small*, the linearity of *material behaviour* may no longer hold. This phenomenon is seen in cases when the material undergoes some internal shape change process (e.g., straightening of molecular chains in polymers), which may result in behaviour that is no longer linear, although it may still be reversible.
- Examples of irreversible behaviour include time-dependent deformation (viscous flow or creep) or time-independent irreversible deformation, such as plasticity. The study of nonlinear material behaviour is a vast subject that we do not touch upon in this book.
- The strains are *not small*, so that large (finite) deformations must be considered. This opens up a large domain of study that has interesting applications for materials capable of sustaining large elastic strains, such as rubber. An introduction to the continuum mechanics of this subject is given, for example, by Ogden (1997).
- Material behaviour *cannot be linearised*. This is generally the case around singular points in the solution. For solid bodies an example of this situation is *buckling*. For an introduction to buckling in relation to the linearization of constitutive behaviour see Ogden (1997); Ballard and Millard (2005).

Leaving the exploration of these domains to other texts, let us only introduce some classical bounds on the domain of linear elasticity imposed by the requirement that stresses and strains must not exceed certain values at which nonlinear or irreversible behaviour sets in. Traditionally the bounds on the domain of applicability of elasticity are defined in terms of stresses, by requiring that the corresponding points in the stress space lie inside a convex set. More precisely, the condition is defined in terms of the stress state $\boldsymbol{\sigma}$ as follows:

$$f(\boldsymbol{\sigma}) \leq 0.$$

The function f is referred to as the yield function, in reference to the plastic yield phenomenon that may occur when $f(\boldsymbol{\sigma}) = 0$.

Two yield functions, or yield criteria, are most commonly used:

- The *Tresca* criterion,

$$f(\boldsymbol{\sigma}) = \frac{1}{2} \max_{i,j} |\sigma_i - \sigma_j| - k,$$

  where σ_i denote the principal stresses (eigenvalues) of $\boldsymbol{\sigma}$, and k is a material parameter.
- The *von Mises* criterion,

$$f(\boldsymbol{\sigma}) = \sqrt{\frac{3}{2}}\, |\,\mathrm{dev}\boldsymbol{\sigma}\,| - \sigma_Y,$$

  where σ_Y is a material parameter, and $|\,\mathrm{dev}\,\boldsymbol{\sigma}\,|$ denotes the tensor norm of $\mathrm{dev}\,\boldsymbol{\sigma} = \boldsymbol{\sigma} - \frac{1}{3}(\mathrm{tr}\sigma)\boldsymbol{I}$, the deviatoric part of the stress tensor, where $\boldsymbol{I}$ is the appropriate unit tensor.

  The tensor norm $|\,\boldsymbol{T}\,|$ used here is the Frobenius norm equal to the square root of the sum of squares of all elements, normalised so that, for example, $|\,\boldsymbol{I}\,| = 1$. For a symmetric tensor the norm can be conveniently defined as $|\,\boldsymbol{T}\,| = [\mathrm{tr}(\boldsymbol{T} \cdot \boldsymbol{T})]^{1/2}$.

The material parameters introduced above, k for the Tresca criterion and σ_Y for the von Mises criterion, define the condition under which inelastic deformation may first take place, and the linear elastic constitutive law for the relationship between stresses and strains becomes no longer valid. Practical questions arise regarding how these material parameters are related to applied loads in deformation experiments, such as uniaxial tension and simple shear experiments, and in other more complex (multiaxial) loading cases. It is then possible to define the shape of the domain within which the linear elasticity description is valid, which is bounded by the *yield surface*. The visualisation of these bounding surfaces in the three-dimensional space of principal stresses is a task that can readily be tackled using MATHEMATICA.

SUMMARY

In this chapter the fundamental linear elastic constitutive relationship between stresses and strains was introduced. The choice of notation for recording this relationship was discussed, and the effect of material symmetry on the number of elastic constants needed to describe the properties of elastic materials was considered. The case of isotropic elasticity was treated in some detail. The limitations of the linear elastic theory were discussed.

EXERCISES

1. Extension of cylindrical bar

Consider a cylindrical bar $\Omega = S \times [-L, L]$ of arbitrary section S, made out of an anisotropic elastic material (see Figure 3.6 (left)). The bar is subjected on its end sections $z = \pm L$ to a distribution of traction vectors, $\boldsymbol{t} = \pm \frac{P}{S} \boldsymbol{e}_z$, with the lateral surface remaining traction-free. Using a cylindrical coordinate system:

(a) Show that the stress field,

$$\sigma = \frac{P}{S} \boldsymbol{e}_z \otimes \boldsymbol{e}_z,$$

is statically admissible, that is, satisfies the balance equation and the boundary conditions.

(b) Compute the corresponding strain field and prove that the compatibility equation demands that $C_{34} = 0$.

(c) Compute the corresponding displacement field, neglecting the infinitesimal rigid displacement. Remark that the solution demands that $C_{13} = -C_{23}$ in order to eliminate the multiplicity of the solution imposed by θ.

Hint: See notebook `C03_cylinder_anisotropic_extension.nb`.

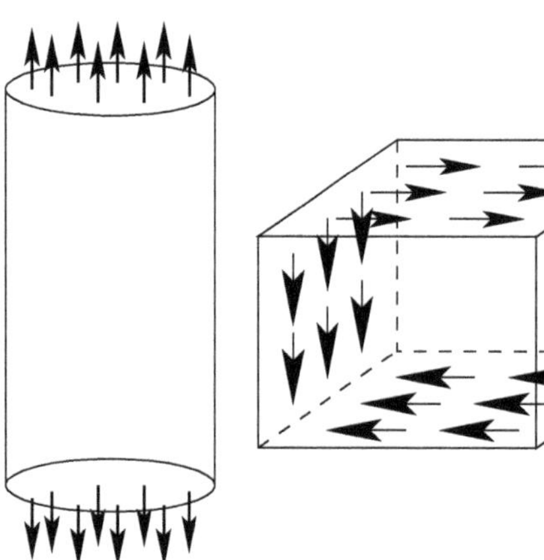

Figure 3.6. Deformation experiments for a cylinder submitted to uniaxial extension (left) and a rectangular parallelepiped subjected to shear (right).

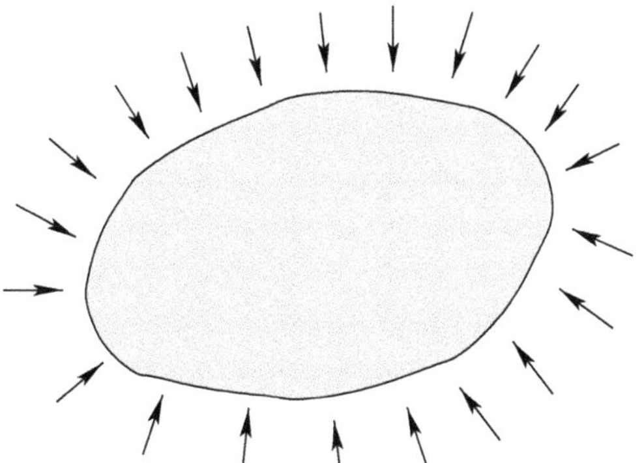

Figure 3.7. Deformation experiment for a body submitted to uniform pressure.

2. Shear of a rectangular parallelepiped

Consider a rectangular parallelepiped $\Omega = [-L_x, L_x] \times [-L_y, L_y] \times [-L_z, L_z]$ of arbitrary section S made out of an anisotropic elastic material (see Figure 3.6 (right)). The stress state in the body is expressed in a cartesian coordinate system as

$$\boldsymbol{\sigma} = \sigma_{xy}\,(\boldsymbol{e}_x \otimes \boldsymbol{e}_y + \boldsymbol{e}_y \otimes \boldsymbol{e}_x).$$

(a) Determine the system of external surface tractions and body forces that balance the given stress field. Determine the linear and angular momenta of the external forces acting on the faces of the plate.
(b) Compute the corresponding strain field and check its compatibility.
(c) Compute the displacement field by neglecting the infinitesimal rigid displacement field.
(d) Determine the extensions of the segments oriented along the coordinate axes.
(e) Show that all faces and plane sections remain plane, but that faces rotate, forming an oblique parallelipiped.

Hint: See notebook `C03_shear_anisotropic.nb`.

3. Uniform compression

Consider a body Ω of arbitrary shape made out of an anisotropic elastic material. The stress state in the body is expressed in a cartesian coordinate system as

$$\boldsymbol{\sigma} = -p\boldsymbol{I}.$$

(a) Determine the system of external surface tractions and body forces that balance the given stress field. Show that the linear and angular momenta of the external forces are both zero.
(b) Compute the corresponding strain field and check its compatibility.
(c) Compute the displacement field by neglecting the infinitesimal rigid displacement field.
(d) Starting now from a uniform compressive strain field

$$\boldsymbol{\varepsilon} = \varepsilon\boldsymbol{I},$$

compute the stress and the displacement fields. Remark the difference with respect to the previous solution. See Figure 3.7.

Hint: See notebook `C03_uniform_compression_anisotropic.nb`.

Figure 3.8. A beam subjected to bending.

4. Bending of a beam

Consider a cylindrical beam $\Omega = S \times [-L, L]$ of arbitrary section S made out of an anisotropic elastic material. The beam is subjected on its end sections $z = \pm L$ to a distribution of traction vectors, $\boldsymbol{t} = \pm\frac{M_x}{I_x} y \boldsymbol{e}_z$, and has traction-free lateral surfaces (see Figure 3.8). Using a cartesian coordinate system:

(a) Show that the stress field

$$\sigma = \frac{M_x}{I_x} y \boldsymbol{e}_z \otimes \boldsymbol{e}_z$$

is statically admissible, that is, satisfies the balance equations and the boundary conditions.

(b) Compute the corresponding strain field and prove that the compatibility equation demands that $C_{34} = 0$.

(c) Compute the corresponding displacement field neglecting the infinitesimal rigid displacement. Show that
   - imposing $C_{35} = C_{34} = 0$ would prevent warping of the cross section;
   - imposing C_{35} would prevent twisting of the cross section.

(d) Compute the displacement field by neglecting the infinitesimal rigid displacement field.

(e) By adding an infinitesimal rigid displacement, compute the constants such that the boundary conditions correspond to bending of the beam supported at its ends. The conditions will use five out of the six constants. What is the role of the last one?

(f) Compute the shape of the deflected axis.

Hint: See notebook `C03_cylinder_anisotropic_bending.nb`.

5. Bending of a plate

Consider a rectangular plate $\Omega = [-L_x, L_x] \times [-L_x, L_x] \times [-H, H]$ of arbitrary cross section S made out of an anisotropic elastic material (see Figure 3.9). The plate is subject to tractions on its lateral surfaces $x = \pm L_x$ and $y = \pm L_y$ that can be equilibrated by the stress field

$$\boldsymbol{\sigma} = \frac{12Z}{H^3} \left(M_{xx} \boldsymbol{e}_x \otimes \boldsymbol{e}_x + M_{xy} (\boldsymbol{e}_x \otimes \boldsymbol{e}_y + \boldsymbol{e}_y \otimes \boldsymbol{e}_x) + M_{yy} z \boldsymbol{e}_y \otimes \boldsymbol{e}_y \right).$$

(a) Determine the system of external surface tractions and body forces that balance the given stress field. Determine the resultants and moments of the external forces acting on the faces of the plate.

(b) Compute the corresponding strain field and check its compatibility.

(c) Compute the displacement field by neglecting the infinitesimal rigid displacement field.

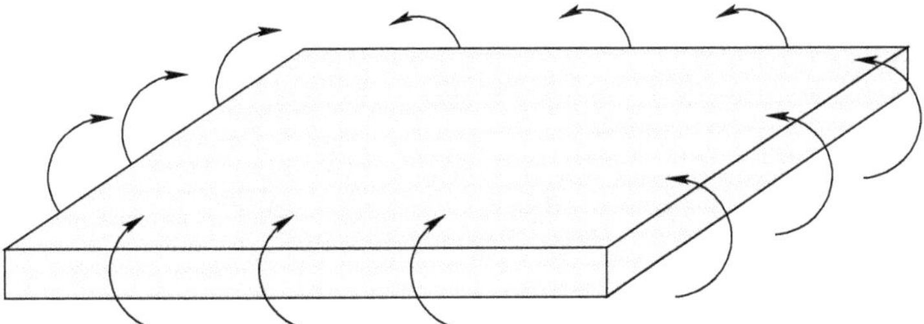

Figure 3.9. Deformation experiment for a plate under bending.

(d) Compute the deformed shape of the middle surface.
(e) Show that
- straight material segments become curved during deformation;
- *orthogonality* between material segments and the middle surface is not preserved through the deformation.

Analysing the preceeding results, explain why this hypothesis could be accepted in the case of thin plates.
(f) Determine the anisotropic rigidities D_{ijkl}, $i, j, k, l = x, y$ relating the bending moments:

$$\boldsymbol{M} = M_{xx}\boldsymbol{e}_x \otimes \boldsymbol{e}_x + M_{xy}(\boldsymbol{e}_x \otimes \boldsymbol{e}_y + \boldsymbol{e}_y \otimes \boldsymbol{e}_x) + M_{yy}\boldsymbol{e}_y \otimes \boldsymbol{e}_y,$$

to the components of the second gradient of the deflection of the middle surface,

$$\boldsymbol{W} = \widehat{\text{grad}}\,\widehat{\text{grad}}\,u_z(x, y, 0).$$

Hint: `C03_plate_anisotropic_bending.nb`.

6. Orientational variation of elastic properties for different materials

For materials presented in Table 3.3 perform the following calculations:
(a) Compute the values of material constants for the fourth and second rank tensor notation of the Hooke tensor.
(b) Compute the orientational variation of the axial extension for the materials.

7. Orientational variation in isotropic elasticity

Show that the surface representing orientational variation of the Young modulus in isotropic elasticity is a sphere.

8. Elastic moduli in isotropic elasticity

Consider an isotropic elastic body with material constants represented in terms of the constants (λ, ν) and (E, ν), respectively. Obtain the formulae in equation (3.49) using the following procedures:
- by direct inversion of the second-order tensor form of Hooke's tensor,
- by transforming equation (3.51) into equation (3.50), and then reciprocally.

Table 3.3. *Elastic compliances (in units of* 10^{-11} Pa^{-1}) in the Voigt notation for a series of crystals at room temperature (*Nye*, 1985)

	Symmetry						
Crystal	s^V_{11}	s^V_{12}	s^V_{44}	s^V_{33}	s^V_{13}	s^V_{14}	s^V_{66}
Sodium chloride	2.21	−0.45	7.83	0	0	0	0
Aluminium	1.59	−0.58	3.52	0	0	0	0
Copper	1.49	−0.63	1.33	0	0	0	0
Nickel	7.99	−0.312	8.44	0	0	0	0
Tungsten	0.257	−0.073	0.660	0	0	0	0
Sodium chloride	2.2	−0.6	8.6	0	0	0	0
Tin	1.85	0.99	5.70	1.18	−0.25	0	13.5
ADP	1.8	0.7	11.3	4.3	−1.1	0	16.2
Zinc	0.84	0.11	2.64	2.87	−0.78	0	0
Cadmium	1.23	−0.15	5.40	5.55	0.93	0	0
Quartz	1.27	−0.17	2.01	0.97	−0.15	−0.43	0
Tourmaline	0.40	−0.10	1.51	0.63	−0.016	−0.058	0

Hint: Take the `Tr` of both sides of equations to obtain the compressibility modulus and the relation between volume change (spherical part of strain) and pressure (related to the spherical part of stress).

9. Positive definiteness of elastic moduli in isotropic elasticity

Consider an isotropic elastic body with material constants represented in terms of the constants (λ, ν) and (E, ν), respectively. Show that

(a) The positive definiteness of the Hooke tensor is eqivalent to the requirements

$$3\lambda + 2\mu > 0, \qquad 2\mu > 0$$

or

$$E > 0, \qquad \frac{1}{2} \geq \nu > -1.$$

(b) The incompressibility condition for the material is equivalent to $\nu = \frac{1}{2}$.

(c) The spherical and the deviatoric parts of strain, and also of stress, are orthogonal with respect to the double dot product, and they represent the eigenvectors of the constitutive equation in isotropic elasticity. Determine the corresponding eigenvalues.

10. Special strain–stress states in isotropic elasticity

Consider an isotropic elastic body with material constants represented in terms of the constants (λ, ν) and (E, ν), respectively. For each of the following cases:

(a) extension $\boldsymbol{\sigma} = \sigma \boldsymbol{e}_x \otimes \boldsymbol{e}_x$

(b) shear $\boldsymbol{\sigma} = \sigma(\boldsymbol{e}_x \otimes \boldsymbol{e}_y + \boldsymbol{e}_y \otimes \boldsymbol{e}_x)$

(c) uniform compression $\boldsymbol{\sigma} = -p\boldsymbol{I}$

perform the following operations:

- determine complete strain states
- check the compatibility and balance equations (in the absence of body and inertial forces)
- compute the spherical and deviatoric parts of the tensors and interpret the results in terms of compression and shear
- determine the traction vectors acting on surfaces of an infinitesimal volume element and interpret physically the conditions of positive definiteness of elastic moduli.

4 General principles in problems of elasticity

OUTLINE

This chapter is devoted to the formulation of the complete elasticity problem. It begins with the formulation of the regular problem of thermoelasticity. Displacement (Navier) and stress (Beltrami–Michell) formulations are introduced, and the one-dimensional problem of a spherical vessel under internal and external pressure is solved as an illustration.

The general principles applicable in linear elasticity are treated next. The superposition principle and the virtual work theorem are introduced, allowing the conditions for the uniqueness of elastic solution to be established. The existence of the strain energy potential and the complementary energy potential is proven, and reciprocity theorems are presented. Saint Venant torsion is considered in detail, and the more general Saint Venant principle is introduced, together with Hoff's counterexample and the von Mises–Sternberg formulation.

4.1 THE COMPLETE ELASTICITY PROBLEM

The complete system of equations of elasticity consists of the equations of kinematics and dynamics, together with the linear elastic constitutive relations introduced in the previous chapter. Solution of the complete system must be found in the form of three field quantities:

vector field of displacements, $\boldsymbol{u}$;
tensor field of small strains, $\boldsymbol{\varepsilon}$;
tensor field of stresses, $\boldsymbol{\sigma}$.

Within domain Ω the following ***system of equations of linear thermoelasticity*** must be satisfied:

- *Kinematic equations*

$$\boldsymbol{\varepsilon} = \frac{1}{2}(\nabla + \nabla^T)\boldsymbol{u}. \tag{4.1}$$

- *Constitutive equations of linear thermoelasticity:*

$$\boldsymbol{\sigma} = \boldsymbol{\sigma}_0 + \boldsymbol{C} : (\boldsymbol{\varepsilon} - \boldsymbol{A}\theta), \tag{4.2}$$

  where $\boldsymbol{\sigma}_0$ is a tensor of initial stresses, $\boldsymbol{C}$ is the tensor of elastic moduli, $\boldsymbol{A}$ is a tensor of linear thermal expansion coefficients, and θ is a scalar field of temperature changes.

The particular case when $\theta = 0$ will be referred to as *isothermal*, and when $\boldsymbol{\sigma}_0 = \mathbf{0}$ will be referred to as the *natural state*.

- *Equations of static equilibrium*

$$\operatorname{div}\boldsymbol{\sigma} + \boldsymbol{f} = 0, \tag{4.3}$$

where $\boldsymbol{f}$ is the vector of body forces.

A set of necessary and sufficient conditions must be imposed on the problem formulation in order to guarantee the existence and uniqueness of the solution. A problem formulation that satisfies such conditions, following Hadamard, will be called a *well-posed*, or *regular* linear thermoelastic problem.

The regular linear thermoelastic problem must be formulated by the above set of equations together with the provision of the following information:

- At every point within the domain Ω:
  - the tensor of elastic moduli $\boldsymbol{C}$ that is symmetric, positive definite, and bounded
  - the tensor of linear thermal expansion coefficients $\boldsymbol{A}$ that is symmetric
  - the tensor of initial stresses $\boldsymbol{\sigma}_0$
  - the vector of body forces $\boldsymbol{f}$
  - the scalar field of temperature changes θ.
- At every point on the domain boundary $\partial\Omega$ and *for each direction* $\boldsymbol{e}_i$, $i = 1, 2, 3$:
  - either a component of the displacement vector,

$$u_i = u_i^B, \tag{4.4}$$

  where u_i^B is a given function,
  - or a component of the traction vector,

$$(\boldsymbol{\sigma} \cdot \boldsymbol{n})_i = t_i^B, \tag{4.5}$$

  where t_i^B is a given function.

Let $\partial\Omega_i^d$ denote the part of the boundary on which the boundary condition is imposed in terms of displacements u_i^B, and respectively let $\partial\Omega_i^t$ denote the part of the boundary on which the boundary condition is imposed in terms of tractions t_i^B. The complementarity of the regions on which boundary conditions are imposed in terms of displacements and tractions leads to the following partition of the boundary $\partial\Omega$:

$$\partial\Omega_i^d \cup \partial\Omega_i^t = \partial\Omega \qquad \partial\Omega_i^d \cap \partial\Omega_i^t = \emptyset \qquad \forall i = 1, 2, 3. \tag{4.6}$$

The above requirement of complementary partition means that at each boundary point one of the factors must be prescribed in each of the terms in the expression for the mechanical work of boundary traction, namely

$$\int_{\partial\Omega} \boldsymbol{u} \cdot \boldsymbol{\sigma} \cdot \boldsymbol{n}\, ds = \int_{\partial\Omega} u_i \sigma_{ij} n_j \, ds \tag{4.7}$$

$$= \int_{\partial\Omega} \left[u_1 \sigma_{1j} n_j + u_2 \sigma_{2j} n_j + u_3 \sigma_{3j} n_j \right] ds. \tag{4.8}$$

Regular, or well-posed problems form only a subset of all problems encountered in the mechanics of linear thermoleastic solids. Examples of nonregular problems, also known as *ill-posed problems*, include nondestructive identification of defects, such as cracks or of inclusions, or constitutive parameters, and tomographic reconstruction problems (Bonnet and Constantinescu, 2005).

Some classical boundary conditions are

- *Encastre*:

$$\boldsymbol{u} = \boldsymbol{0}, \qquad \text{that is,} \quad u_i = 0 \quad \forall i. \tag{4.9}$$

- *Traction-free surface*:

$$\boldsymbol{\sigma} \cdot \boldsymbol{n} = \boldsymbol{0} \qquad \sigma_{ij} n_j = 0 \quad \forall i. \tag{4.10}$$

- *Frictionless sliding contact*:
  Displacement normal to the boundary is prescribed, but lateral sliding is unrestrained, so that shear tractions vanish,

$$\boldsymbol{u} \cdot \boldsymbol{n} = u^d, \quad \boldsymbol{t}_1 \cdot \boldsymbol{\sigma} \cdot \boldsymbol{n} = 0, \quad \text{and} \quad \boldsymbol{t}_2 \cdot \boldsymbol{\sigma} \cdot \boldsymbol{n} = 0, \tag{4.11}$$

  where $\boldsymbol{t}_1$ and $\boldsymbol{t}_2$ are two surface tangent vectors that are perpendicular to each other and to the normal $\boldsymbol{n}$.
- *Prescribed normal and shear tractions*:

$$\boldsymbol{\sigma} \cdot \boldsymbol{n} = -p\,\boldsymbol{n} + q_1 \boldsymbol{t}_1 + q_2 \boldsymbol{t}_2 \tag{4.12}$$

$$\boldsymbol{n} \cdot \boldsymbol{\sigma} \cdot \boldsymbol{n} = -p \quad \text{and} \quad \boldsymbol{t}_1 \cdot \boldsymbol{\sigma} \cdot \boldsymbol{n} = q_1 \quad \text{and} \quad \boldsymbol{t}_2 \cdot \boldsymbol{\sigma} \cdot \boldsymbol{n} = q_2. \tag{4.13}$$

4.2 DISPLACEMENT FORMULATION

The regular linear thermoelastic problem can be solved in the displacement formulation, that is, by assuming that the vector displacement field is the principal unknown variable. This technique is associated with the names of Lamé and Clapeyron. Substituting equation (4.1) into (4.2) and then into (4.3), and using the symmetry of the elastic stiffness tensor $\boldsymbol{C}$, one obtains the *displacement equations of equilibrium*:

$$\operatorname{div}(\boldsymbol{\sigma}_0 + \boldsymbol{C} : (\operatorname{grad}\boldsymbol{u} - \boldsymbol{A}\theta)) + \boldsymbol{f} = \boldsymbol{0}. \tag{4.14}$$

In the displacement formulation the following solution procedure may be adopted:

An admissible displacement field satisfying the displacement boundary conditions (4.4) is assumed and substituted into kinematic equations (4.1) and linear thermoelastic constitutive equations (4.2). It is then verified whether the stresses obtained in this way satisfy the static equilibrium equations (4.3) and the tractions satisfy the boundary conditions (4.5). If not, an improved trial admissible displacement field is selected and the procedure is repeated.

Conversely, once a displacement vector field $\boldsymbol{u}$ is found that satisfies the displacement equations of equilibrium (4.14), with the strain defined by the kinematic equations (4.1), and stress defined by the linear thermoelastic constitutive equation (4.2), then the stress equation of equilibrium (4.3) is satisfied.

In the isothermal case, $\theta = 0$, and in the absence of initial stresses $\boldsymbol{\sigma}_0$ (i.e., in the natural state), equation (4.14) reduces to

$$\operatorname{div}(\boldsymbol{C} : \operatorname{grad}\boldsymbol{u}) + \boldsymbol{f} = \boldsymbol{0}. \tag{4.15}$$

Equation (4.15) for isotropic material reduces to the *Navier equation*,

$$\mu \Delta \boldsymbol{u} + (\lambda + \mu) \operatorname{grad} \operatorname{div} \boldsymbol{u} + \boldsymbol{f} = \boldsymbol{0}, \tag{4.16}$$

or, equivalently, in terms of engineering elastic constants,

$$\Delta \boldsymbol{u} + \frac{1}{1-2\nu} \operatorname{grad} \operatorname{div} \boldsymbol{u} + \frac{2(1+\nu)}{E} \boldsymbol{f} = \boldsymbol{0}. \tag{4.17}$$

Because

$$\Delta \boldsymbol{u} = \operatorname{grad} \operatorname{div} \boldsymbol{u} - \operatorname{curl} \operatorname{curl} \boldsymbol{u}, \tag{4.18}$$

equation (4.16) can be rewritten as

$$(\lambda + 2\mu) \operatorname{grad} \operatorname{div} \boldsymbol{u} - \mu \operatorname{curl} \operatorname{curl} \boldsymbol{u} + \boldsymbol{f} = \boldsymbol{0}. \tag{4.19}$$

The application to the above equation of the operators div and curl respectively leads to the following results:

$$(\lambda + 2\mu) \Delta \operatorname{div} \boldsymbol{u} = -\operatorname{div} \boldsymbol{f}, \tag{4.20}$$

$$\mu \Delta \operatorname{curl} \boldsymbol{u} = -\operatorname{curl} \boldsymbol{f}. \tag{4.21}$$

If the body force field is such that $\operatorname{div} \boldsymbol{f}$ and $\operatorname{curl} \boldsymbol{f}$ both vanish, then both $\operatorname{div} \boldsymbol{u}$ and $\operatorname{curl} \boldsymbol{u}$ are harmonic fields. Furthermore, changing the order of operators yields

$$\operatorname{div} \Delta \boldsymbol{u} = 0 \quad \text{and} \quad \operatorname{curl} \Delta \boldsymbol{u} = 0.$$

Hence, by equation (4.18), $\Delta \boldsymbol{u}$ is a vector field that is harmonic; that is,

$$\Delta \Delta \boldsymbol{u} = \boldsymbol{0}. \tag{4.22}$$

Thus, if the body force field $\boldsymbol{f}$ is divergence-free and curl-free, then the displacement field is *biharmonic*.

From the kinematic equations (4.1), it follows that

$$\Delta \Delta \boldsymbol{\varepsilon} = \boldsymbol{0}; \tag{4.23}$$

that is, the strain tensor is also biharmonic.

4.3 STRESS FORMULATION

In the stress formulation, the tensor stress field is used as the principal unknown variable. This technique is associated with the names of Beltrami and Michell. The strains are determined from the stresses using the compliance form of the linear thermoelasticity constitutive equations (4.2), namely,

$$\boldsymbol{\varepsilon} = \boldsymbol{A}\theta + \boldsymbol{C}^{-1} : (\boldsymbol{\sigma} - \boldsymbol{\sigma}_0). \tag{4.24}$$

As demonstrated in the discussion of kinematics, in order for a given tensor strain field to correspond to a compatible displacement field, the strain compatibility equation must be satisfied:

$$\operatorname{inc} \boldsymbol{\varepsilon} = -(\operatorname{curl} (\operatorname{curl} \boldsymbol{\varepsilon}^T)^T = \boldsymbol{0}. \tag{4.25}$$

Substitution of (4.24) into the above equation leads to the ***stress equation of compatibility***:

$$\mathrm{inc}\,[\boldsymbol{A}\theta + \boldsymbol{C}^{-1} : (\boldsymbol{\sigma} - \boldsymbol{\sigma}_0)] = \mathbf{0}. \tag{4.26}$$

This equation ensures that a unique vector displacement field $\boldsymbol{u}$ can be constructed. Provided the resulting displacements satisfy boundary conditions on parts $\partial\Omega_i^d$ of the boundary, a complete solution of the linear thermoelastic problem is obtained.

The stress formulation solution procedure may be adopted as follows. First, a statically admissible tensor stress field is selected, that is, a field that satisfies stress equilibrium conditions (4.3) and traction boundary on parts $\partial\Omega_i^t$ of the boundary. Strains are then determined in terms of stresses using equation (4.24) and substituted into the compatibility equation (4.25). If these equations are not satisfied, another trial admissible stress field must be selected. If compatibility is verified, displacements can be obtained by back integration, and displacement boundary conditions on parts $\partial\Omega_i^d$ of the boundary enforced.

In the isothermal case, $\theta = 0$, and in the absence of initial stress $\boldsymbol{\sigma}_0$, equation (4.26) reduces to

$$\mathrm{inc}\,(\boldsymbol{C}^{-1} : \boldsymbol{\sigma}) = \mathbf{0}. \tag{4.27}$$

For isotropic linearly elastic material the compliance form of the constitutive linear elastic equations is

$$\boldsymbol{\varepsilon} = \frac{1+\nu}{E}\boldsymbol{\sigma} - \frac{\nu}{E}(\mathrm{tr}\,\boldsymbol{\sigma})\mathbf{1}. \tag{4.28}$$

Useful relationships between stress and strain can be obtained from the above equation by applying trace and divergence operators, respectively:

$$\mathrm{tr}\,\boldsymbol{\varepsilon} = \frac{1-2\nu}{E}(\mathrm{tr}\,\boldsymbol{\sigma}) \tag{4.29}$$

$$\mathrm{div}\,\boldsymbol{\varepsilon} = -\frac{\nu}{E}\mathrm{grad}\,\boldsymbol{\sigma} + \frac{1+\nu}{E}\mathrm{div}\,\boldsymbol{\sigma}. \tag{4.30}$$

To obtain the compatibility equation of stress for isotropic material, expression (4.28) is substituted into the strain compatibility equation in the form

$$\triangle\,\boldsymbol{\varepsilon} + \mathrm{grad}\,\mathrm{grad}\,(\mathrm{tr}\,\boldsymbol{\varepsilon}) - (\nabla + \nabla^T)\boldsymbol{\varepsilon} = \mathbf{0}. \tag{4.31}$$

In conjunction with the stress equilibrium equation (4.3), the following result is obtained:

$$\Delta\boldsymbol{\sigma} + \frac{1}{1+\nu}\mathrm{grad}\,\mathrm{grad}\,(\mathrm{tr}\,\boldsymbol{\sigma}) - \frac{\nu}{1+\nu}\Delta(\mathrm{tr}\,\boldsymbol{\sigma})\mathbf{1} + (\nabla + \nabla^T)\boldsymbol{f} = \mathbf{0}. \tag{4.32}$$

The above equation is known as the *Beltrami–Michell equation*.

Taking the trace of the above equation, it is found that

$$\Delta(\mathrm{tr}\,\boldsymbol{\sigma}) = -\frac{1+\nu}{1-\nu}\mathrm{div}\boldsymbol{f}. \tag{4.33}$$

Substituting this result back into the Beltrami–Michell equation, it is found that

$$\Delta\boldsymbol{\sigma} + \frac{1}{1+\nu}\mathrm{grad}\,\mathrm{grad}\,(\mathrm{tr}\,\boldsymbol{\sigma}) + \frac{\nu}{1-\nu}\mathrm{div}\boldsymbol{f}\,\mathbf{1} + (\nabla + \nabla^T)\boldsymbol{f} = \mathbf{0}. \tag{4.34}$$

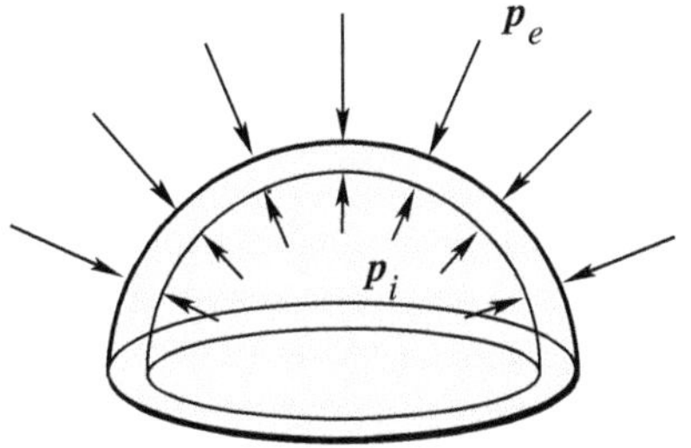

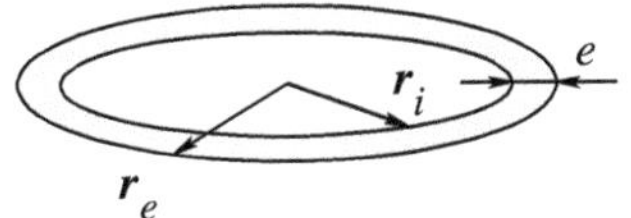

Figure 4.1. The upper half of a spherical reservoir, a spherical shell under pressure.

In the absence of body forces, $\boldsymbol{f} = \mathbf{0}$, the stress compatibility equation takes the simple form

$$\triangle\boldsymbol{\sigma} + \frac{1}{1+\nu}\operatorname{grad}\operatorname{grad}(\operatorname{tr}\boldsymbol{\sigma}) = \mathbf{0}. \tag{4.35}$$

If the volume forces are constant, both $\operatorname{div}\boldsymbol{f}$ and $\operatorname{grad}\boldsymbol{f}$ vanish, and the following results are obtained. It is found from equation (4.33) that

$$\triangle(\operatorname{tr}\boldsymbol{\sigma}) = 0. \tag{4.36}$$

Taking the Laplacian of the Beltrami–Michell equation and using the commutativity of differential operators, it is established that the tensor field of stresses is biharmonic:

$$\triangle\triangle\boldsymbol{\sigma} = \mathbf{0}. \tag{4.37}$$

Equations (4.22), (4.23), and (4.37) demonstrate that for an isothermal natural state under the action of a constant body force, the elastic fields of displacements, strains, and stresses are all biharmonic. Therefore a close relationship exists between the family of solutions of linear thermoelastic problems and the solutions of the biharmonic equation.

The Mathematica package supplied with this book provides an efficient means of evaluating differential operators of arbitrary tensor fields in various orthogonal coordinate systems, and thus of verifying biharmonicity of these fields and their suitability as elastic solutions. Some tools are provided to establish equivalence between differential forms, as well as methods of writing analytical expression for the general solutions of the biharmonic equation in different coordinate systems.

4.4 EXAMPLE: SPHERICAL SHELL UNDER PRESSURE

Let us consider the problem of a spherical reservoir under pressure (see Figure 4.1). The reservoir is a spherical shell with internal and external radii of r_i and r_e, respectively, made from a linear elastic material with moduli (λ, μ). We shall seek to compute the complete solution to this problem, assuming that the internal and the external pressure fields are uniform and equal to p_i and p_e, respectively. Gravity and thermal effects are neglected.

In order to solve the problem we first set the coordinate system to spherical and then start to search for a general form of solution under the assumption of spherical symmetry, with a displacement

$$\boldsymbol{u}(r, \theta, \varphi) = u_r(r)\boldsymbol{e}_r.$$

Strains and stresses are computed next.

```
SetCoordinates[Spherical[r, t, p]]
CoordinatesToCartesian[{r, t, p}]

u[r_, t_, p_] := {ur[r], 0, 0}
eps = Strain[ u [r, t, p]]
sig = Lambda Tr[eps ] IdentityMatrix[3 ] + 2 Mu eps
```

The equation to be satisfied is the stress field equilibrium. Note that only the divergence component along $\boldsymbol{e}_r$ is nonzero. This leads to the following equation:

$$(\lambda + 2\mu)\frac{\partial}{\partial r}\left(\frac{1}{r^2}\frac{\partial}{\partial r}\left(r^2 u_r(r)\right)\right) = 0.$$

The general solution of this equation is computed using **DSolve**. The option **GeneratedParameters** sets the name of the constants of integration.

```
(divsig = Simplify[Div[ sig ]] ) // MatrixForm
solur = DSolve[ divsig[[1]] == 0, ur, r ,
            GeneratedParameters -> CR][[1, 1]]

uu = u [r, t, p] /. solur
 eps = Strain[ uu ]
 sig = \[Lambda] Tr[eps ] IdentityMatrix[3 ] + 2 \ [Mu] eps
```

The general form of displacement, strain, and stress fields satisfying the balance equations in spherical coordinates with spherical symmetry is

$$\boldsymbol{u}(r, \theta, \varphi) = \left(c_1 r + \frac{c_2}{r^2}\right)\boldsymbol{e}_r$$

$$\boldsymbol{\varepsilon}(r, \theta, \varphi) = \left(c_1 - 2\frac{c_2}{r^3}\right)\boldsymbol{e}_r \otimes \boldsymbol{e}_r + \left(c_1 + \frac{c_2}{r^3}\right)(\boldsymbol{e}_\theta \otimes \boldsymbol{e}_\theta + \boldsymbol{e}_\varphi \otimes \boldsymbol{e}_\varphi)$$

$$\boldsymbol{\sigma}(r, \theta, \varphi) = \left((3\lambda + 2\mu)c_1 - 4\mu\frac{c_2}{r^3}\right)\boldsymbol{e}_r \otimes \boldsymbol{e}_r + \left((3\lambda + 2\mu)c_1 + 2\mu\frac{c_2}{r^3}\right)(\boldsymbol{e}_\theta \otimes \boldsymbol{e}_\theta + \boldsymbol{e}_\varphi \otimes \boldsymbol{e}_\varphi),$$

where the constants c_1 and c_2 will be determined from the boundary conditions.

The boundary conditions are defined by the internal and the external pressure as

$$\sigma(r, \theta, \phi) \cdot \boldsymbol{n} = -p(r)\boldsymbol{n}$$

with

$$r = r_i \quad \boldsymbol{n} = -\boldsymbol{e}_r$$

and respectively

$$r = r_e \quad \boldsymbol{n} = \boldsymbol{e}_r.$$

```
er = {1, 0, 0}
normal = - er
eqi = ( Thread [ (sig /. r -> ri) . normal == - pi normal ] )

normal =  er
eqe = (  Thread[ (sig /. r -> re) . normal == - pe normal ] )

sol = Solve[ {eqi[[1]], eqe[[1]]} ,   {C[1], C[2]} ][[1]]

uf = Simplify[ uu /. sol ]
epsf = Simplify[ eps /. sol ]
sigf = Simplify[ sig /. sol ]
```

The constants obtained are

$$c_1 = -\frac{1}{3\lambda + 2\mu} \frac{p_e r_e^3 - p_i r_i^3}{r_e^3 - r_i^3} \qquad c_2 = -\frac{1}{4\mu} \frac{(p_e - p_i) r_e^3 r_i^3}{r_e^3 - r_i^3}.$$

Let us now simplify the previous solution, using the assumption that the shell forming the reservoir is thin, that is,

$$r_i = \left(R - \frac{e}{2}\right) \quad r_e = \left(R + \frac{e}{2}\right) \qquad \frac{e}{R} \ll 1,$$

where e represents the thickness of the shell.

In order to carry out the series expansion we have to introduce *two* small parameters,

$$\eta = \frac{e}{R} \ll 1, \qquad \delta = \left(\frac{r}{R} - 1\right) \ll 1,$$

defining the rule that makes it possible to move forth and back between the different variables.

Using the `Series` MATHEMATICA operator we proceed to the series expansion. The two steps ensure that the combinations of δ and η are eliminated.

We can finally check the divergence of the obtained stress field and note that it is of order $o(\delta)$.

```
r2eta = {ri-> R(1-eta/2), re-> R(1 + eta/2) , r-> R(1 + delta)}
delta2r = {delta ->  r / R - 1, eta -> ee / R}
```

```
Expand[ Normal[
   Series[sigf /. r2eta, {eta,0,1}, {delta,0,1}]] ]

ssig = Normal[Series[%, {eta, 0, 0}]]
rsig = Simplify[ ssig /. delta2r ]

Simplify[ Div[ rsig ] ]
Series[% /. r2eta, {delta, 0, 1}]
```

By introducing the notation $\bar{p} = (p_i + p_e)/2$, $\Delta p = p_i - p_e$, the final stress field for a thin spherical shell is obtained as

$$\boldsymbol{\sigma}(r,\theta,\varphi) = \left(-\bar{p} - \Delta p\,\frac{r-R}{e}\right)\boldsymbol{e}_r \otimes \boldsymbol{e}_r + \left(-\bar{p} - \Delta p\,\frac{r-2R}{2e}\right)(\boldsymbol{e}_\theta \otimes \boldsymbol{e}_\theta + \boldsymbol{e}_\varphi \otimes \boldsymbol{e}_\varphi)\,. \tag{4.38}$$

The stress fields of a thin spherical shell can equally be obtained by simply computing the balance of forces of a hemisphere under the hypothesis that $\sigma_{\theta\theta}$ is constant through the thickness. We obtain

$$2\pi\, Re\sigma_{\theta\theta} = \pi R^2(p_i - p_e) \tag{4.39}$$

and

$$\sigma_{\theta\theta} = (p_e - p_i)\frac{R}{2e},$$

which is an expression of order η^{-1}.

4.5 SUPERPOSITION PRINCIPLE

Linearity of the system of equations of thermoelasticity has certain strong implications for the properties of solutions. An equivalent formulation is known as the ***superposition principle***.

Let there exist two deformation states of a linear thermoelastic solid Ω that satisfy the system of equations of elasticity and are given by

- *State 1*: solution $(\boldsymbol{u}_1, \boldsymbol{\varepsilon}_1, \boldsymbol{\sigma}_1)$ corresponding to the following fields:
  initial stresses $\boldsymbol{\sigma}_1^0$
  temperature changes θ_1
  body forces $\boldsymbol{f}_1$ in Ω
  surface tractions $\boldsymbol{t}_1$ on $\partial\Omega_i^t$
  boundary displacements $\boldsymbol{u}_1$ on $\partial\Omega_i^d$.
- *State 2*: solution $(\boldsymbol{u}_2, \boldsymbol{\varepsilon}_2, \boldsymbol{\sigma}_2)$ corresponding to the following fields:
  initial stresses $\boldsymbol{\sigma}_2^0$
  temperature changes θ_2
  body forces $\boldsymbol{f}_2$ in Ω
  surface tractions $\boldsymbol{t}_2$ on $\partial\Omega_i^t$

boundary displacements $\boldsymbol{u}_2$ on $\partial\Omega_i^d$.

Then the following state **is also a solution.**

- *State 3*: linear combination of *State 1* and *State 2* with real coefficients a_1 and a_2, given by $(a_1\boldsymbol{u}_1 + a_2\boldsymbol{u}_2, a_1\boldsymbol{\varepsilon}_1 + a_2\boldsymbol{\varepsilon}_2, a_1\boldsymbol{\sigma}_1 + a_2\boldsymbol{\sigma}_2)$. This solution corresponds to
  initial stresses $a_1\boldsymbol{\sigma}_1^0 + a_2\boldsymbol{\sigma}_2^0$
  temperature changes $a_1\theta_1 + a_2\theta_2$
  body forces $a_1\boldsymbol{f}_1 + a_2\boldsymbol{f}_2$ in Ω
  surface tractions $a_1\boldsymbol{t}_1 + a_2\boldsymbol{t}_2$ on $\partial\Omega_i^t$
  boundary displacements $a_1\boldsymbol{u}_1 + a_2\boldsymbol{u}_2$ on $\partial\Omega_i^d$.

4.6 QUASISTATIC DEFORMATION AND THE VIRTUAL WORK THEOREM

Quasistatic deformation implies that all deformation states between the initial and final configuration are in a state of equilibrium. Physically this assumption means that the force and work terms associated with mass acceleration are negligible compared to the terms associated with internal and external forces acting on the body. In other words, it is assumed that under quasistatic deformation the virtual power of inertial terms is equal to zero.

Consider a sufficiently smooth virtual displacement field in Ω, denoted $\boldsymbol{u}^*$. Under quasistatic deformation, $\boldsymbol{u}^*$ evolves with the corresponding virtual velocity field $\boldsymbol{v}^*$ in Ω. Under the assumption of zero power of inertial forces, the total virtual power of external and internal forces must be equal to zero,

$$\mathcal{P}_e^*(\boldsymbol{v}^*) + \mathcal{P}_i^*(\boldsymbol{v}^*) = 0, \quad \forall \boldsymbol{v}^*, \tag{4.40}$$

where $\mathcal{P}_e^*(\boldsymbol{v}^*)$ and $\mathcal{P}_e^*(\boldsymbol{v}^*)$ denote respectively the power of external and internal forces on the virtual velocity field $\boldsymbol{v}^*$, given by

$$\mathcal{P}_e^*(\boldsymbol{v}^*) = \int_\Omega \boldsymbol{f} \cdot \boldsymbol{v}^*\, dv + \int_{\partial\Omega} \boldsymbol{t} \cdot \boldsymbol{v}^*\, ds \tag{4.41}$$

and

$$\mathcal{P}_i^*(\boldsymbol{v}^*) = -\int_\Omega \boldsymbol{\sigma} : \boldsymbol{D}^*(\boldsymbol{v}^*)\, dv, \tag{4.42}$$

where $\boldsymbol{D}^*(\boldsymbol{v}^*)$ denotes the deformation rate corresponding to the virtual velocity field $\boldsymbol{v}^*$,

$$D^*(v^*) = \frac{1}{2}(\text{grad}\, v^* + \text{grad}\,^T v^*) \tag{4.43}$$

4.7 UNIQUENESS OF SOLUTION

Consider two equlibrium states of the same body Ω:

- **State 1**: solution $(\boldsymbol{u}_1, \boldsymbol{\varepsilon}_1, \boldsymbol{\sigma}_1)$ for body force $\boldsymbol{f}_1$ in Ω, tractions $\boldsymbol{t}_1$ on $\partial\Omega_i^t$, and displacements $\boldsymbol{u}_1$ on $\partial\Omega_i^d$.
- **State 2**: solution $(\boldsymbol{u}_2, \boldsymbol{\varepsilon}_2, \boldsymbol{\sigma}_2)$ for body force $\boldsymbol{f}_2$ in Ω, tractions $\boldsymbol{t}_2$ on $\partial\Omega_i^t$, and displacements $\boldsymbol{u}_2$ on $\partial\Omega_i^d$.

Assume that there are no initial stresses, $\boldsymbol{\sigma}_0 = \mathbf{0}$, and the temperature state is isothermal, $\theta = 0$. Now let

$$\boldsymbol{f}_1 = \boldsymbol{f}_2 \qquad \boldsymbol{t}_1 = \boldsymbol{t}_2 \qquad \boldsymbol{u}_1 = \boldsymbol{u}_2.$$

Then the two solutions may at most differ only by a rigid body displacement

$$\boldsymbol{u}_1 = \boldsymbol{u}_2 + \boldsymbol{a} + \boldsymbol{b} \times \boldsymbol{x} \tag{4.44}$$

$$\boldsymbol{\varepsilon}_1 = \boldsymbol{\varepsilon}_2 \tag{4.45}$$

$$\boldsymbol{\sigma}_1 = \boldsymbol{\sigma}_2, \tag{4.46}$$

where $\boldsymbol{a}, \boldsymbol{b}$ are two real vectors representing rigid body translation and rigid body rotation, respectively.

The proof of uniqueness is obtained by considering the difference between the two states. The solution $(\boldsymbol{\xi}, \boldsymbol{\varepsilon}, \boldsymbol{\sigma}) = (\boldsymbol{u}_1, \boldsymbol{\varepsilon}_1, \boldsymbol{\sigma}_1) - (\boldsymbol{u}_2, \boldsymbol{\varepsilon}_2, \boldsymbol{\sigma}_2)$ corresponds to the body force $\mathbf{0}$ in Ω, tractions $\mathbf{0}$ on $\partial\Omega_i^t$, and displacements $\mathbf{0}$ on $\partial\Omega_i^d$. Because

In the absence of initial stresses and temperatures changes, the application of Clapeyron's theorem (see below) leads to the following equality:

$$2\mathcal{U}^I(\boldsymbol{\xi}) = \int_\Omega \boldsymbol{\varepsilon} : \boldsymbol{C} : \boldsymbol{\varepsilon}\, dv = 0. \tag{4.47}$$

The positive definiteness of the tensor of elastic moduli, $\boldsymbol{C}$, implies from the above equation that

$$\boldsymbol{\varepsilon} = 0 \quad \text{and therefore} \quad \boldsymbol{\sigma} = 0.$$

The integration of the strain field to obtain the displacement field leads to the desired conclusion that the two solutions may only differ by the displacements due to rigid body motion:

$$\boldsymbol{\xi} = \boldsymbol{a} + \boldsymbol{b} \times \boldsymbol{x}, \quad \boldsymbol{a}, \boldsymbol{b} \in \mathbb{R}^3.$$

4.8 ENERGY POTENTIALS

Existence of the strain energy potential

External work done on an elastic body is stored in the form of the potential energy of elastic deformation, or elastic strain energy, and can be recovered provided the heat tranfer is negligible. The elastic strain energy is a scalar field. Total strain energy of the body can be expressed as

$$\mathcal{U}^I = \int_\Omega u^I\, dv.$$

In a deformed state described by the strain field $\boldsymbol{\varepsilon}$, strain energy u^I is a function of $\boldsymbol{\varepsilon}$. Strain energy serves as the potential function for the corresponding stress field; that is,

$$\boldsymbol{\sigma} = \frac{\partial u^I}{\partial \boldsymbol{\varepsilon}} \qquad \sigma_{ij} = \frac{\partial u^I}{\partial \varepsilon_{ij}} \quad \forall i, j = 1, 2, 3. \tag{4.48}$$

For linear thermoelastic material the strain potential $u^I(\boldsymbol{\xi}^*)$ corresponding to the displacement field $\boldsymbol{\xi}^*$ and strain field $\boldsymbol{\varepsilon}^*$ is given by the expression

$$u^I(\boldsymbol{\xi}^*) = \boldsymbol{\sigma}_0 : \boldsymbol{\varepsilon}^* + \frac{1}{2}\boldsymbol{\varepsilon}^* : \boldsymbol{C} : \boldsymbol{\varepsilon}^* + \boldsymbol{\varepsilon}^* : \boldsymbol{A}\theta. \tag{4.49}$$

The scalar potential function u^I does not represent the complete strain energy, since it does not take into account all energy terms, for example, those representing heat transfer.

The complete strain energy potential of the body Ω is defined as

$$\mathcal{U}^I(\boldsymbol{\xi}^*) = \int_\Omega \left(\boldsymbol{\sigma}_0 : \boldsymbol{\varepsilon}^* + \frac{1}{2}\boldsymbol{\varepsilon}^* : \boldsymbol{C} : \boldsymbol{\varepsilon}^* + \boldsymbol{\varepsilon}^* : \boldsymbol{A}\theta\right) dv. \tag{4.50}$$

In the absence of initial stresses and temperature changes ($\boldsymbol{\sigma}_0 = \mathbf{0}$, $\theta = O$), the strain energy potential takes the simplified form

$$\mathcal{U}^I = \frac{1}{2}\int_\Omega \boldsymbol{\varepsilon}^* : \boldsymbol{C} : \boldsymbol{\varepsilon}^* \, dv.$$

It can be verified by a derivation showing that the definition of the strain energy potential (4.49) satisfies (4.48). It is worth noting the key role played by the symmetry of the elastic tensor $\boldsymbol{C}$, $C_{ijkl} = C_{klij}$.

The construction of the strain energy potential offers some useful insight into the properties of potential functions and the phyical meaning of deformation energy. We start with the definition of the power of internal forces associated with material particle at $\boldsymbol{x} \in \Omega$:

$$\boldsymbol{\sigma}(\boldsymbol{x}) : \dot{\boldsymbol{\varepsilon}}(\boldsymbol{x}).$$

The mechanical work done to transform the particle from State 1 determined by $(\boldsymbol{u}_1, \boldsymbol{\varepsilon}_1, \boldsymbol{\sigma}_1)$ into State 2 determined by $(\boldsymbol{u}_2, \boldsymbol{\varepsilon}_2, \boldsymbol{\sigma}_2)$ is obtained by integrating the mechanical power of internal forces over the path of equlibrium states $(\boldsymbol{u}(t), \boldsymbol{\varepsilon}(t), \boldsymbol{\sigma}(t))$ $(t \in [t_1, t_2])$:

$$\int_{t_0}^{t_1} \boldsymbol{\sigma} : \dot{\boldsymbol{\varepsilon}} \, dt = \int_{\varepsilon_0}^{\varepsilon_1} \boldsymbol{\sigma} : d\boldsymbol{\varepsilon}. \tag{4.51}$$

The existence of a strain energy potential is ensured by the path independence of the above integral; that is, a function $u^I(\boldsymbol{\varepsilon})$ exists such that

$$\boldsymbol{\sigma} = \frac{\partial u^I}{\partial \boldsymbol{\varepsilon}}.$$

In the mathematical formalism of differential forms this is expressed as

$$\boldsymbol{\sigma} : d\boldsymbol{\varepsilon} = d\mathcal{F}$$

and leads to

$$\int_{\varepsilon_0}^{\varepsilon_1} \boldsymbol{\sigma} : d\boldsymbol{\varepsilon} = \int_{\varepsilon_0}^{\varepsilon_1} d\mathcal{F} = \mathcal{F}(\boldsymbol{\varepsilon}_1) - \mathcal{F}(\boldsymbol{\varepsilon}_0).$$

By the Poincaré lemma (Spivak, 1965) (also known as the Cauchy integration formula), the existence of the potential $u^I(\boldsymbol{\varepsilon})$ is ensured if and only if

$$\frac{\partial \sigma_{ij}}{\partial \varepsilon_{kl}} = \frac{\partial \sigma_{kl}}{\partial \varepsilon_{ij}} \qquad i, j, k, l = 1\text{–}3,$$

which is equivalent to the previously cited symmetry of the elastic moduli,

$$C_{ijkl} = C_{klij}\,.$$

The potential function is then defined by equation (4.49).

Existence of the complementary energy potential

Theorems: Virtual work theorem For every quasistatic deformation process (i.e., in the absence of inertial forces) and for every virtual displacement field $\boldsymbol{v}$ the work of internal and the work of external forces are equal,

$$\int_\Omega \boldsymbol{\sigma} : \boldsymbol{\varepsilon}[\boldsymbol{v}]\, dv = \int_\Omega \boldsymbol{f} \cdot \boldsymbol{v}\, dv + \int_{\partial\Omega} \boldsymbol{t} \cdot \boldsymbol{v}\, ds, \tag{4.52}$$

where $\boldsymbol{\varepsilon}[\boldsymbol{v}] = \frac{1}{2}\left(\nabla \boldsymbol{v} + \nabla^T \boldsymbol{v}\right)$ and $\boldsymbol{t} = \boldsymbol{\sigma} \cdot \boldsymbol{n}$.

The equality is derived directly from the principle of virtual work by replacing the virtual velocity field with a virtual displacement field, or from the integration of the quasistatic equilibrium equations using the Stokes theorem.

The transition from the virtual velocity field to the virtual displacement field amounts only to a change of physical dimensions in volume integrals, thereby transforming power into work. The virtual displacement field considered here is arbitrary and independent of the actual deformation of the solid.

Theorems: The Clapeyron theorem In the absence of initial stresses ($\boldsymbol{\sigma}_0 = \mathbf{0}$), and of temperature changes ($\theta = 0$), the virtual work of external and internal forces computed on the actual displacement field is equal to twice the value of the strain energy potential:

$$\int_\Omega \boldsymbol{\sigma} : \boldsymbol{\varepsilon}\, dv = \int_\Omega \boldsymbol{f} \cdot \boldsymbol{u}\, dv + \int_{\partial\Omega} \boldsymbol{t} \cdot \boldsymbol{u}\, dv = 2\mathcal{U}_i(\boldsymbol{u}). \tag{4.53}$$

The proof follows directly from the virtual work theorem (4.52) and the definition of the strain energy potential (4.49).

The Clapeyron theorem provides a physical interpretation of the positive definiteness of the tensor of elastic moduli $\boldsymbol{C}$. More precisely, the positive definiteness of $\boldsymbol{C}$ ensures that exterior loading must supply a positive finite quantity of mechanical work in order to deform the body:

$$0 < \int_\Omega \boldsymbol{\varepsilon} : \boldsymbol{C} : \boldsymbol{\varepsilon}\, dv < +\infty \tag{4.54}$$

$$0 < \int_\Omega \boldsymbol{f} \cdot \boldsymbol{v}\, dv + \int_{\partial\Omega} \boldsymbol{t} \cdot \boldsymbol{v}\, ds < +\infty. \tag{4.55}$$

A negative definite tensor of elastic moduli $\boldsymbol{C}$ would imply that the body may supply energy to the exterior, leading to the lack of stability of the deformation process.

We also recall simple mechanical tests illustrating the positive definiteness of $\mathbf{C}$ for case of isotropic elasticity presented in Section 4.3.

It is important to note the difference between the *virtual* work of a force computed on the real displacement as a particular realisation of the field of virtual displacement, for example, for the body force

$$\int_{\Omega} \boldsymbol{f} \cdot \boldsymbol{u} \, dv,$$

and the *actual* mechanical work of external surface tractions or body forces done on the body between the initial and actual configurations. The latter in the case of *linear* elasticity is equal to

$$\frac{1}{2} \int_{\Omega} \boldsymbol{f} \cdot \boldsymbol{u} \, dv.$$

The above expression can be obtained by integration of instantaneous power over time.

4.9 RECIPROCITY THEOREMS

Theorems: The Maxwell–Betti reciprocity theorem In the absence of initial stresses $\boldsymbol{\sigma}_0 = \mathbf{0}$ and temperature changes $\theta = 0$, consider two equilibrium states of the same body Ω:

- *State 1*: solution $(\boldsymbol{u}_1, \boldsymbol{\varepsilon}_1, \boldsymbol{\sigma}_1)$ for body force $\boldsymbol{f}_1$ in Ω, tractions $\boldsymbol{t}_1$ on $\partial\Omega_i^t$ and displacements $\boldsymbol{u}_1$ on $\partial\Omega_i^d$.
- *State 2*: solution $(\boldsymbol{u}_2, \boldsymbol{\varepsilon}_2, \boldsymbol{\sigma}_2)$ for body force $\boldsymbol{f}_2$ in Ω, tractions $\boldsymbol{t}_2$ on $\partial\Omega_i^t$ and displacements $\boldsymbol{u}_2$ on $\partial\Omega_i^d$.

Then the virtual mechanical work of external forces of State 1 computed on the displacements of State 2 is equal to the virtual mechanical work of external forces of State 2 computed on the displacements of State 1:

$$\int_{\Omega} \boldsymbol{f}_1 \cdot \boldsymbol{u}_2 \, dv + \int_{\partial\Omega} \boldsymbol{t}_1 \cdot \boldsymbol{u}_2 \, ds = \int_{\Omega} \boldsymbol{f}_2 \cdot \boldsymbol{u}_1 \, dv + \int_{\partial\Omega} \boldsymbol{t}_2 \cdot \boldsymbol{u}_1 \, ds.$$

The proof is obtained as a consequence of the virtual work theorem and the symmetry of $\boldsymbol{C}$. Indeed, the virtual work theorem implies the following equality between the virtual mechanical work of internal forces:

$$\int_{\Omega} \boldsymbol{\varepsilon}_1 : \boldsymbol{C} : \boldsymbol{\varepsilon}_2 \, dv = \int_{\Omega} \boldsymbol{\varepsilon}_2 : \boldsymbol{C} : \boldsymbol{\varepsilon}_1 \, dv.$$

This equality is satisfied if and only if $\boldsymbol{C}$ is symmetric,

$$C_{ijkl} = C_{klij}, \qquad \forall i, j, k, l = 1\text{–}3.$$

This is a fundamental property of $\boldsymbol{C}$ and has already been pointed out as a key requirement in the proof of the existence of strain energy potential.

It is therefore possible to check the symmetry of the tensor of elastic moduli and implicitly for the existence of the strain energy potential by performing loading tests on elastic bodies and by computing the 'reciprocal work' from experimental measurements. This kind of experiment was first invented and preformed by Faraday in 1834 on elastic rods (Figure 4.2).

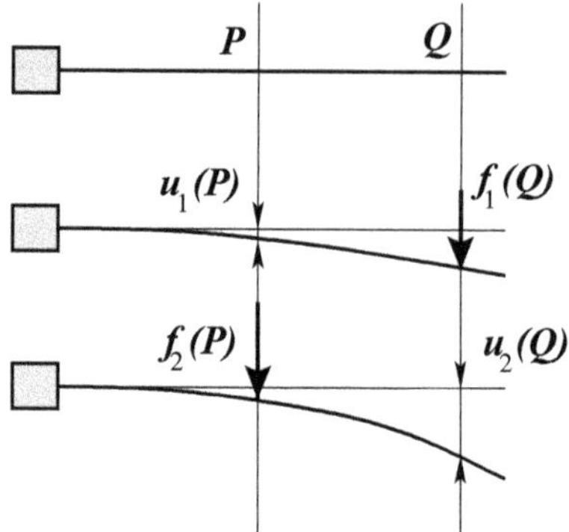

Figure 4.2. Faraday's experiments on elastic rods providing an illustration of the Maxwell–Betti reciprocity principle.

The two states considered in the Maxwell–Betti reciprocity theorem are defined by the application of forces $f_1(P)$ at P and $f_2(Q)$ at Q. The reciprocity theorem in this case is expressed by

$$f_1(P) \cdot u_2(P) = f_2(Q) \cdot u_1(Q),$$

where $u_1(Q)$ is the displacement in State 1 measured at Q and respectively $u_2(P)$ is the displacement in State 2 measured at P.

A simple proof of the reciprocity theorem can be obtained for this case if the work done on the elastic system is compared for two loading sequences leading to the same final state. Sequence 1 involves the simultaneous proportional application of forces $f_1(P)$ and $f_2(Q)$. Sequence 2 involves the application of force $f_1(P)$ followed by the application of force $f_2(Q)$. The external work for each sequence is stored in the form of elastic strain energy of the final state, and therefore

$$\begin{aligned}&\frac{1}{2}f_1(P)(u_1(P)+u_2(P))+\frac{1}{2}f_2(Q)(u_1(Q)+u_2(Q))\\&\quad=\frac{1}{2}f_1(P)u_1(P)+\frac{1}{2}f_2(Q)u_2(Q)+f_1(P)u_2(P),\end{aligned}$$

leading directly to the required relationship.

Within the framework of linear elasticity the forces and displacements are related through a linear matrix equation:

$$\begin{bmatrix} f_1(P) \\ f_2(Q) \end{bmatrix} = \begin{bmatrix} L_{PP} & L_{PQ} \\ L_{QP} & L_{QQ} \end{bmatrix} \begin{bmatrix} u_1(P) \\ u_2(Q) \end{bmatrix}. \tag{4.56}$$

A simple algebraic computation that is left to the reader as an exercise permits to show that the Maxwell–Betti theorem for this case is verified if and only if the matrix L is symmetric.

Another exercise related to the analysis of Faraday's experiments with elastic rods is to to show that the work done between the initial state at time $t = 0$ and the final state at time $t = 1$ computed over two different integration paths is equal if and only if the matrix L is symmetric (see Figure 4.3).

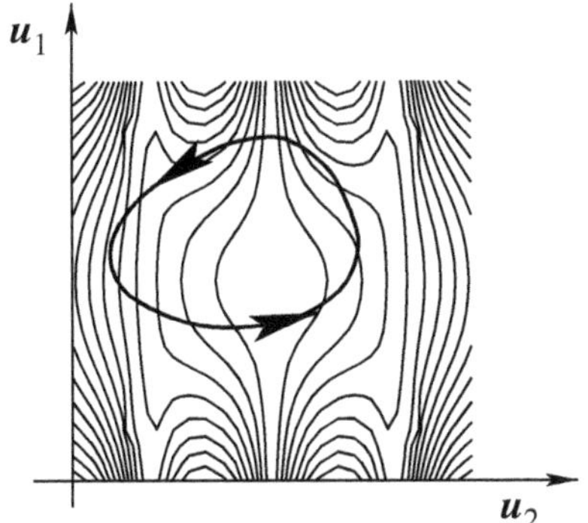

Figure 4.3. A closed loading path over a potential energy landscape.

Hint: Consider two different paths and compute the actual work from power using the integral

$$\int_{t_0}^{t_1} \boldsymbol{f}(t) \cdot \dot{\boldsymbol{u}}(t) dt = \int_{t_0}^{t_1} f_1(P,t)\dot{u}(P,t) + f_2(Q,t)\dot{u}(Q,t) dt.$$

4.10 THE SAINT VENANT PRINCIPLE

The formulation of elastic problems introduced in the beginning of this chapter emphasises the crucial role played by the boundary conditions in determining the solution. A natural question therefore arises: how sensitive are the solutions for the elastic stress and strain fields to the precise details of the displacement and traction boundary conditions? What boundary conditions are most appropriate, for example, when considering bending of a beam or torsion of a shaft?

The fundamental idea that provides an answer to this question was first published in 1855 by J. C. Adhémar (also known as Barré de Saint Venant, 1855). Saint Venant continued earlier work by Navier on the bending of beams and put forward the following conjecture:

> **The Saint Venant principle**
> Consider the elastic solution for a shaft subjected to a torque applied at its end. Far from the ends the elastic fields describing this solution are independent of the exact distribution of surface tractions that generate the end torque loading.

In this section we consider a cylindrical rod, illustrated in Figure 4.4, and assume that it is subject to loading only at its extremes. We shall then construct a closed form solution for this problem that will depend only on the force and moment resultants of the end traction distribution. The demonstration of the Saint Venant conjecture in application to this case would amount to showing that the difference between this approximate solution and the full solution becomes negligible far from the shaft ends.

The extension of the conjecture in the general case, that is, bodies of arbitrary shape, as well as the existing mathematical proofs for these results will be discussed briefly in the next section.

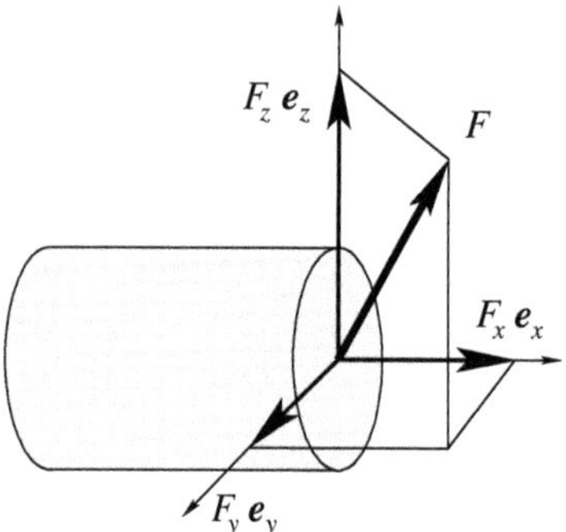

Figure 4.4. The resultant traction field and its components with respect to the coordinate system associated with the end section of a cylindrical shaft.

Consider an elastic shaft of length L with a cylindrical section S, $\Omega = [0, L] \times S$. Coordinate axes are chosen to represent the geometric axes of inertia of section S:

$$\int_S y\, ds = 0, \qquad \int_S z\, ds = 0, \qquad \int_S y\, z\, ds = 0. \tag{4.57}$$

The shaft is in equilibrium under two distributions of surface tractions applied to its end sections. The force and moment resultants for the section $x = 0$, see Figure 4.4, are given by

$$\begin{aligned} \boldsymbol{F} &= F_x \boldsymbol{e}_x + F_y \boldsymbol{e}_y + F_z \boldsymbol{e}_z = \int_{S(x=0)} \boldsymbol{\sigma} \cdot \boldsymbol{n}\, ds, \\ \boldsymbol{M} &= M_x \boldsymbol{e}_x + M_y \boldsymbol{e}_y + M_z \boldsymbol{e}_z = \int_{S(x=0)} \boldsymbol{x} \times (\boldsymbol{\sigma} \cdot \boldsymbol{n})\, ds. \end{aligned} \tag{4.58}$$

Below we construct the Saint Venant solution of this problem using MATHEMATICA following the reasoning presented in Ballard and Millard (2005) and Bamberger (1997). We recall the relation between elastic moduli:

$$\lambda = \frac{\nu E}{(1+\nu)(1-2\nu)}, \qquad \mu = \frac{E}{1+\nu}.$$

The necessary packages must be loaded and the coordinate system set to cartesian in this example.

```
<< Tensor2Analysis.m

SetCoordinates[Cartesian[x, y, z]]
```

We begin with the assumption that the stress tensor corresponds to antiplane shear and tension,

$$\boldsymbol{\sigma} = \begin{pmatrix} \sigma_{xx} & \sigma_{xy} & \sigma_{xz} \\ \sigma_{xy} & 0 & 0 \\ \sigma_{xz} & 0 & 0 \end{pmatrix}.$$

Consideration of the equilibrium equation $\operatorname{div} \boldsymbol{\sigma} = 0$ shows that it requires that σ_{xy}, σ_{xz} be independent of x.

```
sigma[x_, y_, z_] :=
   {{sxx[x, y, z], sxy[x, y, z], sxz[x, y, z]},
    {sxy[x, y, z], 0, 0},
    {sxz[x, y, z], 0, 0}}
sigma[x, y, z] // MatrixForm

div := Thread[ Div[ sigma[x, y, z]] == {0, 0, 0} ]
div // MatrixForm

sxy[x, y, z] = sxy[y, z]
sxz[x, y, z] = sxz[y, z]

eqdiv := Thread[ Div[ sigma[x, y, z]] == {0, 0, 0} ]
eqdiv // MatrixForm
```

We now calculate the strain field using the isotropic linear elastic constitutive law.

```
eps[x, y, z] :=
  (1+nu)/EE sigma[x, y, z] -
   nu/EE Tr[ sigma[x, y, z]] IdentityMatrix[3]
eps[x, y, z]  // MatrixForm
```

The application of the strain compatibility equation leads to the result that

$$\sigma_{xx,xx} = \sigma_{xx,yy} = \sigma_{xx,zz} = \sigma_{xx,yz} = 0.$$

This implies that stress σ_{xx} has the form of a polynomial function,

$$\sigma_{xx} = a_1\, xy + a_2\, xz + a_3\, x + a_4\, y + a_5\, z + a_6,$$

where $a_i \in \mathbb{R}$ are constants.

Before seeking the solution, we also check the form of boundary conditions involving σ_{xx}. The axial (x) component of the resultant applied to the rod should be equilibrated. Taking into account the expressions for moments of inertia of cross sections given in equation (4.57), this leads to

$$a_3 = 0.$$

```
inc := Thread[Flatten[Inc[eps[x,y,z]] ] == Array[0 &, 9]]
inc  // MatrixForm

sxx[x, y, z] = a1 x y + a2 x z + a3 x + a4 y + a5 z + a6

(traction = sigma[x, y, z] . {-1, 0, 0} ) // MatrixForm
(traction  /. x -> 0 ) // MatrixForm
(traction = sigma[x, y, z] . {1, 0, 0} ) // MatrixForm
```

```
(traction /. x -> L ) // MatrixForm

a3 = 0

inc // MatrixForm
```

The remaining equations (one equilibrium equation and two strain compatibility conditions) provide a series of three PDEs for $\sigma_{xy}(y, z)$ and $\sigma_{xz}(y, z)$,

$$\sigma_{xy,y} + \sigma_{xz,z} + a_1\, y + a_2\, z + a_3 = 0 \tag{4.59}$$

$$\Delta_2 \sigma_{xy}(y, z) = \frac{a_1}{1+\nu} \tag{4.60}$$

$$\Delta_2 \sigma_{xz}(y, z) = \frac{a_2}{1+\nu}, \tag{4.61}$$

where Δ_2 denotes the Laplacian operator in the (y, z) plane.

```
eq1 = div[[1, 1]]

sol = Solve[ D[ div[[1]], y ], D[sxz[y, z], y, z]]
eq2 = Simplify[ inc[[2]] /. sol[[1]]]

sol = Solve[ D[ div[[1]], z ], D[sxy[y, z], y, z]]
eq3 = Simplify[ inc[[3]] /. sol[[1]]]
```

The displacements can be obtained by integration of normal strains,

$$\varepsilon_{xx} = u_{x,x} \quad \varepsilon_{yy} = u_{y,y} \quad \varepsilon_{zz} = u_{z,z},$$

with the addition of a vector of unknown functions:

$$\frac{1}{E} K_x(y, z)\boldsymbol{e}_x - \frac{\nu}{E} K_y(x, z)\boldsymbol{e}_x - \frac{\nu}{E} K_z(y, z)\boldsymbol{e}_x.$$

The use of the '*set delayed*' operator $:=$ for defining the displacement vector $\boldsymbol{u}$ permits automatical update of expression during manipulation.

```
u := Map[
   Integrate[ strain[[#, #]] , pp[[#]] ]  +
       If[ # == 1, 1 / EE, - nu / EE]
    ToExpression["K" <> ToString[pp[[#]]] ]
       @@ Drop[pp, {#}] & , Range[3]]

u
```

Expressions for unknown functions $K_x(y, z)$, $K_y(x, z)$, $K_z(y, z)$ allow additional simplification by observing that

$$\varepsilon_{yz} = \frac{1+\nu}{E} \sigma_{yz} = 0.$$

```
epsyz = D[ u[[3]], y ] + D[ u[[2]], z ]

sky = DSolve[ D[epsyz, z] == 0 , Ky[x, z], z]
Ky[x, z] = sky[[1, 1, 2]] /. C[1] -> Ky[x] /. C[2] -> -  K[x]

skz = DSolve[ D[epsyz, y] == 0 , Kz[x, y], y]
Kz[x, y] = skz[[1, 1, 2]] /. C[1] -> Kz[x] /. C[2] ->  K[x]
```

Substituting the results into the expressions for strains and stresses and integrating to obtain the equilibrium equation, expressions for unknown functions $K(x)$, $K_y(x)$, and $K_z(x)$ are found.

Equilibrium also provides an equation for $K_z(y, z)$:

$$\Delta_2 K_z(y, z) = -2(a_1\, y + a_2\, z). \tag{4.62}$$

The solution is now complete after the following function is introduced:

$$\phi(y, z) = K_x(y, z) + E\, r_z\, z - E\, r_y\, y.$$

```
strain = Simplify[ 1/2 (Grad[u] + Transpose[Grad[u]] )]
sig = Simplify[
 EE nu /(1+nu)/(1-2 nu) Tr[strain] IdentityMatrix[3]+
      EE / (1 + nu) strain]


newdiv := Simplify[Div[sig]]
newdiv
newdiv //  MatrixForm


K[x] = c x - EE / nu rx
newdiv = newdiv /. K''[x] -> 0

DSolve[ newdiv[[2]] == 0  , Ky''[x], x]
Ky[x] =%[[1, 1, 2]] /. C[1] -> - EE ly  /. C[2] -> - EE ry

DSolve[ newdiv[[3]] == 0, Kz''[x], x]
Kz[x] = %[[1, 1, 2]] /. C[1] -> - EE lz /. C[2] -> - EE rz

FullSimplify[ u /. Kx[y, z] -> phi[y, z] - EE rz  z + EE ry y]
strain = FullSimplify[ 1/2 (Grad[u] + Transpose[Grad[u]] )]
(sig = FullSimplify[
  EE  nu /(1 + nu) / (1 - 2 nu) Tr[strain] IdentityMatrix[3]+
  EE / (1 + nu) strain] ) // MatrixForm
```

The computed closed form elastic solution is defined by the following *displacement field*,

$$u_x = \frac{1}{E}\left(\frac{a_1}{2}x^2 y + \frac{a}{2}x^2 z + a_4\, xy + a_5\, xz + a_6 x\right) + \frac{1}{E}\phi(y,z) + (r_y\, y - \ r_z\, z) \qquad (4.63)$$

$$\begin{aligned} u_y = &-\frac{1}{E}\left(\frac{a_1}{6}\left(-\left(x\left(x^2+3\,\nu\, y^2\right)\right)+3\,\nu\, x\, z^2\right) - a_2\,\nu\, x\, y\, z \right.\\ &\left. -\frac{a_4}{2}\left(x^2+\nu\,(y-z)\,(y+z)\right) - a_5\,\nu\, y\, z + c\,\nu\, x\, z - a_6\,\nu\, y\right)\\ &+(l_y + r_z\, x - r_x\, z) \end{aligned} \qquad (4.64)$$

$$\begin{aligned} u_z = &-\frac{1}{E}\left(a_1\,\nu\, x\, y\, z - \frac{a_2}{6}\left(-x^3+3\,\nu\, x\, y^2 - 3\,\nu\, x\, z^2\right) + a_4\,\nu\, y\, z\right.\\ &\left. -\frac{a_5}{2}\left(-3\,x^2+3\,\nu\, y^2-3\,\nu\, z^2\right) + a_6\,\nu\, z + c\,\nu\, x\, y\right)\\ &+(l_z + r_y\, x + r_x\, y), \end{aligned} \qquad (4.65)$$

and the corresponding components of the *stress field*,

$$\sigma_{xx} = a_6 + a_4\, y + a_1\, x\, y + a_5\, z + a_2\, x\, z \qquad (4.66)$$

$$\sigma_{xy} = \frac{1}{4\,(1+\nu)}\left(\nu\,\left(2\,(c-a_2\, y)\, z + a_1\,\left(z^2-y^2\right)\right) + 2\,\phi^{(1,0)}(y,z)\right) \qquad (4.67)$$

$$\sigma_{xz} = \frac{1}{4\,(1+\nu)}\left(-\left(\nu\,\left(2\,c\,y - a_2\,y^2 + 2\,a_1\,y\,z + a_2\,z^2\right)\right) + 2\,\phi^{(0,1)}(y,z)\right) \qquad (4.68)$$

$$\sigma_{yy} = \sigma_{zz} = \sigma_{yz} = 0. \qquad (4.69)$$

We can now verify that the stress field satisfies the equilibrium equation

$$\operatorname{div}\boldsymbol{\sigma} = 0.$$

We also verify that the boundary condition on the cylindrical surface with the normal

$$\boldsymbol{n} = n_y \boldsymbol{e}_y + n_z \boldsymbol{e}_z = \cos\theta \boldsymbol{e}_y + \sin\theta \boldsymbol{e}_z$$

is satisfied.

```
Simplify[Div[sig]]

(traction = sig . 0, Cos[t], Sin[t] ) // MatrixForm
```

The deplanation (warping) function ϕ(y, z)

Function $\phi(y,z)$, which appeared in the derivation, is referred to as the deplanation, or the warping function. It is the solution of the equation

$$\Delta_2\phi(y,z) = -2(a_1 y + a_2 z) \qquad (4.70)$$

with Neumann boundary condition arising from the traction-free requirement imposed on the cylindrical surface in the form

$$\frac{\partial\phi}{\partial\boldsymbol{n}} = \nu\left\{\frac{a_1}{2}(y^2-z^2) + a_2\, y\, z + \nu\, c\, z\right\} n_y + \nu\left\{a_1\, y\, z + \frac{a_2}{2}(z^2-y^2) - \nu\, c\, y\right\} n_z.$$

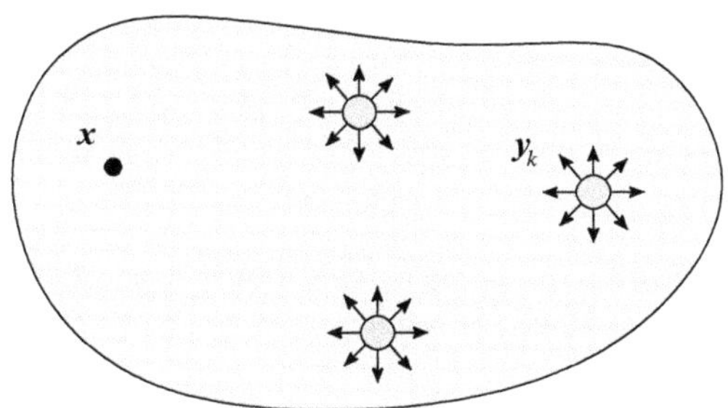

Figure 4.5. The elastic body and applied surface tractions considered in the von Mises–Sternberg formulation of the Saint Venant principle.

Function $\phi(y, z)/E$ describes axial displacements of points lying on planar sections normal to the longitudinal axis of the rod.

From the form of the displacement field we can readily deduce that constants l_x, l_y, l_z and r_x, r_y, r_z define the components of translation and rotation of the rigid body displacement field:

$$\boldsymbol{l} + \boldsymbol{r} \times \boldsymbol{x}.$$

Constants $a_1, a_2, a_4, a_5, a_6, c$ are defined from the boundary conditions at the loaded ends $x = 0$ or $x = L$.

The solution of the problem about torsion of a shaft has been established for the case when loading is described only in terms of the force and moment resultants acting on the end sections of the rod, rather than by imposing the detailed distribution of the vector field of tractions, as one might expect for a well-posed boundary value problem of elasticity. It is apparent that, according to Saint Venant, far from the region of application of surface tractions the stress solution depends only on the resultant force and moment of applied surface tractions.

In spite of its apparent simplicity, precise mathematical proof of this conjecture turns out to be challenging.

An important result was obtained by Boussinesq (1885) in the form of the solution for an elastic half-space subjected to a concentrated load (Section 7.1). The resulting stress field decays as $1/r^2$ with distance r from the point of application of the concentrated load. This confirms, albeit indirectly, the Saint Venant conjecture: only the force resultant affects the stress distribution. A similar conclusion can also be drawn from the Cerruti solution (Section 7.1) for a tangential concentrated force applied at the surface of an elastic half plane.

General formulations for bodies of arbitrary shape were proposed by von Mises and later revisited by Sternberg 1954. (See Figure 4.5.)

The Saint Venant principle in the von Mises–Sternberg formulation

Consider Ω to be a regular linear elastic domain with boundary $\partial\Omega$.

Let us assume that the distribution of surface tractions $\boldsymbol{t}$ vanishes outside small nonintersecting neighbourhoods $B_{\boldsymbol{y}_k}(r)$ of the point $\boldsymbol{y}_k$ k, 1, K of radius r.

Then for each $\boldsymbol{x} \in \Omega$ we have for $r \longrightarrow 0$ the relations for the solution

$$\boldsymbol{\sigma}(\boldsymbol{x}) = o(r^{\alpha}) \qquad \boldsymbol{\xi}(\boldsymbol{x}) = o(r^{\alpha})$$

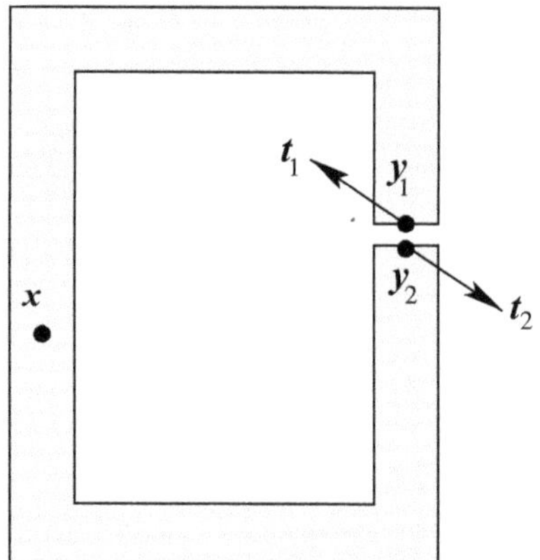

Figure 4.6. A structure proposed by Hoff (1952) as a counter example to the Saint Venant principle.

with

- $\alpha = 2$ if at least one of the resultant forces is not vanishing, and if there exist k such that

$$\int_{B_y(r)} \boldsymbol{t}\, ds \neq 0.$$

- $\alpha = 3$ if all the resultant forces are zero:

$$\int_{B_y(r)} \boldsymbol{t}\, ds = 0.$$

- $\alpha = 4$ if the resultant forces and moments are zero:

$$\int_{B_y(r)} \boldsymbol{t}\, ds = 0 \qquad \int_{B_y(r)} \boldsymbol{x} \wedge \boldsymbol{t}\, ds = 0.$$

Counter-examples proposed by Hoff

Hoff was an aerospace engineer working on thin-walled and slender structures. Hoff (1952) identified a number of cases where the Saint Venant principle does not apply. As an example, consider a rod of the shape shown in Figure 4.7. If this structure is subject to loads represented by the opposite forces $\boldsymbol{t}_1$ in $\boldsymbol{y}_1$ and $\boldsymbol{t}_2$ in $\boldsymbol{y}_2$, then both force and moment

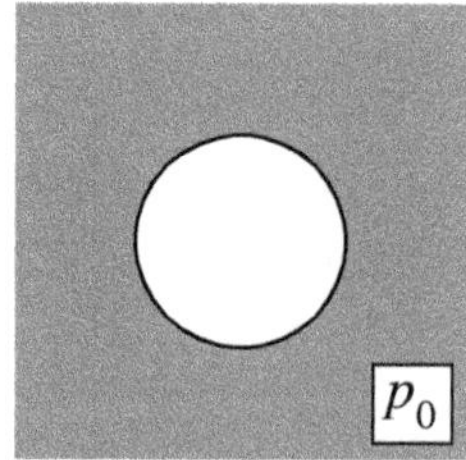

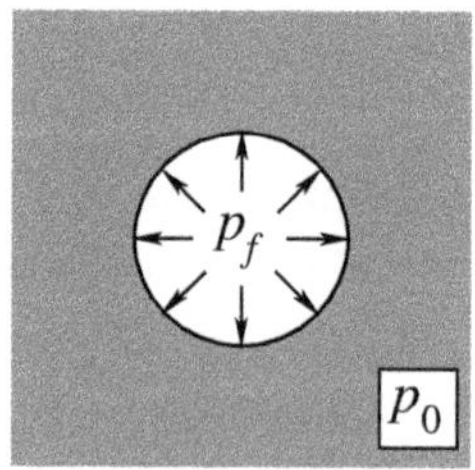

Figure 4.7. Undisturbed rock mass (left), excavated cavity (middle) and filled cavity (right).

resultants vanish as points $\boldsymbol{y}_1$ and $\boldsymbol{y}_2$ approach each other. However, the application of the original Saint Venant formulation would lead one to conlcude falsely that stress at $\sigma(\boldsymbol{x})$ at point $\boldsymbol{x}$ remote from $\boldsymbol{y}_1$ and $\boldsymbol{y}_2$ becomes negligibly small. The more precise von Mises–Sternberg formulation shows that if one of the two independent neighbourhoods of points $\boldsymbol{y}_1$ and $\boldsymbol{y}_2$ are considered, then the resultant force in each of them taken separately does not vanish in this case. The Hoff example therefore does not refute the Saint Venant principle, but highlights the importance of using its correct formulation.

SUMMARY

In this chapter the formulation of the complete elasticity problem is addressed. Displacement and stress formulations are introduced. The superposition priniciple is then presented, and the fundamental virtual work theorem is used to prove uniqueness of the elastic solution, followed by other fundamental theorems (Clapeyron and Maxwell–Betti). Finally, the Saint Venant principle is discussed.

EXERCISES

1. Isotropic elastic spherical shell under a temperature gradient

Consider a spherical shell made from an isotropic elastic material with constants (λ, μ). The shell is subjected to a temperature gradient defined by Θ_i and Θ_e, the temperatures of the internal $(r = r_i)$ and external $(r = r_e)$ surfaces, respectively. Using a spherical coordinate system under the assumption of spherical symmetry:

(a) Compute the temperature distribution, assuming a time-independent state. We recall that the heat equation for an isotropic homogenous material is defined as

$$\operatorname{div} k \operatorname{grad} \Theta(\boldsymbol{x}, t) = c \frac{\partial}{\partial t} \Theta(\boldsymbol{x}, t).$$

(b) Compute the displacement field using spherically symmetric solution of the balance equation.

(c) Compute the displacement, strain, and stress field solutions of the problem, assuming that internal and external surfaces of the shell are traction-free.

(d) Compute the solution in the case of a thin shell, that is, if

$$r_i = \left(R - \frac{e}{2}\right) \quad r_e = \left(R + \frac{e}{2}\right) \quad \frac{e}{R} \ll 1,$$

where e represents the shell thickness.

(e) Can the principal terms of the thin shell solution be obtained through reasoning similar to that used in Section 4.4 ?

Hint: See notebook `C04_sphere_temperature.nb`.

2. Isotropic elastic spherical shell subjected to pressure and temperature loading

Consider a spherical shell subjected to both a pressure and a temperature loading,

- pressures p_i and p_e, and
- temperatures Θ_i and Θ_e,

at the internal ($r = r_i$) and external ($r = r_e$) surfaces, respectively. Solve the problem:
(a) directly, using the method of solution of the preceeding problem,
(b) using the superposition principle.

Hint: Combine notebooks `C04_sphere_pressure.nb` and `C04_sphere_temperature.nb`.

3. Isotropic solid elastic sphere subjected to pressure and temperature loading

Consider the case of a solid sphere of radius r_e subject to external pressure p_e and temperature Θ_e.
(a) Express explicitly the spherical symmetry, and write the boundary conditions.
(b) Compute the complete solution.

4. Excavation of an underground cavity (G.Rousset, J.Salençon, in Polytechnique, 1990–2005)

Consider exacavation of an underground cavity of spherical shape with a radius $r = r_i$ (see Figure 4.7). The cavity is to be situated at a depth h, $h \gg r$, below the surface. The rock mass is considered to be a homogenous isotropic elastic body with moduli (λ, μ). We shall model this situation as an infinite elastic body subject to an initial pressure distribution p_0. Note that a spherical coordinate system centred in the middle of the cavity will be used.
(a) Compute the initial stress field $\boldsymbol{\sigma}_0$ and verify that this field satifies balance equations.
(b) Suppose that the cavity is excavated instantaneously and that the surface of the cavity becomes traction-free. Present explicitly the boundary conditions in this case after excavation.
(c) Compute the displacement, strain, and stress fields in the underground rock mass after the excavation.
(d) Taking into account the elastic displacement of the rock mass, determine the volume change of the cavity after exacavation. Interpret the result from the physical point of view.

Hint: See notebook `C04_underground_reservoir.nb.`

5. Filling of an underground cavity with liquified gas (Polytechnique, 1990–2005)

In the second stage of this analysis we fill the underground cavity of the preceeding exercise with liquified gas at pressure p_f.
(a) Suppose that the cavity is filled instantaneously. Write explicitly the boundary conditions in this case after filling.
(b) Compute the displacement, strain, and stress fields of the rock mass arising between the excavation phase and the filling phase.
(c) Discuss the application of the superposition principle in this case.

Hint: See notebook `C04_underground_reservoir.nb.`

6. Transversely isotropic elastic sphere under pressure (Lehnitski, 1981)

Consider a spherical shell of internal and external radii r_i and r_e, respectively. The shell is made of material that possesses transversely isotropic material symmetry oriented along the $\boldsymbol{e}_r$ direction, meaning that $\boldsymbol{\theta}$ and $\boldsymbol{\varphi}$ are equivalent directions everywhere.

The shell is subjected to both pressure and temperature loading,

- pressures p_i and p_e, and
- temperatures Θ_i and Θ_e,

at the internal ($r = r_i$) and external ($r = r_e$) surfaces, respectively.

(a) Explain why one can consider the problem under the assumptions of spherical symmetry.
(b) Compute the displacement, strain, and stress fields for the shell subjected to external pressure in the absence of thermal loading.
(c) Compute the temperature field in the shell, assuming that the stationary heat conduction equation can be written in the form

$$\operatorname{div} \boldsymbol{k} \operatorname{grad} \Theta = 0,$$

with the transversely isotropic thermal conductivity given by the second-order tensor

$$\boldsymbol{k} = k_{rr} \boldsymbol{e}_r \otimes \boldsymbol{e}_r + k_{tt} (\boldsymbol{e}_\theta \otimes \boldsymbol{e}_\theta + \boldsymbol{e}_\varphi \otimes \boldsymbol{e}_\varphi).$$

(d) Compute the displacement, strain, and stress fields for the shell subjected to thermal loading, assuming that the dilatation tensor has the following form:

$$\boldsymbol{A} = a_{rr} \boldsymbol{e}_r \otimes \boldsymbol{e}_r + a_{tt} (\boldsymbol{e}_\theta \otimes \boldsymbol{e}_\theta + \boldsymbol{e}_\varphi \otimes \boldsymbol{e}_\varphi).$$

Hint: See notebook **C04_sphere_anisotropic.nb.**

Remark for the interested reader: The complete transient thermal problem can also be solved. For a complete reference on the subject see Eason and Ogden (1964). A brief draft of the solution is given in notebook **C04_sphere_anisotropic_heat.nb.**

7. Finite cylindrical tube subjected to pressure loading

Consider a cylindrical tube made from isotropic elastic material with moduli (λ, μ). The internal and external radii and the height of the cylinder will be denoted by r_i, r_e, h, respectively.

The cylinder is subjected to internal and external pressures p_i and p_e on its internal and external cylindrical surfaces, respectively. The end sections of the tube are subject to a pressure p_z.

(a) Compute the displacement, strain, and stress field for the shell subject to external pressures in the absence of thermal loading.
(b) How can the solution be obtained using the superposition principle?
(c) Compute the solution in the case of a thin cylinder, that is,

$$r_i = \left(R - \frac{e}{2}\right) \quad r_e = \left(R + \frac{e}{2}\right) \quad \frac{e}{R} \ll 1,$$

where e represents the shell thickness. Does the height of the cylinder play any role in the solution?

8. Finite cylindrical tube subjected to thermal loading

Consider the same cylinder as in the preceding exercise.

The cylinder this time has uniform temperatures Θ_i and Θ_e on its internal and external cylindrical surfaces, respectively. The end sections are assumed to prevent axial displacement of the tube without restricting transverse displacements, while at the same time ensuring zero thermal flux through the boundary.

(a) Propose a kinematically admissible displacement field that obeys cylindrical symmetry.
(b) Compute the temperature distribution within the cylinder.
(c) Compute displacement, strain, and stress fields for the shell due to thermal dilatation.

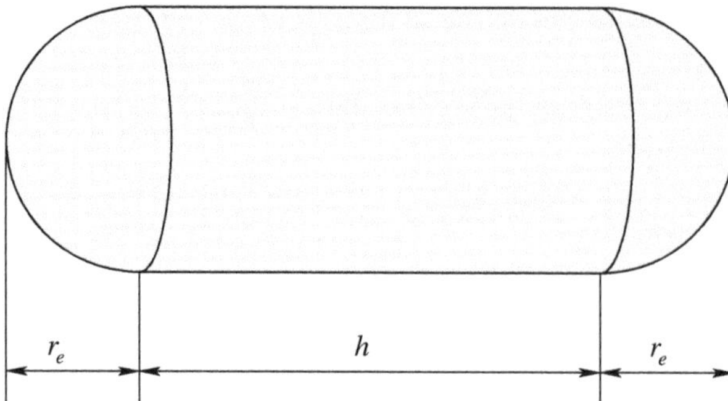

Figure 4.8. A 'diving bottle' shape formed out two hemispherical shells and one cylindrical shell.

(d) Compute the solution in the case of a thin cylinder, that is,

$$r_i = \left(R - \frac{e}{2}\right) \quad r_e = \left(R + \frac{e}{2}\right) \qquad \frac{e}{R} \ll 1,$$

where e represents the shell thickness.

(e) How does the solution change if the tube is replaced with a bar with the same external diameter ?

9. Finite elastically anisotropic cylindrical tube (Lehnitski, 1981)

Determine the solution for a cylindrical tube subject to the loading of the preceding two exercises considering that the tube is made from an anisotropic material, for the following cases of material symmetry:

(a) transverse isotropy
(b) cubic symmetry

10. A diving bottle under pressure loading

Diving bottles are manufactered by welding two hemispherical caps to a length of cylindrical tube of the same diameter. We (see Figure 4.8) that the spherical caps and the cylindrical tube have the same internal and external radii and are manufactured from the same linear elastic material. The bottle is subjected to pressures p_i and p_e on its internal and external surfaces, respectively.

(a) Compute the complete stress field for the case of materials with the following material symmetry:
   - isotropic
   - transversely isotropic.

(b) Compute the solution in the case of a thin spherical shell.
(c) Is it possible to obtain a closed form solution for the diving bottle subject to thermal loading by imposing uniform temperatures Θ_i and Θ_e on its internal and external surfaces, respectively?

11. Thermal assembly ('frettage') of cylindrical tubes (Ballard, in Polytechnique 1990–2005).

'Frettage' is a technique that permits the assembly of prestressed cylindrical tubes. An inner and an outer tube used are characterised by the radii r_i, r_e and R_i, R_e, respectively.

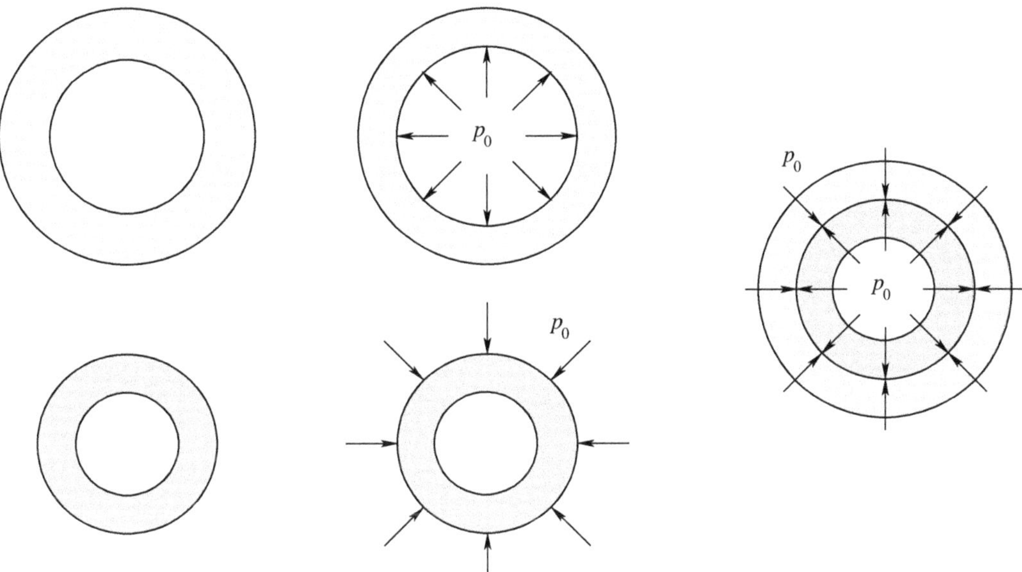

Figure 4.9. Tubes before and after 'assembly.'

The outer tube has a smaller interior radius than the exterior radius of the inner tube, that is, $R_i < r_e$. In order to make the assembly possible, the inner tube is cooled down (or the outer heated) to create a temparature difference θ. Once assembled, the tubes reform to the equilibrium state characterised by the interface pressure p_0 and a residual stress field $\boldsymbol{\sigma}_0$. The procedure is illustrated in Figure 4.9.

Applications of this method of assembly include, for example, fixing of valve seats in motor engine cylinder heads and creation of residual stress distributions in tubular structures in order to improve service performance.

In the following we assume that both tubes are made from isotropic linear elastic material with moduli (λ, μ) and that the contact between the tubes is frictionless. The assumption of frictionless contact is somewhat counterintuitive, as it means that the inner tube can be extracted from the assembly without effort; however, we shall preserve it as the first approximation in order to simplify computations.

(a) Specify the complete set of continuity conditions at the interface for displacements and tractions.
(b) Determine the minimal temperature change during cooling to allow assembly.
(c) Determine the frettage pressure p_0 and specify the stress distributions in both tubes.
(d) Now assume that the initial configuration of the structure corresponds to the already assembled frettage tube. The inner, interface, and outer radii are r_i, r_0, r_e, and the frettage pressure is p_0.
   - Compute the initial stress state $\boldsymbol{\sigma}_0$ in the assembly.
   - Compute the final displacement and stress state in the assembly when the frettage tube is subjected to service loading characterised by p_i and p_e, the inner and the outer pressure.

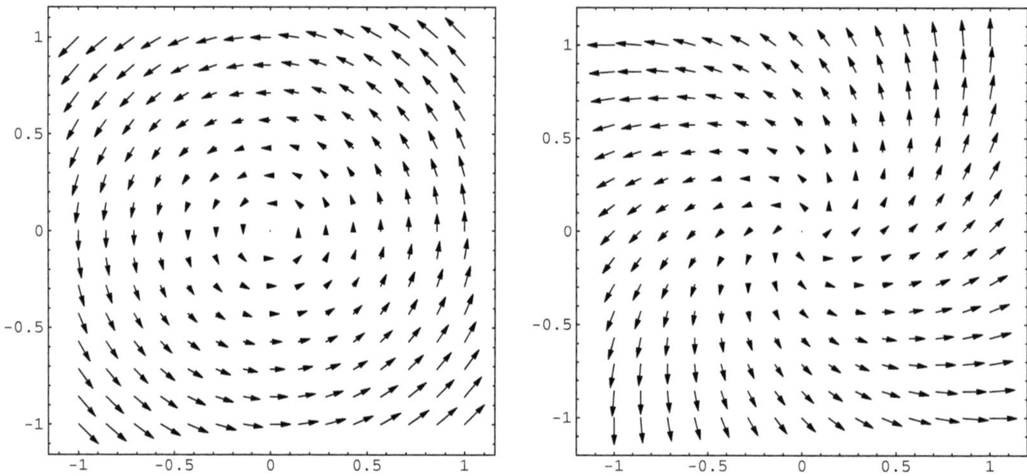

Figure 4.10. Two traction fields creating an equivalent torsion in the sense of Saint Venant.

12. Traction vector fields

Show that the following two vector fields, illustrated in Figure 4.10, are equivalent in the sense of Saint Venant when applied to a shaft with a sqare cross section $[-1, 1] \times [-1, 1]$:

- $\boldsymbol{t}_1 = -y\,\boldsymbol{e}_x + x\,\boldsymbol{e}_y$
- $\boldsymbol{t}_1 = (x\,\boldsymbol{e}_x + y\,\boldsymbol{e}_y) + \boldsymbol{t}_1$.

13. Cylindrical rod of arbitrary cross section under torsion

Consider an isotropic elastic cylindrical rod with arbitrary cross section under torsional loading. The body occupies in the initial reference frame the domain $\Omega = S \times [0, L]$ (see Figure 4.11). The relative angle of rotation between end sections by an angle α is prescribed. The lateral surface of the cylinder is traction-free.

(a) *Circular cross section S.* Verify that displacement field of the form

$$\boldsymbol{u} = \alpha \frac{z}{L} \boldsymbol{e}_z \times (x\,\boldsymbol{e}_x + y\,\boldsymbol{e}_y)$$

defines a complete solution of the torsion problem. Compute surface tractions $\boldsymbol{t} = \boldsymbol{\sigma} \cdot \boldsymbol{n}$ on the end sections, and show that their linear and angular momenta are

$$\boldsymbol{F} = \int_S \boldsymbol{t}\,ds = 0, \qquad \boldsymbol{M} = \int_S (x\,\boldsymbol{e}_x + y\,\boldsymbol{e}_y) \times \boldsymbol{t}\,ds = \frac{1}{2}\frac{\alpha}{L}\mu R^4,$$

where R is the radius of the cross section.

Under what conditions is the small strain hypothesis still valid?

(b) *Arbitrary cross section S.* Show that the previous solution is valid if and only if the cross section is circular.

Hint: Consider a parameterised description of the boundary of the cross section:

$$\boldsymbol{p} = r(\theta)\boldsymbol{e}_r = r(\theta)(\cos\theta\,\boldsymbol{e}_x + \sin\theta\,\boldsymbol{e}_y) \qquad \boldsymbol{p} \in \partial S.$$

Compute the tangent and normal vectors to ∂S, and examine the traction vector of the free surface:

$$\boldsymbol{t} = \boldsymbol{\sigma} \cdot \boldsymbol{n}.$$

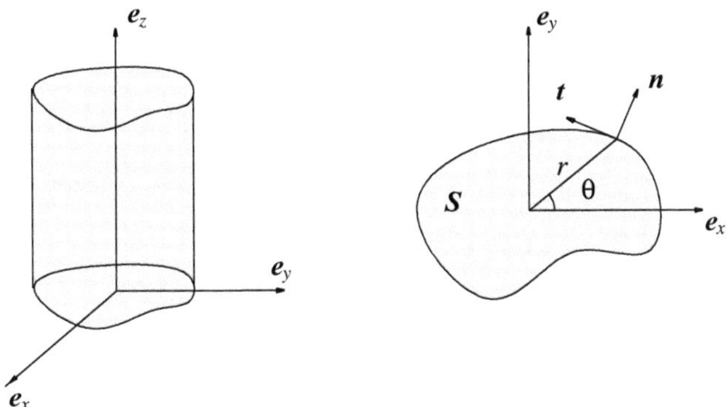

Figure 4.11. Cylindrical rod of arbitrary cross section.

(c) *Arbitrary cross section S; introduction of the warping function.* Show that the displacement field of the form

$$\boldsymbol{u} = \frac{\alpha z}{L}\boldsymbol{e}_z \times (x\,\boldsymbol{e}_x + y\,\boldsymbol{e}_y) + \frac{\alpha z}{L}\varphi(x, y)\boldsymbol{e}_z$$

defines a complete solution of the torsion problem, provided the warping function φ is a solution of the Neumann problem,

$$\Delta\varphi(x, y) = 0 \qquad (x, y) \in S$$

$$\frac{\partial\varphi}{\partial\boldsymbol{n}}(x, y) = \operatorname{grad}\varphi \cdot \boldsymbol{n} = yn_x - xn_y \qquad (x, y) \in \partial S,$$

where $\boldsymbol{n} = (n_x\,\boldsymbol{e}_x + n_y\,\boldsymbol{e}_y)$ is the normal unit vector to ∂S.

(d) *Arbitrary cross section S, solution with the warping function.* Using the solution defined in the previous question, compute the traction vector acting on the end sections. Show using the properties on the warping function and integration by parts that the linear and angular momenta on the end section are in this case

$$\begin{aligned}\boldsymbol{F} &= \int_S \boldsymbol{t}\,ds \\ &= \mu\frac{\alpha}{L}\int_{\partial S}\left(\frac{\partial\varphi}{\partial n} - yn_x + xn_y\right)(x\,\boldsymbol{e}_x + y\,\boldsymbol{e}_y)\,ds + \mu\frac{\alpha}{L}\int_{\partial S}(\Delta\varphi)\,(x\,\boldsymbol{e}_x + y\,\boldsymbol{e}_y)\,ds,\\ \boldsymbol{M} &= \int_S (x\,\boldsymbol{e}_x + y\,\boldsymbol{e}_y) \times \boldsymbol{t}\,ds = M_z\boldsymbol{e}_z = \frac{\alpha\,\mu}{L}J_0,\\ J_0 &= \int_S x\left(\frac{\partial\varphi}{\partial y} + x\right) - y\left(\frac{\partial\varphi}{\partial y} - y\right) = \int_S (x^2 + y^2)\,ds - \int_S |\operatorname{grad}\varphi|^2\,ds,\end{aligned}$$

where J_0 denotes the geometric torsional moment of cross section S.

5 Stress functions

OUTLINE

This chapter is devoted to the solution of elastic problems using the stress function approach. The Beltrami potential has already been introduced previously as a convenient form of representation for self-equilibrated stress fields. However, the main emphasis in the chapter is placed on the analysis of the Airy stress function formulation, even though it represents only a particular case of the Beltrami representation. The reason for this is the particular importance of this approach in the context of plane elasticity.

The Airy stress function approach is introduced taking particular care to ensure that conditions of strain compatibility are properly satisfied. The approximate nature of the plane stress formulation is elucidated.

The properties of Airy stress functions in cylindrical polar coordinates are then addressed. Particular care is taken to analyse some important fundamental solutions that serve as *nuclei of strain* within the elasticity theory, namely the solutions for a disclination, dislocations and dislocation dipoles, and concentrated forces.

Williams eigenfunction analysis of the stress state in an elastic wedge under homogeneous loading is presented next, and the elastic stress fields found around the tip of a sharp crack subjected either to opening or shear mode loading. Finally, two further important problems are treated, namely the Kirsch problem of remote loading of a circular hole in an infinite plate, and the Inglis problem of remote loading of an elliptical hole in an infinite plate.

5.1 PLANE STRESS

The previously introduced expression for the stress tensor in terms of the Beltrami potential is

$$\boldsymbol{\sigma} = \operatorname{inc} \boldsymbol{B}. \tag{5.1}$$

Because div inc $\boldsymbol{B} = 0$, the stress tensor defined in this way automatically satisfies equilibrium.

The Airy stress function solution corresponds to a special form of the Beltrami potential, namely

$$\boldsymbol{B} = A(x, y, z)\, \boldsymbol{e}_z \otimes \boldsymbol{e}_z. \tag{5.2}$$

In matrix form with respect to cartesian coordinates this Beltrami tensor is given by

$$\boldsymbol{B} = \begin{pmatrix} 0 & 0 & 0 \\ 0 & 0 & 0 \\ 0 & 0 & A(x, y, z) \end{pmatrix}. \tag{5.3}$$

Introducing the stress using the Airy form of the Beltrami potential ensures that the stress tensor satisfies the equilibrium equation $\operatorname{div}\boldsymbol{\sigma} = 0$ and that the corresponding stress state is planar:

$$\boldsymbol{\sigma} = \operatorname{inc}\boldsymbol{B} = \operatorname{inc}\, A(x, y, z)\,\boldsymbol{e}_z \otimes \boldsymbol{e}_z = \begin{pmatrix} \dfrac{\partial^2 A}{\partial y^2} & -\dfrac{\partial^2 A}{\partial x \partial y} & 0 \\ -\dfrac{\partial^2 A}{\partial x \partial y} & \dfrac{\partial^2 A}{\partial x^2} & 0 \\ 0 & 0 & 0 \end{pmatrix}. \tag{5.4}$$

The derivation is illustrated in the notebook **CQS_airy1.nb**:

```
<< Tensor2Analysis.m

SetCoordinates[Cartesian[x, y, z]]

B = {{0, 0, 0}, {0, 0, 0}, {0, 0, A[x, y, z]}}

(Stress = Inc[B]) // MatrixForm

(Div[Stress]) // MatrixForm
```

From now on we use the MATHEMATICA notation for partial differentiation with respect to the arguments, so that, for example

$$\mathbf{A}^{(\mathtt{i},\mathtt{j},\mathtt{k})} = \frac{\partial^{i+j+k} A}{\partial x^i \partial y^j \partial z^k}. \tag{5.5}$$

For a stress state to give a solution of the complete system of equations of elasticity, the corresponding strain state must also satisfy the compatibility equations. To compute the strain tensor we require the isotropic compliance tensor, which is constructed as follows:

$$S_{ijkl} = \frac{-\nu}{E}\delta_{ij}\delta_{kl} + \frac{(1+\nu)}{2E}(\delta_{ik}\delta_{jl} + \delta_{il}\delta_{jk}). \tag{5.6}$$

This allows the strain to be computed from the stress by the double dot product:

$$\epsilon = \boldsymbol{S} : \boldsymbol{\sigma}. \tag{5.7}$$

We implement the calculations in the form of the MATHEMATICA functions **DDot** (double dot product) and **IsotropicCompliance**. For the latter we include an alternative definition of the isotropic compliance tensor as a function of one parameter, **nu**. This is done for convenience so that Young's modulus **EE** can be set to unity.

```
DDot[T4_, t2_] := GTr[GDot[T4, t2, 1, 1], 1, 4]

IsotropicCompliance[EE_, nu_] := Array[
 -nu/EE KroneckerDelta[#1,#2] KroneckerDelta[#3,#4]+
 (1+nu)/2/EE(KroneckerDelta[#1,#3] KroneckerDelta[#2,#4]+
  KroneckerDelta[#1,#4]KroneckerDelta[#2,#3])&,{3,3,3,3}]

IsotropicCompliance[nu_] := IsotropicCompliance[1, nu]

SS = IsotropicCompliance[nu]

(Strain = DDot[SS, Stress]) // MatrixForm
```

The strain tensor has the form

$$\begin{pmatrix} \mathbf{A}^{(0,2,0)} - 2\,\mathbf{nu}\,\mathbf{A}^{(2,0,0)} & -(1+\mathbf{nu})\mathbf{A}^{(1,1,0)} & 0 \\ -(1+\mathbf{nu})\mathbf{A}^{(1,1,0)} & \mathbf{A}^{(2,0,0)} - 2\,\mathbf{nu}\,\mathbf{A}^{(0,2,0)} & 0 \\ 0 & 0 & -\mathbf{nu}\,(\mathbf{A}^{(0,2,0)} + \mathbf{A}^{(2,0,0)}) \end{pmatrix}. \tag{5.8}$$

Here the dependence of **A** on the arguments (**x**, **y**, **z**) is implied, but has been omitted.

The incompatibility tensor is computed by applying the operator inc to the strain tensor.

```
(Inc[Strain]) // MatrixForm
```

The resulting incompatibility tensor is a symmetric 3×3 tensor. The requirement of strain compatibility therefore leads to six equations which must be satisfied by the function $A(x, y, z)$.

In the context of plane elasticity, compliances are often expressed in terms of the Lamé coefficient μ and the Kolosov constant $\kappa = (3 - \nu)/(1 + \nu)$ for plane stress; for plane strain $\kappa = 3 - 4\nu$. For convenience we therefore also introduce another definition of compliance tensor given below.

IsotropicComplianceK is defined as a function of one or two parameters.

The one-parameter version assumes for simplicity that $2\mu = 1$.

```
IsotropicComplianceK[K_, mu_] :=
 Array[-(3-K)KroneckerDelta[#1,#2] KroneckerDelta[#3,#4]+
  2(KroneckerDelta[#1,#3] KroneckerDelta[#2,#4]+
   KroneckerDelta[#1,#4] KroneckerDelta[#2,#3])&,
          {3,3,3,3}]/(8 mu);

IsotropicComplianceK[K_] := IsotropicComplianceK[K, 1/2];
```

5.2 AIRY STRESS FUNCTION OF THE FORM A_0(x, y)

Conventionally the Airy stress function is used to solve plane problems of elasticity, for which the principal dependence of the stress and strain tensors is on the plane coordinates x and y. Starting the consideration with this case, in equation (5.3) we use a function $A_0(x, y)$ that is independent of the z coordinate, instead of the function $A(x, y, z)$.

The strain tensor is represented by the matrix

$$\begin{pmatrix} \mathtt{A0}^{(0,2)} - \mathtt{nu}\ \mathtt{A0}^{(2,0)} & -(\mathtt{1}+\mathtt{nu})\mathtt{A0}^{(1,1)} & 0 \\ -(\mathtt{1}+\mathtt{nu})\mathtt{A0}^{(1,1)} & -\mathtt{nu}\mathtt{A0}^{(0,2)} + \mathtt{A0}^{(2,0)} & 0 \\ 0 & 0 & -\mathtt{nu}(\mathtt{A0}^{(0,2)} + \mathtt{A0}^{(2,0)}) \end{pmatrix}. \tag{5.9}$$

The corresponding strain incompatibility matrix has the form

$$\mathtt{II} = \begin{pmatrix} -\mathtt{nu}(\mathtt{A0}^{(0,4)} + \mathtt{A0}^{(2,2)}) & \mathtt{nu}(\mathtt{A0}^{(1,3)} + \mathtt{A0}^{(3,1)}) & 0 \\ \mathtt{nu}(\mathtt{A0}^{(1,3)} + \mathtt{A0}^{(3,1)}) & -\mathtt{nu}(\mathtt{A0}^{(2,2)} + \mathtt{A0}^{(4,0)}) & 0 \\ 0 & 0 & \mathtt{A0}^{(0,4)} + 2\,\mathtt{A0}^{(2,2)} + \mathtt{A0}^{(4,0)} \end{pmatrix}. \tag{5.10}$$

It is now possible to identify the conditions on the Airy stress function $A_0(x, y)$ that ensure that the components of the above incompatibility tensor vanish.

Strain incompatibility in plane stress

We note that the component `II[[3, 3]]` in equation (5.10) is the result of applying the biharmonic operator to the Airy stress function, $\Delta\Delta A_0(x, y)$.

Recall that the corresponding compatibility equation in terms of strains has the form

$$I_{33} = \frac{\partial^2 \epsilon_{xx}}{\partial y^2} + \frac{\partial^2 \epsilon_{yy}}{\partial x^2} - 2\frac{\partial^2 \epsilon_{xy}}{\partial x \partial y} = 0. \tag{5.11}$$

This equation is special in that it only relates to the strain components in the xy plane.

In many approximate treatments found in the literature this strain compatibility equation is incorrectly identified as the only one that needs to be satisfied in the plane problem, on the basis that it is the only one relating the in-plane strains alone. However, it is clear from the form of the incompatibility tensor (5.10) that this leaves other compatibility equations unsatisfied.

If only the requirement of biharmonicity of the Airy stress function is imposed, an *approximate* solution of a plane problem results. The corresponding state of stress is planar, whereas the strain tensor contains nonzero terms ϵ_{xx}, ϵ_{yy}, ϵ_{xy}, ϵ_{zz}. Compatibility of strains is satisfied only partially.

Harmonic Airy stress function: $\Delta A_0(x, y) = 0$

Choosing the Airy stress function to be harmonic, $\Delta A_0(x, y) = 0$, ensures full compatibility of strains. Indeed, inspection reveals that *all* of the components of the incompatibility tensor `II` can be obtained by differentiation of the laplacian of the Airy stress function, $\Delta A_0(x, y)$.

Reducibility of differential operators

Although in the present case the above statement can be verified fairly easily, in more complex computations it is often difficult to recognise higher order derivatives of the laplacian and their combinations. One of the purposes of this book is to demonstrate how some of the laborious analytical calculations in classical elasticity can be elucidated and simplified with the help of MATHEMATICA . Here we develop a useful technique that can be applied in the general case.

Coefficients of the terms $A_0^{(i,j)}$ do not depend on the variables x and y. We therefore establish a one-to-one mapping between the differential operators and polynomials in auxiliary variables v_x and v_y, as follows:

$$A_0^{(i,j)} = \frac{\partial^{(i+j)} A_0}{\partial x^i \partial y^j} \quad \longleftrightarrow \quad v_x^i\, v_y^j. \tag{5.12}$$

We now specify an appropriate MATHEMATICA rule which implements this mapping. The rule can be applied to the components of the incompatibility tensor `II` and the laplacian of $A_0(x, y)$.

```
deriv2poly = Derivative[px_,py_][A0][x,y]->vx^px vy^py
```

```
(incpoly = II /. deriv2poly) // MatrixForm
```

```
lappoly = (Laplacian[A0[x, y]]) /. deriv2poly
```

Now, in order to verify the reducibility of components of tensor `II` to the derivatives of the laplacian of $A_0(x, y)$, we need to verify the divisibility of the corresponding polynomials.

```
(reducelap = Table[ PolynomialReduce[incpoly[[i, j]],
    lappoly, {x, y} ], {i, 3}, {j,3}])
                                    // MatrixForm
```

According to the definition of `PolynomialReduce`, the second part of the resulting list contains the residue. We remark that all residues can be extracted to show that they all indeed evaluate to zero.

```
(residuelap = Map[ #[[2]] &, reducelap, {2} ] )
                                    // MatrixForm
```

The result signifies that each component of the incompatibility tensor in polynomial form can be factorized into a product of the ‘laplacian polynomial’ and a quotient polynomial. Using the one-to-one mapping between the differential operators and the polynomials, we can establish the following equivalence:

$$\left(a_0 + \ldots + a_{ij}\,\frac{\partial^{(i+j)}}{\partial x^i\,\partial y^j} + \ldots\right)\,\Delta A_0 \quad\longleftrightarrow\quad \left(a_0 + \ldots + a_{ij}\,v_x^i\,v_y^j + \ldots\right)\left(v_x^2 + 2\,v_x\,v_y + v_y^2\right). \tag{5.13}$$

As a consequence, we can obtain explicitly the differential operator that needs to be applied to $\Delta\,A_0$ to construct each of the components of the incompatibility tensor.

For completeness of presentation we now define the procedures which perform this validation exercise.

First we construct the quotient tensor.

```
(quotientlap = Map[ #[[1, 1]] &, reducelap, {2}] )
                                        // MatrixForm
```

Let us now exemplify the procedure using one of the components, say, `II[[3,3]]`.

We first build the coefficient list and multiply each coefficient by the appropriate derivative of the laplacian. We then add together all the terms. Because the initial table of coefficients was two-dimensional, we flatten this table and apply the plus operator to the resulting one-dimensional list. Finally, we verify that the result is equal to the component `II[[3,3]]`.

```
(coeflist = Simplify[
   CoefficientList[quotientlap[[3, 3]], {dx, dy}]])
                                     // MatrixForm
(comblist =
   Table[ coeflist[[i, j]]  D[lap, {x, i - 1}, {y, j - 1}],
      {i, 1, Dimensions[coeflist] [[1]]},
      {j, 1, Dimensions[coeflist] [[2]]} ])
                                          // MatrixForm

Simplify[Plus @@ Flatten[ comblist]]
Simplify[Plus @@ Flatten[ comblist] - II[[3, 3]]]
```

All operations can be grouped into a module as shown here.

```
ComputeReduction[quotientpoly_, divisorderiv_, vars_] :=
  Module
    [{coeflist},
    coeflist = CoefficientList[quotientpoly, vars];
    Simplify[
      Plus @@ Flatten[
        Table[ coeflist[[i, j]]  D[
          divisorderiv, {x, i - 1}, {y, j - 1}],
```

```
            {i, 1, Dimensions[coeflist][[1]]},
            {j,1,Dimensions[coeflist][[2]]} ]
         ]
      ]
   ]
```

It now becomes a trivial task to verify the original assumption.

```
Table[ Simplify[
  ComputeReduction[quotientlap[[i, j]], lap, {dx, dy}]
      - II[[i, j]]],{i, 3}, {j, 3}]
                                          // MatrixForm
```

Remarks on the approximate nature of plane stress

The foregoing discussion has identified several deficiencies of the formulation using the Beltrami potential tensor in the form

$$B = \begin{pmatrix} 0 & 0 & 0 \\ 0 & 0 & 0 \\ 0 & 0 & A_0(x,y) \end{pmatrix}. \tag{5.14}$$

- Requiring *biharmonicity* of $A_0(x, y)$ does not satisfy strain compatibility, except for one equation out of six. This approximate formulation is nevertheless widely used in applications. In the next section we shall demonstrate the nature of the approximations involved.
- Requiring *harmonicity* of $A_0(x, y)$ satisfies strain compatibility in full. However, practice has shown that this requirement is restrictive in terms of the variety of solutions that can be obtained, and therefore it is rarely used.

5.3 AIRY STRESS FUNCTION WITH A CORRECTIVE TERM: $A_0(x,y) - z^2A_1(x,y)$

We begin again with the Beltrami potential tensor given in (5.3). Our aim is to establish a form of the function $A(x, y, z)$ which leads to a compatible strain field. A recipe due to Clebsch (see Love, 1944) is to consider $A(x, y, z)$ in the form

$$A(x, y, z) = A_0(x, y) - z^2 A_1(x, y) \tag{5.15}$$

and to set

$$A_1(x, y) = \frac{\nu}{2(1+\nu)} \Delta A_0(x, y). \tag{5.16}$$

The Clebsch form of the Airy stress function is introduced.

```
B := {{0, 0, 0}, {0, 0, 0}, {0, 0, A[x, y, z]}}
A[x_,y_,z_] = A0[x, y] -nu/2/(1 + nu)Laplacian[A0[x,y]] z^2;
(Stress = Inc[B]) // MatrixForm
```

The resulting stress components have the form

$$\sigma_{ij} = \sigma_{ij}^0 - \frac{\nu}{2(1+\nu)} z^2 \Delta\sigma_{ij}^0, \tag{5.17}$$

where σ_{ij}^0 denotes the stress components derived using only the biharmonic function $A_0(x, y)$ of the previous section.

We use MATHEMATICA to demonstrate that the resulting strain field is indeed compatible, provided $A_0(x, y)$ is biharmonic.

```
SS = IsotropicCompliance[nu]

(Strain = Simplify[Collect[DDot[SS,Stress],z]])// MatrixForm

( Incstrain = Simplify[ Inc[Strain] ] ) // MatrixForm
```

The strain incompatibility tensor is represented by a large matrix. It is necessary to demonstrate that all components vanish if A_0 is biharmonic, $\Delta\Delta A_0(x, y) = 0$. This task is similar to that tackled in the preceding section: we use the rule **deriv2poly** and command **ComputeReduction** defined previously.

```
deriv2poly = Derivative[a_, c_][A0][x, y] -> dx^a dy^c

(incpoly = II /. deriv2poly ) // MatrixForm
bih = Biharmonic[A0[x, y]]
bihpoly = (bih) /. deriv2poly

(reducebih =
      Table[ PolynomialReduce[incpoly[[i, j]],
                  bihpoly, {x, y} ], {i, 3}, {j, 3}])
                                          // MatrixForm

(quotientbih = Map[ #[[1, 1]] &, reducebih, {2} ]  )
                                          // MatrixForm

(coeflist =
    Simplify[
      CoefficientList[quotientbih[[3,3]],{dx,dy}]])
                                          // MatrixForm

Table[ Simplify[
      ComputeReduction[quotientbih[[i,j]],bih,{dx,dy}]
          - II[[i, j]]], {i, 3}, {j, 3}]
                                          // MatrixForm
```

The proof of compatibility of the strain field is complete. On this basis a complete set of expressions for the displacements can be written (see Love, 1944).

Remark on the plane stress approximation

Referring to the stresses given by (5.17), we note that the corrective stress terms are proportional to z^2. These can be made as small as one likes, provided that the extent of the body in the z direction is kept small compared to its other dimensions. As a consequence, the plane stress approximation can be applied to thin plates, with the penalty of violating strain compatibility conditions. The difference between the two formulations amounts to a parabolic stress variation through the thickness of the plate.

5.4 PLANE STRAIN

The incompatibility of strain that arises in the plane potential formulation involving the function $A_0(x, y)$ can be satisfied by adopting a different approach.

Consider the Beltrami potential tensor in the form

$$\boldsymbol{B} = \begin{pmatrix} A_1(x,y) & 0 & 0 \\ 0 & A_2(x,y) & 0 \\ 0 & 0 & A_3(x,y) \end{pmatrix}. \tag{5.18}$$

The Beltrami potential formulation of this form is sometimes referred to as the Maxwell stress potential.

Compute the stress tensor and note that the resulting stress state is no longer planar.

```
B := {{A1[x, y], 0, 0}, {0, A2[x, y], 0}, {0, 0, A3[x, y]}}
B // MatrixForm
Stress := Inc[B]
Stress // MatrixForm
```

Find the strain tensor and compute the incompatibility tensor `II`.

```
SS = IsotropicCompliance[nu]
Strain := Simplify[DDot[SS, Stress]]
(Strain) // MatrixForm

II := Inc[Strain]
(II) // MatrixForm
```

Let us now consider the structure of the nonzero components of this tensor:

$$II(1,1) = A_1^{(0,4)} - \nu A_3^{(0,4)} + A_2^{(2,2)} - \nu A_3^{(2,2)} \tag{5.19}$$

$$II(1,2) = -A_1^{(1,3)} + \nu A_3^{(1,3)} - A_2^{(3,1)} + \nu A_3^{(3,1)} \tag{5.20}$$

$$II(2,1) = -A_1^{(1,3)} + \nu A_3^{(1,3)} - A_2^{(3,1)} + \nu A_3^{(3,1)} \tag{5.21}$$

$$II(2,2) = A_1^{(2,2)} - \nu A_3^{(2,2)} + A_2^{(4,0)} - \nu A_3^{(4,0)} \tag{5.22}$$

$$II(3,3) = -\nu A_1^{(0,4)} + A_3^{(0,4)} - \nu A_1^{(2,2)} - \nu A_2^{(2,2)} + 2A_3^{(2,2)} - \nu A_2^{(4,0)} + A_3^{(4,0)}. \tag{5.23}$$

We note that the first four components can be made identically zero by setting

$$A_1 = \nu A_3, \quad A_2 = \nu A_3. \tag{5.24}$$

Under the same substitution, component `II[[3,3]]` assumes the form

$$II(3,3) = -(1-\nu^2)\left(A_3^{(0,4)} + 2A_3^{(2,2)} + A_3^{(4,0)}\right) = -(1-\nu^2)\Delta\Delta A_3. \tag{5.25}$$

We conclude that biharmonicity of $A_3(x, y)$ ensures strain compatibility under this formulation.

The stress tensor is now found in the form

```
Stress // MatrixForm
```

$$\begin{pmatrix} \mathrm{A3}^{(0,2)} & -\mathrm{A3}^{(1,1)} & 0 \\ -\mathrm{A3}^{(1,1)} & \mathrm{A3}^{(2,0)} & 0 \\ 0 & 0 & -\mathrm{nu}(\mathrm{A3}^{(0,2)} + \mathrm{A3}^{(2,0)}) \end{pmatrix}. \tag{5.26}$$

Note that the out-of-plane stress component is now present.

The strain state is now found.

```
Simplify[Strain] // MatrixForm
```

$$-(1+\mathrm{nu})\begin{pmatrix} (-1+\mathrm{nu})\mathrm{A3}^{(0,2)} + \mathrm{nu}\ \mathrm{A3}^{(2,0)} & \mathrm{A3}^{(1,1)} & 0 \\ \mathrm{A3}^{(1,1)} & \mathrm{nu}\ \mathrm{A3}^{(0,2)} + (-1+\mathrm{nu})\mathrm{A3}^{(2,0)} & 0 \\ 0 & 0 & 0 \end{pmatrix} \tag{5.27}$$

We note that this compatible strain state is indeed planar. Moreover, the above equation can be rewritten in the form

$$(1-\mathrm{nu}^2)\begin{pmatrix} \mathrm{A3}^{(0,2)} - \mathrm{A3}^{(2,0)}\mathrm{nu}/(1-\mathrm{nu}) & -\mathrm{A3}^{(1,1)}/(1-\mathrm{nu}) & 0 \\ -\mathrm{A3}^{(1,1)}/(1-\mathrm{nu}) & \mathrm{A3}^{(0,2)} - \mathrm{A3}^{(2,0)}\mathrm{nu}/(1-\mathrm{nu}) & 0 \\ 0 & 0 & 0 \end{pmatrix}. \tag{5.28}$$

By comparison with equation (5.9), we note that the two expressions for strain in terms of the potential function (and hence stresses) differ only in terms of coefficients. In fact, the definition

$$E^* = \frac{E}{1-\nu^2}, \qquad \nu^* = \frac{\nu}{1-\nu} \tag{5.29}$$

ensures that the equations for planar strain become identical with those for planar stress provided Young's modulus and Poisson's ratio are replaced with the starred symbols throughout.

5.5 AIRY STRESS FUNCTION OF THE FORM $A_0(r,\theta)$

The change of coordinates from cartesian (x, y, z) to cylindrical polar (r, θ, z) preserves all of the properties introduced in the previous sections. In fact, any coordinate transformation within the plane perpendicular to the z axis can be performed with the help of the `TensorAnalysis` package, provided the resulting coordinate system remains orthogonal.

For cylindrical polar coordinates the result has the form

$$\sigma = \mathrm{inc}\,(A_0(r,\theta)\,\boldsymbol{e}_z \otimes \boldsymbol{e}_z) = \begin{pmatrix} \frac{1}{r^2}\frac{\partial^2 A_0}{\partial\theta^2} + \frac{1}{r}\frac{\partial A_0}{\partial r} & \frac{1}{r}\frac{\partial A_0}{\partial\theta} - \frac{1}{r^2}\frac{\partial^2 A_0}{\partial r\partial\theta} & 0 \\ \frac{1}{r}\frac{\partial A_0}{\partial\theta} - \frac{1}{r^2}\frac{\partial^2 A_0}{\partial r\partial\theta} & \frac{\partial^2 A_0}{\partial r^2} & 0 \\ 0 & 0 & 0 \end{pmatrix}. \tag{5.30}$$

Using MATHEMATICA, the derivation is performed in a few lines.

```
<< Tensor2Analysis.m
SetCoordinates[Cylindrical[r, t, z]]
B = {{0, 0, 0}, {0, 0, 0}, {0, 0, Psi[r, t]}}
(Stress1 = Inc[B]) // MatrixForm
```

Airy stress function

`B := {{0, 0, 0}, {0, 0, 0}, {0, 0, A0[x, y]}}`	Airy stress function form of the Beltrami–Maxwell tensor potential `B`

5.6 BIHARMONIC FUNCTIONS

The above analysis of the plane problem demonstrates the important role played by biharmonic functions in the solution of elastic plane problems.

The general form of the solution of the biharmonic equation in two dimensions has been established by Goursat using the apparatus of functions of the complex variable $\zeta = x + \mathrm{i}y = r\exp \mathrm{i}\theta$. The general solution is found in the forms

$$A_1(r,\theta) = \mathrm{Re}(\bar{\zeta}\phi(\zeta) + \chi(\zeta)), \tag{5.31}$$

$$A_2(r,\theta) = \mathrm{Im}(\bar{\zeta}\phi(\zeta) + \chi(\zeta)), \tag{5.32}$$

where ϕ and χ are arbitrary analytic (and therefore harmonic) functions of ζ. The function χ above therefore describes the subset of biharmonic functions that are also harmonic, whereas the form $\bar{\zeta}\phi(\zeta)$ represents the set of functions that in this context could be termed 'essentially biharmonic.'

Without constructing a rigorous proof (which can be found, for example, in Muskhelishvili (1953)) one may remark that in the complex plane ζ

$$\Delta A(\zeta) = \left(\frac{\partial^2}{\partial x^2} + \frac{\partial^2}{\partial y^2}\right) A(\zeta) = 4\frac{\partial^2}{\partial\zeta\partial\bar{\zeta}}A(\zeta).$$

Therefore the biharmonic equation is

$$\frac{\partial^4}{\partial\bar{\zeta}^2\partial\zeta^2}A(\zeta) = 0.$$

Considering ζ and $\bar{\zeta}$ as two independent variables and integrating twice with respect to $\bar{\zeta}$ results in

$$\frac{\partial^2}{\partial\zeta^2}A(\zeta) = \bar{\zeta}\Phi(\zeta) + \Phi_1(\zeta),$$

where Φ, Φ_1 are analytic functions of ζ. Further integration in ζ preserves harmonicity and leads to the solution $A(\zeta) = A_1(\zeta) + \mathrm{i}A_2(\zeta)$ in the form of equation (5.31).

The convenience offered by the Goursat form (5.31) is that the complex variable ζ can be expressed in terms of an arbitrary pair of coordinates in the complex plane, leading to a great variety of forms of solution.

The solution of the biharmonic equation in polar coordinates is particularly relevant to problems with boundaries defined as segments of the r or θ coordinate lines. Exploration of solutions in the cylindrical polar coordinate representation may begin by noting that essentially biharmonic terms can be readily obtained by considering analytic functions in the ζ plane, which on their own give rise to harmonic solutions as their real and imaginary parts. With the multiplier $\bar{\zeta}$ these functions generate 'essentially biharmonic' solutions.

One natural choice for a basis family of functions is the powers ζ^n. These give rise to the harmonic soutions

$$\zeta^n = r^n \cos n\theta + ir^n \sin n\theta.$$

Essentially biharmonic solutions are obtained in the form

$$\bar{\zeta}\zeta^{n+1} = r^{n+2} \cos n\theta + ir^n \sin n\theta.$$

This representation stands in an obvious relationship with the series expansions, namely the Fourier series in θ and the power law series in r. In the general real form one can write

$$A(r,\theta) = \sum_{n=-\infty}^{\infty} r^n \left(a_{n1} \cos n\theta + a_{n2} \cos(n-2)\theta + b_{n1} \sin n\theta + a_{n2} \sin(n-2)\theta\right). \quad (5.33)$$

It turns out, however, that this representation alone does not provide a sufficient variety of solutions. It needs to be enhanced by the logarithmic function and its combinations with powers, because of a particularly important role in plane elasticity played by this family of solutions.

Two harmonic solutions are generated by the real and imaginary parts of the logarithmic function,

$$\log\zeta = \log r + i\theta = \frac{1}{2}\log(x^2+y^2) + i\,arctan(y/x).$$

These two terms give rise to stresses and strains that decay as $1/r^2$ with distance from the origin, and therefore to displacements varying as $1/r$. The application of `IntegrateStrain` utility from Chapter 1 reveals that the displacements are purely radial for $A(r,\theta) = \log r$ and purely tangential for $A(r,\theta) = \theta$.

A set of important essentially biharmonic solutions is generated by considering equations (5.31) and (5.32) with the functions

$$\bar{\zeta}\phi(\zeta) + \chi(\zeta) = \frac{1}{2}(\zeta \pm \bar{\zeta}) \log \zeta.$$

The real solutions arising in this case are

$$r \log r \cos\theta, \qquad r\theta\cos\theta, \qquad r\log r \sin\theta, \qquad r\theta\sin\theta.$$

The stresses and strains due to these solutions decay as $1/r$ with distance from the origin. The displacements may therefore be found to vary logarithmically with r or to be a linear function of angle θ. The elastic fields implied by these functions are of particular interest in problems involving forces concentrated at a point, and dislocations, as discussed below.

Two more solutions that are essentially biharmonic arise from the function

$$\bar{\zeta}\zeta \log \zeta = r^2 \log r + ir^2\theta.$$

These solutions give rise to displacement fields of the form $r\theta$ either in the radial or tangential components, and thus cannot be continuous in a body that spans the full range of polar angles. A displacement discontinuity that grows linearly with the distance r from the origin must be thought to arise at a line that corresponds to the branch cut of the function $\theta = \operatorname{atan}(y/x)$ in the ζ plane. The solutions thus correspond to 'disclinations' that might be created in the material by inserting or removing a wedge of material.

This method of generating biharmonic solutions can be used to derive families of functions of arbitrary order n based on $\zeta^n \log \zeta$ and $\bar{\zeta}\zeta^{n-1} \log \zeta$ and their combinations. For example, with the help with the operator **Biharmonic**, one can readily verify the validity, for all values of n, of the following biharmonic solutions:

$$r^n(\cos n\theta \log r - \theta \sin n\theta), \qquad r^n(\sin n\theta \log r - \theta\cos n\theta), \tag{5.34}$$

$$r^n(\cos(n-2)\theta \log r - \theta\sin(n-2)\theta), \qquad r^n(\sin(n-2)\theta\log r - \theta\cos(n-2)\theta). \tag{5.35}$$

These are, of course, neither even nor odd in θ. The analysis of displacements reveals that these solutions contain the term θ in the expressions for stress components. Although this observation has been used as the basis for excluding these solutions from further consideration, it seems appropriate to catalogue them here, because they may turn out to be useful in the solution of some boundary value problems. We note in passing that, for example, the Airy function term given by the first expression in equation (5.35) for $n = 0$,

$$\cos 2\theta \log r + \theta \sin 2\theta,$$

gives rise to stresses varying as $\cos 2\theta \log r/r^2$, and as $\theta\cos 2\theta/r^2$, not obtainable from the set of solutions usually considered.

Stress fields corresponding to any test solution can be readily computed and plotted. The necessary packages **Tensor2Analysys.m, Displacement.m, IntegrateStrain.m** must be loaded and the coordinate system defined. For any chosen function **Airy[r,t]** its biharmonicity is checked first.

For convenience, we define the function **AiryStress**, which allows the stress components to be computed in one line.

The strain is then calculated using `IsotropicCompliance[nu]` and `DDot` operators, and displacements obtained using `IntegrateStrain`.

```
<< Tensor2Analysis.m
<< Displacement.m
<< IntegrateStrain.m

SetCoordinates[Cylindrical[r, t, z]]

Airy[r_,t_]=r^2 Log[r];
Biharmonic[Airy[r,t]]

AiryStress[Airy_] := Inc[{{0,0,0}, {0,0,0}, {0,0,Airy}}]
Stress = AiryStress[Airy[r,t]]

Strain = Simplify[DDot[IsotropicCompliance[nu], Stress]];
uint = IntegrateStrain[Strain]
```

A note of caution must be added, to draw attention to the fact that correct integration of displacements must take account of rigid body displacement and rotation effects. Furthermore, the definition of the module `IntegrateStrain` introduced in Chapter 1 includes the verification of strain compatibility. As discussed above, the majority of plane elasticity solutions do not satisfy this requirement. If the general procedure for strain integration is 'forced' for a given plane strain tensor, the resulting displacements may contain terms depending on coordinate z. These terms may be discarded and strain recalculated to confirm if the integration has been correct. This is illustrated below using the solution for disclination.

It is apparent from this discussion that the variety of forms of biharmonic functions can be spanned by the full family of analytic functions in the ζ plane. The truly complete solution for the two-dimensional case is thus given by the Goursat forms (5.31) and (5.32).

Alternative and less general approaches to exploration of the forms of solution in polar coordinates in real form have been widely employed in the literature. One method relies on the use of Fourier series expansion in the polar angle θ, seeking $A(r, \theta)$ in the form

$$A(r,\theta) = \sum_{n=0}^{\infty} f_n(r)\cos n\theta + \sum_{n=0}^{\infty} g_n(r)\sin n\theta. \tag{5.36}$$

The requirement that this function be biharmonic leads to the governing equation for the functions $f_n(r)$, $g_n(r)$ in the form

$$\left(\frac{d^2}{dr^2} + \frac{1}{r}\frac{d}{dr} - \frac{n^2}{r^2}\right)^2 f(r) = 0. \tag{5.37}$$

The difficulty associated with this approach lies in the fact that it does not lead to the most general form of solution, and somewhat artificial methods have to be employed to consider the so-called degenerate cases. Michell was the first to present, without proof, a form of 'general' solution in polar coordinates. It is very important to note, however, that the solution is not complete. Following Barber (2002) we record the Michell solution (with the addition of the term $r^2\theta$ for consistency) as

$$\begin{aligned} A(r,\theta) = {} & A_{01}r^2 + A_{02}r^2\log r + A_{03}\log r + A_{04}\theta + A_{05}r^2\theta \\ & + (A_{11}r^3 + A_{12}r\log r + A_{14}r^{-1})\cos\theta + A_{13}r\theta\sin\theta \\ & + (B_{11}r^3 + B_{12}r\log r + B_{14}r^{-1})\sin\theta + B_{13}r\theta\cos\theta \\ & + \sum_{n=2}^{\infty}(A_{n1}r^{n+2} + A_{n2}r^{-n+2} + A_{n3}r^n + A_{n4}r^{-n})\cos n\theta \\ & + \sum_{n=2}^{\infty}(B_{n1}r^{n+2} + B_{n2}r^{-n+2} + B_{n3}r^n + B_{n4}r^{-n})\sin n\theta\,. \end{aligned} \tag{5.38}$$

Following the existing convention, functions (5.34) and (5.35), which arise for $n \geq 2$, have been omitted from the Michell solution.

5.7 THE DISCLINATION, DISLOCATIONS, AND ASSOCIATED SOLUTIONS

In this section we consider a particular solution for the Airy stress function that allows some important properties of displacement fields to be discussed, and relationships between solutions to be explored.

Choosing

$$A(r,t) = Dr^2\log r$$

gives rise to a particularly simple stress field given by $\sigma_{rr} = D(1+2\log r)$, $\sigma_{\theta\theta} = D(3+2\log r)$, $\sigma_{r\theta} = 0$.

Of particular interest are the strain and displacement fields for this solution.

The strain tensor is computed from stress using the isotropic compliance tensor in terms of the Kolosov constant **K**, with 2μ set to unity for simplicity. Application of the **Inc** operator shows, however, that this strain is not compatible. Hence the application of **IntegrateStrain** runs into difficulties. If the steps of this procedure are formally completed, however, a displacement field can be found that contains terms depending on z. If these are discarded, a plane displacement field remains that in fact can be shown to give rise to the correct strain.

```
Stress = AiryStress[D r^2 Log[r]]
Strain=DDot[IsotropicComplianceK[K],Stress];
Inc[Strain]
```

```
theta = IntegrateGrad[-Curl[Strain]];
omega =Table[Sum[Signature[{i, j, k}] theta[[k]],
{k, 3}], {i, 3}, {j, 3}];
Uint = Simplify[IntegrateGrad[Strain + omega]]
```

```
Eint= Simplify[(Grad[uint] + Transpose[Grad[uint]])/2]
```

The displacement field has the form

$$2\mu u_r = D[(\kappa-1)r\log r - r], \qquad 2\mu u_\theta = D(\kappa+1)r\theta, \qquad 2\mu u_z = 0. \tag{5.39}$$

Of primary importance here is the fact that the tangential displacement component u_θ contains the polar angle θ and therefore is multivalued, unless a branch cut is introduced. Taking the location of this cut to be the coordinate line $\theta = 0$, we conclude that the u_θ displacement component is discontinuous and undergoes a jump as if the brach cut were opened to the width $-D\pi(\kappa+1)r/\mu$. This displacement field could be created in an infinite elastic plane by inserting into the branch cut a wedge of material that was infinitely extended along the z axis, and spanned the (small) angle $D\pi(\kappa+1)/\mu$ in the (x, y) plane. This solution will be referred to as the *disclination* solution.

Stress and displacement fields around a disclination possess singularities at the origin. They are an example of an important family of singular solutions in elasticity that serve as 'sources' of deformation fields and are known as *nuclei of strain.*

It is useful to consider such solutions in different coordinate frames. For example, Cartesian coordinates provide a fixed reference frame that is preferable whenever a distribution of sources is considered in order to construct new solutions.

In Cartesian coordinates associated with the same origin, the Airy stress function for a disclination assumes the form

$$A(x, y) = \frac{D}{2}(x^2 + y^2)\log(x^2 + y^2).$$

If instead the disclination is installed at point (ξ, η) then

$$A(x-\xi, y-\eta) = \frac{D}{2}((x-\xi)^2 + (y-\eta)^2)\log((x-\xi)^2 + (y-\eta)^2).$$

We now consider two disclinations of opposite sign located close to each other at (ξ, η) and $(\xi + d\xi, \eta)$ respectively. The physical consequence of forming this *disclination dipole* amounts to the creation of a *dislocation* that corresponds to a uniform layer of material being inserted into the branch cut so that the displacement component u_y undergoes a jump of magnitude $b_y = -\pi(\kappa+1)Dd\xi/\mu$ as the line $y = \eta, x > \xi$ is crossed in the direction of positive y. To form a true dipole in the limit $d\xi \to 0$ it is necessary to assume that the constant D is allowed to vary in the process so that

$$\lim_{d\xi\to 0} Dd\xi = -\mu b_y/\pi(\kappa+1).$$

The displacement discontinuity across the branch cut (in this case b_y in component u_y) is referred to as the Burgers vector.

The mathematical consequence of forming the disclination dipole amounts to the differentiation of the Airy stress function with respect to ξ. Discarding the trivial term that is linear in $x - \xi$ produces a new Airy stress function,

$$A_y(x-\xi, y-\eta) = -\frac{\mu b_y}{\pi(\kappa+1)}(x-\xi)\log((x-\xi)^2 + (y-\eta)^2)).$$

Now, placing the dislocation at the origin $(\xi, \eta) = (0, 0)$ and reverting to the polar coordinates, the solution for this dislocation with Burgers vector b_y can be written as

$$A_y(r, \theta) = -\frac{2\mu b_y}{\pi(\kappa+1)} r\cos\theta\log r. \tag{5.40}$$

The nature of the dipole formed by disclinations of equal and opposite signs located at (ξ, η) and $(\xi, \eta + d\eta)$ is best understood if the branch cut location is given by $x = \xi$, $y > \eta$, or $\theta = \pi/2$. Similar reasoning then shows that the tangential displacement u_θ is now associated with the component u_x in Cartesian coordinates, which undergoes a jump of magnitude $b_x = \pi(\kappa + 1)Dd\eta/\mu$ as the line $x = \xi$, $y > \eta$ is crossed in the direction of positive x. In other words, the solution for dislocation of this type is given by

$$A_x(x - \xi, y - \eta) = \frac{\mu b_x}{\pi(\kappa + 1)}(y - \eta)\log((x - \xi)^2 + (y - \eta)^2)).$$

If the dislocation with Burgers vector b_x is located at the origin, then

$$A_x(r, \theta) = \frac{2\mu b_x}{\pi(\kappa + 1)} r \sin\theta \log r. \tag{5.41}$$

Integration of the strain fields for the dislocation solution A_y to obtain displacements using the `IntegrateStrain` procedure leads to the result

$$2\mu u_r = \frac{\mu b_y}{\pi}\left[-\theta\sin\theta + \frac{1}{\kappa + 1}\cos\theta - \frac{\kappa - 1}{\kappa + 1}\log r\cos\theta\right], \tag{5.42}$$

$$2\mu u_\theta = \frac{\mu b_y}{\pi}\left[-\theta\cos\theta + \frac{1}{\kappa + 1}\sin\theta - \frac{\kappa - 1}{\kappa + 1}\log r\sin\theta\right]. \tag{5.43}$$

It can now be readily verified that, in accordance with the mathematical definition of a dislocation, the displacement component u_r is continuous everywhere, whereas the displacement component u_θ undergoes a positive jump of magnitude b_y if the positive half of the x-axis is crossed in the direction of increasing polar angle θ (or coordinate y).

The two dislocation solutions (5.40) and (5.41) were obtained by applying the two-dimensional gradient operator $(\partial/\partial\xi, \partial/\partial\eta, 0) = -(\partial/\partial x, \partial/\partial y, 0) = -\nabla_{x,y}$ to the scalar field $Dr^2 \log r$, resulting in a vector field.

Further application of $\nabla_{x,y}$ generates a family of other valid solutions, referred to as *dislocation dipoles*. This family of solutions possess the properties of a second rank tensor field.

In an infinite elastic solid, differentiation with respect to ξ, η differs only in sign from differentiation with respect to x, y. It is therefore sufficient to consider the simple definition of the Airy stress function for a disclination at the origin. Single and double application of the gradient operator gives rise to the dislocation and dislocation dipole families of solutions.

```
Disclin = D/2(x^2 + y^2)Log[x^2 + y^2];
```

```
Disloc = -Grad[Disclin]
```

```
DislocDipole= -Grad[Disloc]
```

Dislocation dipole solutions are found to have the forms

$$A_{xx} = 2D(x^2/r^2 + \log r), \quad A_{xy} = 2Dxy/r^2, \quad A_{yy} = 2D(y^2/r^2 + \log r). \tag{5.44}$$

Written in polar coordinates, the term A_{xy} becomes

$$A_{xy} = D \sin 2\theta,$$

and corresponds to the 2D centre of shear (that is, 3D centres of shear, arranged along an infinite line along the z axis).

Consideration of the difference

$$A_{yy} - A_{xx} = 2D(y^2 - x^2)/r^2 = D \cos 2\theta$$

shows that it, similarly, corresponds to a plane centre of shear. Indeed, it can be obtained from A_{xy} by rotation of the coordinate set by the angle $-\pi/4$.

Finally, the combination

$$A_{xx} + A_{yy} = 4D \log r$$

corresponds to the 2D centre of dilatation at the origin. It produces a planar radial displacement field with only one nonzero component, $u_r = -2D/(\mu r)$.

Dislocation dipole solutions can be thought of as true 'nuclei of strain,' because they generate displacement, strain, and stress fields that are continuous everywhere, with the exception of a singularity at the origin. The source installed at the singular point can be thought of as concentrated inelastic strain, or point *eigenstrain*. Solutions A_{xy} and $A_{yy} - A_{xx}$ correspond to shear, or the deviatoric component of such point eigenstrain, whereas solution $A_{xx} + A_{yy}$ corresponds to the dilatational, or isotropic thermal expansion component. This classification may be helpful in the analysis of inelastic deformation and residual stresses.

5.8 A WEDGE LOADED BY A CONCENTRATED FORCE APPLIED AT THE APEX

Consider the wedge illustrated in Figure 5.1 made from isotropic homogeneous elastic material with Young's modulus **EE** and Poisson's ratio **nu**. Introduce a system of cartesian coordinates with the z axis associated with the wedge apex, as shown in the figure, and the x axis along the bisector of the wedge angle. We also introduce the associated system of cylindrical polar coordinates with the same z-axis. In this frame the body occupies the angle 2α and is infinitely extended in the direction of the y-axis.

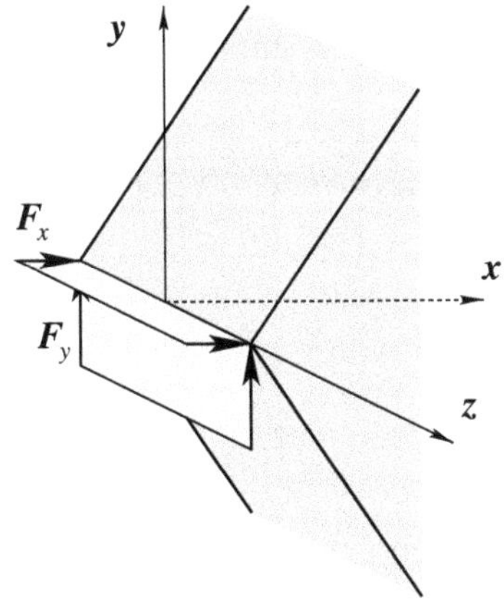

Figure 5.1. Elastic wedge.

Let the wedge be loaded by a concentrated line force $\boldsymbol{F} = F_x \boldsymbol{e}_x + F_y \boldsymbol{e}_y$ applied along the edge, as shown in Figure 5.1. The faces of the body defined by $\theta = \pm\alpha$ are free from tractions, and stresses in the wedge vanish at infinite values of r. The faces of the body defined by $z = \pm L$ are subjected to a uniform pressure p.

In the next sections we shall consider separately the cases of loading by the force components $F_x \boldsymbol{e}_x$ and $F_y \boldsymbol{e}_y$. The complete solution follows from these two analyses by superposition.

It is clear from the description that the problem possesses a certain translational invariance with respect to the z-axis, and the main dependence of the elastic fields is on the coordinates in the x–y plane. Therefore we shall seek the solution in terms of functions depending only on these coordinates. In the cylindrical polar coordinate system this means that the functions will depend on the r–θ coordinates.

Following the analysis of the previous chapter, we shall seek the solution of the plane problem in terms of a biharmonic function $A_0(r, \theta)$. The corresponding solution can then be viewed as an appropriate plane stress solution. Alternatively, it can be corrected using parabolic dependence on the z coordinate (although the boundary conditions may then become violated). Finally, a plane strain solution can also be obtained.

Axial force F_x

For the analysis of the problem about axial loading of a wedge (concentrated force at the apex applied along the wedge bisector) we use a single term from the Michell solution (5.38):

$$A_1(r, \theta) = kF_x \, r\theta \sin\theta.$$

Note that this form of solution can also be obtained from the Goursat formulation (5.31) as follows,

$$\phi(z) = \frac{1}{2} kF_x \log(z), \qquad \chi(z) = -\frac{1}{2} kF_x z \log(z), \tag{5.45}$$

and the biharmonic function

$$A_1(r, \theta) = \frac{1}{2} kF_x \, \mathrm{Re}(\bar{z} \log(z) - z \log(z)) = kF_x \, r\theta \sin\theta. \tag{5.46}$$

The form of the potentials is chosen from the terms in the Michell solution exhibiting linear behaviour in r. This selection is appropriate on the basis of self-similarity Barber (2002).

The plane stress solution can now be constructed using this biharmonic function as the Airy stress function according to the recipe given by equation (5.3). This leads to the result

$$\boldsymbol{\sigma} = \mathrm{inc}\,\boldsymbol{B} = \mathrm{inc}\,(A_1(x, y, z)\,\boldsymbol{e}_z \otimes \boldsymbol{e}_z) = \begin{pmatrix} \dfrac{2F_x k \cos\theta}{r} & 0 & 0 \\ 0 & 0 & 0 \\ 0 & 0 & 0 \end{pmatrix}. \tag{5.47}$$

Note that the representation of the resulting stress state is very simple in polar coordinates: the only nonzero stress component present is σ_{rr}, which decays in inverse proportion to the distance from the apex. The solution satisfies the stress-free boundary conditions at infinity and also the boundary condition that the wedge sides $\theta = \pm\alpha$ remain traction-free (since $\sigma_{\theta\theta} = \sigma_{r\theta} = 0$ everywhere).

The remaining boundary condition concerns the point force F_x acting along the wedge bisector. This condition can be satisfied as follows. Consider a section through the wedge made along a circular arc around the wedge apex at an arbitrary radial position r (Figure 5.1). Force balance requires that, for any r value, the total force along the x-axis be equal to $-F_x$. The total force is given by the integral

$$\int_{-\alpha}^{\alpha} \sigma_{rr} \cos\theta r d\theta = \int_{-\alpha}^{\alpha} 2F_x k \cos^2\theta d\theta = F_x k(2\alpha + \sin 2\alpha). \tag{5.48}$$

The suitable choice of constant k is

$$k = -\frac{1}{(2\alpha + \sin 2\alpha)}, \tag{5.49}$$

so that

$$\sigma_{rr} = -\frac{2F_x \cos\theta}{(2\alpha + \sin 2\alpha)}. \tag{5.50}$$

The above expression describes the entire family of solutions for apex loading of wedges of arbitrary opening angle.

The Flamant problem

For the purpose of illustration we consider in more detail the important case of a wedge of half-angle $\alpha = \pi/2$. The problem is that of a concentrated force acting perpendicular to the straight edge of a semi-infinite elastic plate (the Flamant problem).

Consider contours of equal stress component σ_{rr} given by the condition

$$-\frac{2\cos\theta}{\pi r} = -\frac{2}{\pi r_0}. \tag{5.51}$$

The locus of points given by

$$r = r_0 \cos\theta \tag{5.52}$$

is a single-parameter (r_0) family of circles (Boussinesq circles). All circles touch at the point of application of the force, where they have the plate edge as the common tangent. The Boussinesq circles are illustrated in Figure 5.2.

Transverse force F_y

The foregoing analysis can be repeated in an entirely analogous manner for the transverse force F_y (acting perpendicular to the wedge bisector). The biharmonic function is chosen as follows:

$$\phi(z) = \frac{1}{2}kF_y \log(z), \qquad \chi(z) = \frac{1}{2}kF_y z \log(z),$$

$$A_2(r,\theta) = \frac{1}{2}kF_y \ \mathrm{Im}(\bar{z}\log(z) + z\log(z)) = kF_y \ r\theta\cos\theta. \tag{5.53}$$

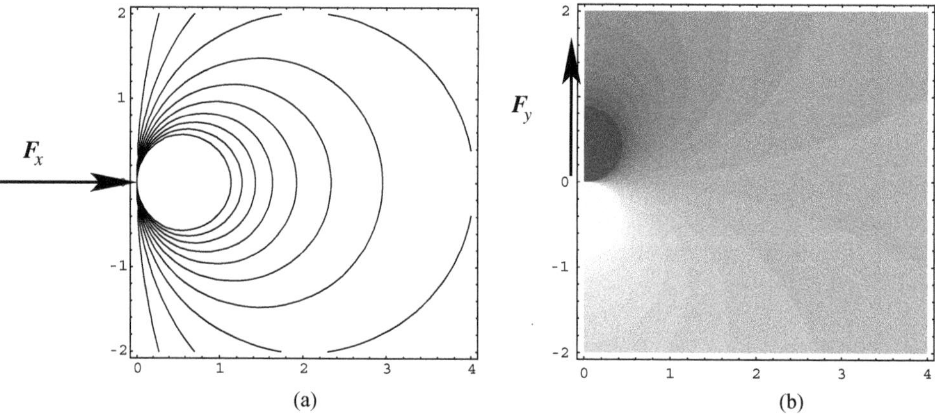

Figure 5.2. Contours of radial stress component for the $\alpha = \pi/2$ 'wedge' problem (semi-infinite region $x \geq 0$) under the action of concentrated forces (a) F_x and (b) F_y.

The plane stress solution is now given by

$$\boldsymbol{\sigma} = \begin{pmatrix} -\frac{2F_y k \sin\theta}{r} & 0 & 0 \\ 0 & 0 & 0 \\ 0 & 0 & 0 \end{pmatrix}. \tag{5.54}$$

The apex boundary condition can again be satisfied by considering a circular arc around the wedge apex and requiring that the total force along the y-axis be equal to $-F_y$:

$$\int_{-\alpha}^{\alpha} \sigma_{rr} \sin\theta r d\theta = \int_{-\alpha}^{\alpha} 2F_y k \sin^2\theta d\theta = -F_y k(2\alpha - \sin 2\alpha) = -F_y. \tag{5.55}$$

Hence

$$k = \frac{1}{(2\alpha - \sin 2\alpha)} \tag{5.56}$$

and

$$\sigma_{rr} = -\frac{2F_y \sin\theta}{(2\alpha - \sin 2\alpha)}. \tag{5.57}$$

Once again, a wedge of half-angle $\alpha = \pi/2$ can be considered for illustration. Contours of equal stress σ_{rr} are families of circles with a common tangent at the load application point.

Load the **Tensor2Analysis** package and define the cylindrical polar coordinate system.

```
<< Tensor2Analysis.m
SetCoordinates[Cylindrical[r, t, z]]
```

Define the Airy stress function $A_1(r, \theta)$ using the Goursat approach by evaluating the real and imaginary parts of a complex function.

```
RePart[z_] := ComplexExpand[Re[z]] /. Arg[r] -> 0;
ImPart[z_] := ComplexExpand[Im[z]] /. Arg[r] -> 0;

Z = r Exp[I t];
zeta = Log[r] + I t;
phi = k Fx zeta; chi = - k Fx Z zeta;

A1 = RePart[Conjugate[Z] phi + chi]
```

The stress tensor is now computed by formulating the Beltrami matrix potential $\boldsymbol{B}$ and applying the incompatibility operator inc:

```
B := {{0, 0, 0}, {0, 0, 0}, {0, 0, A1}}

Stress := Inc[B]
```

The force applied at the apex is balanced by the total force transmitted across a circular arc:

$$\boldsymbol{F} + \int_{-\alpha}^{\alpha} \boldsymbol{\sigma} \cdot \boldsymbol{e}_r \, r \, d\theta = 0.$$

The projection of this equation on the x-axis gives the condition for evaluating k.

```
Forcex = Integrate[Stress[[1,1]] r Cos[t], {t, -a, a}]

ksol = Simplify[Solve[Forcex == -Fx, k]][[1]]

Stress = Simplify[Stress /. ksol]
```

The results can be displayed in the form of a contour plot by converting the coordinates back to the cartesian triple (x, y, z) and evaluating the stress component σ_{rr}:

```
myrule = Thread[{Cos[t], Sin[t], r} ->
{x/Sqrt[x^2 + y^2], y/Sqrt[x^2 + y^2], Sqrt[x^2 + y^2]}]

S11 = Stress[[1, 1]] /. a -> Pi/2 /. Fx -> 1
ContourPlot[N[S11 /. myrule], {x, 0, 4}, {y,-2, 2},
PlotPoints -> 50, ContourShading -> False]
```

5.9 THE KELVIN PROBLEM

The Kelvin problem concerns a point force F_x in an infinite plane. The solution possesses an apparent similarity to the problem of loading of a wedge apex with a concentrated force (and the Flamant problem), because the stresses must vary as $1/r$ to allow the force balance condition to be satisfied on a circle of arbitrary radius.

The first step towards constructing the Kelvin solution is to select the case of axial concentrated force applied at the apex of an elastic wedge with the half-angle $\alpha = \pi$, which conforms to the general Airy stress function form $A = k_1 r\theta \sin\theta$.

We next note that the application of the `IntegrateStrain` procedure to the strain field arising from this solution leads to the displacement field

$$2\mu u_r = \frac{k_1}{2}\left[(\kappa-1)\,\theta\sin\theta - \cos\theta + (\kappa+1)\log r\cos\theta\right], \tag{5.58}$$

$$2\mu u_\theta = \frac{k_1}{2}\left[(\kappa-1)\theta\cos\theta - \sin\theta + (\kappa+1)\log r\sin\theta\right]. \tag{5.59}$$

It is apparent that the displacement field contains a discontinuity in the displacement component u_θ on the line $\theta = 0$, i.e., the positive half of the x-axis. Furthermore, the magnitude of this discontinuity is constant along this half-axis and is equal to

$$2\mu[u_\theta] = k_1(\kappa-1)\pi.$$

We have already encountered a different Airy stress function solution that contained a constant discontinuity of displacement component u_θ along the positive half of the x-axis, namely, the dislocation b_y solution (5.40) that has the general Airy stress function form $k_2 r\log r\cos\theta$. It follows that a superposition of these two solutions can be found such that the displacement is continuous everywhere, and that would correspond to a concentrated force applied at the origin of an infinitely extended elastic plane.

We therefore postulate an Airy stress function in the form

$$A(r,\theta) = F_x(k_1 r\theta\sin\theta + k_2 r\log r\cos\theta). \tag{5.60}$$

The values of the two unknown constants are found from the following two conditions:

- Enforcement of static equilibrium between the external force F_x applied at the origin and the distribution of stresses on a circular contour of arbitrary radius centred on the origin.
- Application of the `IntegrateStrain` procedure and enforcement of continuity of the displacement component u_θ for $\theta = 0$.

Consideration of static equilibrium reveals that, as expected, the dislocation solution does not contain a resultant force at the origin, so that the value of the constant k_1 can be deduced directly from the wedge solution (5.49) as

$$k_1 = -\frac{F_x}{2\pi}. \tag{5.61}$$

Adjusting the constants appropriately in the dislocation solution (5.43) and enforcing displacement continuity by combining it with the expression (5.59) leads to the equation

$$(\kappa-1)k_1 + (\kappa+1)k_2 = 0,$$

so that

$$k_2 = \frac{(\kappa-1)}{(\kappa+1)}\frac{F_x}{2\pi}. \tag{5.62}$$

Airy stress function form of the Kelvin solution for a concentrated force F_x at the origin is

$$A(r, \theta) = \frac{F_x}{2\pi}\left[-r\theta\sin\theta + \frac{(\kappa-1)}{(\kappa+1)} r\log r\cos\theta\right]. \tag{5.63}$$

5.10 THE WILLIAMS EIGENFUNCTION ANALYSIS

The family of solutions associated with the wedge is of special interest in the theory of elasticity. By considering properties of these solutions, it is possible to make valuable deductions about the influence of plane geometry (e.g., the wedge angle) on the stress state in the vicinity of the apex.

Following Williams (1952), we carry out an analysis of the eigenfunctions and eigenvalues for the wedge (plane elastic problem). Wedge geometry $-\alpha < \theta < \alpha$ of Figure 5.1 is once again considered. Solutions must satisfy traction-free boundary conditions $\sigma_{\theta\theta} = \sigma_{r\theta} = 0$ for $\theta = \pm\alpha$. In this section we search for such solutions that have the Airy stress function in the variable-separable form

$$A_0(r, \theta) = r^{\lambda+1} f(\theta). \tag{5.64}$$

The requirement of biharmonicity of $A_0(r, \theta)$ leads to

$$\Delta\Delta A_0 = r^{\lambda-3}\left[(\lambda^2-1)^2 f(\theta) + 2(\lambda^2+1)f''(\theta) + f^{(4)}(\theta)\right]. \tag{5.65}$$

The solution of this equation for the unknown function $A_0(r, \theta)$ has the form

$$A_0(r, \theta) = r^{\lambda+1}\left[a_1\cos(\lambda+1)\theta + a_2\sin(\lambda+1)\theta + a_3\cos(\lambda-1)\theta + a_4\sin(\lambda-1)\theta\right]. \tag{5.66}$$

Now the Beltrami tensor potential can be built and the stress tensor calculated using the familiar procedure of equation (5.3). The stress components $\sigma_{\theta\theta}$ and $\sigma_{r\theta}$ assume the forms

$$\sigma_{\theta\theta} = r^{\lambda-1}\left[a_1\cos(\lambda+1)\theta + a_2\sin(\lambda+1)\theta + a_3\cos(\lambda-1)\theta + a_4\sin(\lambda-1)\theta\right] \tag{5.67}$$

$$\sigma_{r\theta} = r^{\lambda-1}\left[a_1\sin(\lambda+1)\theta - a_2\cos(\lambda+1)\theta + a_3\sin(\lambda-1)\theta - a_4\cos(\lambda-1)\theta\right]. \tag{5.68}$$

These must satisfy traction-free boundary conditions on the edges $\theta = \pm\alpha$, which lead to four linear algebraic equations for the four unknown coefficients a_1, a_2, a_3, a_4. The system matrix has the form

$$\begin{pmatrix} (\lambda+1)\cos(\lambda+1)\alpha & (\lambda+1)\sin(\lambda+1)\alpha & (\lambda+1)\cos(\lambda-1)\alpha & (\lambda+1)\sin(\lambda-1)\alpha \\ (\lambda+1)\cos(\lambda+1)\alpha & -(\lambda+1)\sin(\lambda+1)\alpha & (\lambda+1)\cos(\lambda-1)\alpha & -(\lambda+1)\sin(\lambda-1)\alpha \\ (\lambda+1)\sin(\lambda+1)\alpha & -(\lambda+1)\cos(\lambda+1)\alpha & (\lambda-1)\sin(\lambda-1)\alpha & -(\lambda-1)\cos(\lambda-1)\alpha \\ -(\lambda+1)\sin(\lambda+1)\alpha & -(\lambda+1)\cos(\lambda+1)\alpha & -(\lambda-1)\sin(\lambda-1)\alpha & -(\lambda-1)\cos(\lambda-1)\alpha \end{pmatrix}. \tag{5.69}$$

An eigenfunction of the problem can be found if this system has a nontrivial solution, which happens only if the determinant of the above matrix vanishes. Evaluation leads to the equation

$$(\lambda+1)^2(\lambda\sin 2\alpha - \sin 2\lambda\alpha)(\lambda\sin 2\alpha + \sin 2\lambda\alpha) = 0. \tag{5.70}$$

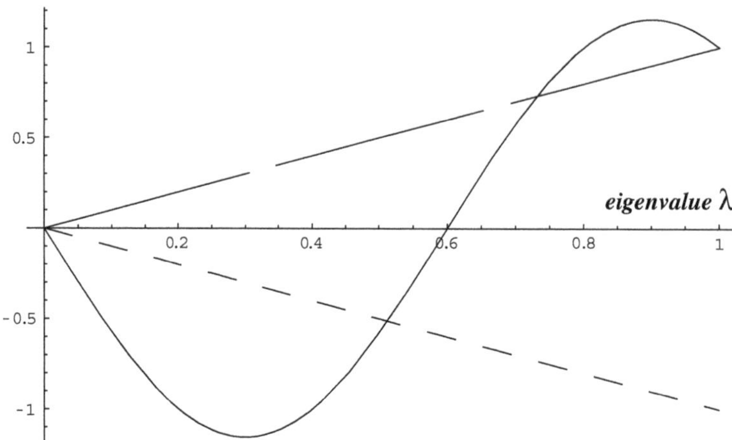

Figure 5.3. An illustration of the root-finding procedure for the transcendental eigenvalue equation (see text).

Apart from the trivial solution $\lambda = -1$, the above equation gives rise to two families of solutions, which can be identified with symmetric and antisymmetric loading of the wedge.

Of particular interest for applications are cases where eigenvalues of λ assume values below unity, because this leads to the stress fields exhibiting singular behaviour ($\sigma = Cr^{\lambda-1}$) in the vicinity of the wedge apex.

Because the characteristic equation (5.70) is transcendental, solutions must be sought numerically. The procedure for finding the roots is illustrated in Figure 5.3: intersections are sought between the curve $\sin 2\lambda\alpha / \sin 2\alpha$ (shown for $\pi/2 < \alpha < \pi$) and the lines λ (antisymmetric case, long dash) and $-\lambda$ (symmetric case, short dash). For the antisymmetric case a solution $\lambda = 1$ is always present, but for wedge angles $2\alpha > 255.4^\circ$ an additional, more singular solution $\lambda < 1$ appears. For the symmetric case singular solutions appear for reentrant wedges, that is, for $2\alpha > 180^\circ$.

The dependence of singular solution eigenvalues λ on the wedge angle 2α is illustrated in Figure 5.4.

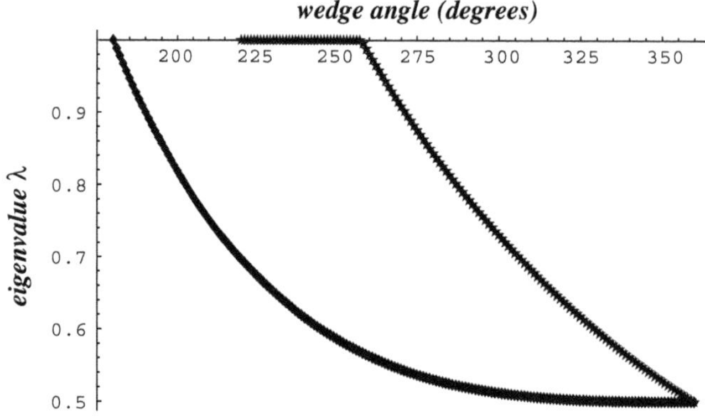

Figure 5.4. The dependence of singular solution eigenvalues λ on the wedge angle 2α (upper curve – antisymmetric solution; lower curve – symmetric solution).

MATHEMATICA programming

The procedure for Williams wedge analysis is implemented in the MATHEMATICA notebook with the help of the **MultipleListPlot** package as follows.

```
Graphics`MultipleListPlot`
Get["Tensor2Analysis.m'']
SetCoordinates[Cylindrical[r, t, z]]
A0[r_, t_] = r^(\[Lambda] + 1) f[t]
eq = Simplify[Biharmonic[A0[r, t]]]
fsol = DSolve[eq == 0, f[t], t][[1]]
f[t_] = f[t] /. fsol; ff = Simplify[f[t]]
A0 = Simplify[ComplexExpand[r^(\[Lambda] + 1)ff],
             Element[\[Lambda], Reals]]
```

The general form of the solution for $A_0(r, \theta)$ can be simplified to the form shown.

```
A0 = (A1 Cos[(\[Lambda] + 1)t] + A2 Sin[([Lambda] + 1)t] +
          A3 Cos[(\[Lambda] - 1)t] +
          A4 Sin[(\[Lambda] - 1)t])r^(\[Lambda] + 1) ;
```

Now the stress tensor can be constructed.

```
(* Build the stress tensor *)
B := {{0, 0, 0}, {0, 0, 0}, {0, 0, A0}}
B // MatrixForm
Stress := Inc[B]
Stress // MatrixForm
```

Components of the stress tensor can be displayed individually.

```
Stress[[1]][[1]]
Stress[[1]][[2]]
Stress[[2]][[2]]
```

The procedure shown can be followed to construct the linear system matrix.

```
line1 = Coefficient[Stress[[2]][[2]]
        /. t -> a, A1,A2,A3,A4]/(r^(\[Lambda]-1)[[1]]
line1 = Coefficient[Stress[[2]][[2]]
        /. t -> -a, {A1,A2,A3,A4}]/(r^(\[Lambda]-1)[[1]]
line1 = Coefficient[Stress[[1]][[2]]
        /. t -> a, {A1,A2,A3,A4}]/(r^(\[Lambda]-1)[[1]]
line1 = Coefficient[Stress[[1]][[2]]
        /. t -> -a, {A1,A2,A3,A4}]/(r^(\[Lambda]-1)[[1]]
```

```
mat = {line1, line2, line3, line4};
MatrixForm[mat/\[Lambda]]
```

Next the determinant can be computed and factored into a form suitable for solving, with the help of an additional substitution rule.

```
eq = Simplify[Det[mat/\[Lambda]]]
Factor[eq /. Cos[4 a] -> (1 - 2 (Sin[2 a])^2) /.
    Cos[4 a \[Lambda]] -> (1 - 2 (Sin[2 a \[Lambda]])^2)]
```

The resulting expression shows the form of the characteristic equation as given in equation (5.70). This equation can be plotted (here for the case $2\alpha = 300°$) as shown.

```
(* Characteristic equation *)
aa = 300 Pi/360;
Plot[{\[Lambda] , -\[Lambda], Sin[2 aa \[Lambda]]/Sin[2 aa ]},
     {\[Lambda], 0, 1},
     PlotStyle -> {Dashing[ {0.15, 0.02} ], Dashing[ {0.02, 0.02} ],
                   Dashing[ {1, 0} ]} ]
```

The roots can now be sought using intrinsic MATHEMATICA functions.

The result is the plot shown in Figure 5.4.

```
asymm = Table[{i, \[Lambda] /. FindRoot[
Sin[2 i Pi \[Lambda]/360] - \[Lambda] Sin[2 i Pi/360] == 0,
[\Lambda], 0.65, 0.25, 1.1}][[1]]}, {i, 220, 360}];
symm1 = Table[{i, \[Lambda] /. FindRoot[
Sin[2 i Pi \[Lambda]/360] + \[Lambda] Sin[2 i Pi/360] == 0,
{\[Lambda], 0.75, 0.25, 1.1}][[1]]}, {i, 180, 320}];
symm2 = Table[{i, \[Lambda] /. FindRoot[
Sin[2 i Pi \[Lambda]/360] + \[Lambda] Sin[2 i Pi/360] == 0,
{\[Lambda], 0.5, 0.25, 1.1}][[1]]}, {i, 320, 360}];
symm = Join[symm1, symm2];
MultipleListPlot[symm, asymm, PlotJoined -> {True, True},
  PlotStyle -> {Dashing [ {1, 0} ], Dashing [ {0.15, 0.15} ]}]
```

The limiting case of the Williams analysis for $2\alpha \to 360°$ is a semi-infinite straight cut in an infinite plane. The asymptotic behaviour of the stress fields with distance from crack tip that arises in this case is governed by the exponent $\lambda - 1 = -0.5$. The angular variation of stresses is governed by the function $f(\theta)$. The antisymmetric and symmetric solutions represent the characteristic near-tip stress fields in Mode II and Mode I loading, respectively.

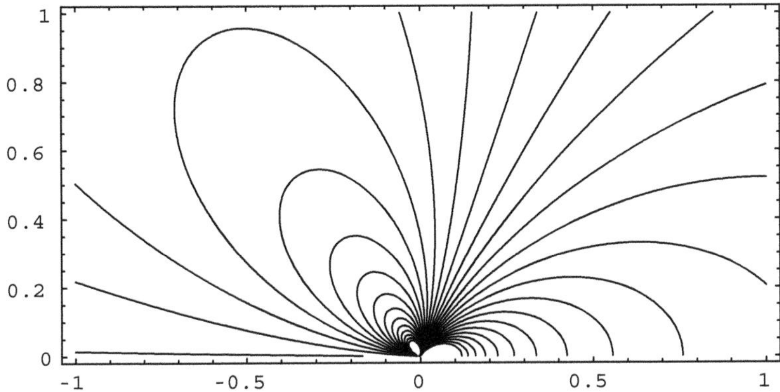

Figure 5.5. Contour plot of stress component $\sigma_{r\theta}$ above an antisymmetrically loaded crack tip at the origin.

The near-tip stress field for a particular case can be plotted as follows. First, define a rule **lrule** defining the geometry and the appropriate value of $\lambda = 0.5$. Then set one of the coefficients a_i to unity and find the eigenvector (a_1, a_2, a_3, a_4) by solving the linear system. For the antisymmetric case this is implemented as shown.

```
lrule = {\[Lambda] -> 0.5, a -> Pi};
mymat = mat /. lrule;
mat3 = mymat [[{1, 2, 3}, {1, 2, 3}]]
b = -mymat [[{1, 2, 3}, {4}]]
s = LinearSolve[mat3, b]
```

The coordinates are now transformed to cartesian, and the shear stress $\sigma_{r\theta}$ is plotted.

```
arule = {A1 -> 0, A2 -> -1., A3 -> 0, A4 -> 1};
backrule =
   Thread[{r,t,z}->CoordinatesFromCartesian[{x,y,z}]];
myrule = Thread[{Cos[t], Sin[t], r} ->
{x/Sqrt[x^2+y^2], y/Sqrt[x^2+y^2], Sqrt[x^2+y^2 ]}];
(* Plot shear stress *)
S11 = Stress[[1, 2]]
        /. arule /. lrule /. backrule /. myrule
ContourPlot[ N [ S11 ], {x, -1, 1}, {y, 0.001, 1},
  PlotPoints -> 50, ContourShading -> False,
  AspectRatio -> Automatic]
```

The symmetric solution can be developed in a similar way by setting $a_1 = 1$, etc. Figures 5.5 and 5.6 show contour plots for stress components $\sigma_{r\theta}$ and $\sigma_{\theta\theta}$ for the antisymmetric and symmetric cases, respectively.

As one final observation on the nature of the crack tip solution we recast the solution in cartesian coordinates.

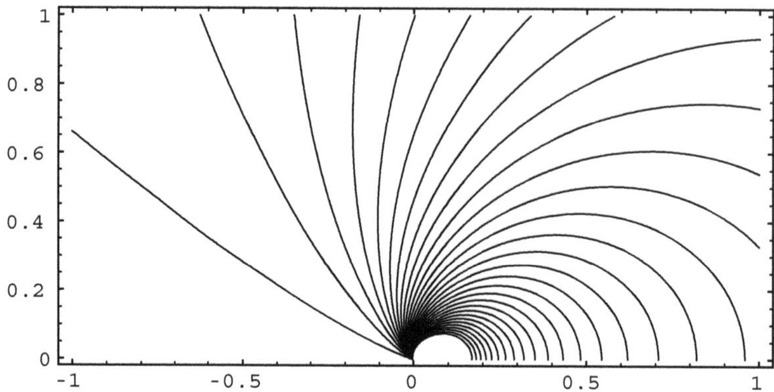

Figure 5.6. Contour plot of stress component $\sigma_{\theta\theta}$ above a symmetrically loaded crack tip at the origin.

We use the MATHEMATICA command **FieldToCartesian**, implementating the procedure as shown.

```
StressCart = Simplify[FieldToCartesian[Stress]]

S22 = StressCart[[2, 2]]
      /.arule /.lrule /.backrule /.myrule
ContourPlot [N [ S22 ], {x, -1, 1}, {y, 0.001, 1},
  PlotPoints -> 250, ContourShading -> True,
  AspectRatio -> Automatic,
  ColorFunction -> ( Hue[2/3(1 - #)]&),
  Contours -> 25, ContourLines -> True]
```

The contour map of the transverse stress component near the tip of a crack loaded in the opening mode is shown in Figure 5.7.

Finally, the von Mises stress can be readily computed.

The result is shown in Figure 5.8.

```
DeviatoricStress =
  Simplify[Stress-Tr[Stress]IdentityMatrix[3]/3];
VonMisesStress =
  Sqrt[Simplify[
3 Coefficient[Det[DeviatoricStress - w IdentityMatrix[3]],w]]
          /. arule /. lrule /. myrule /. backrule];
ContourPlot[N[VonMisesStress], {x, -1, 1}, {y, 0.001, 1},
  PlotPoints -> 250, ContourShading -> True,
  AspectRatio -> Automatic,
  ColorFunction -> (Hue[2/3(1 - #)] &),
  Contours -> 25, ContourLines -> True]
```

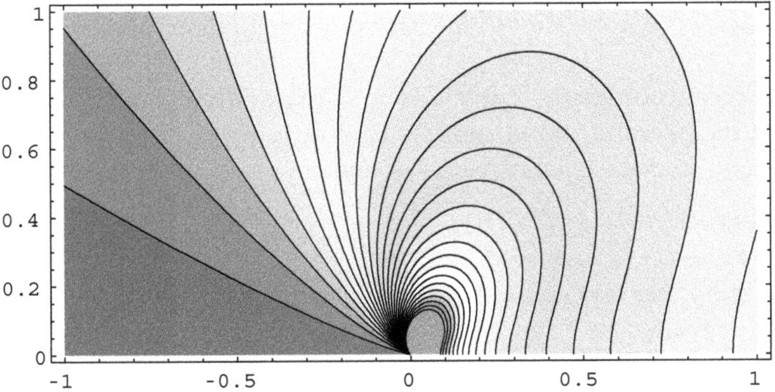

Figure 5.7. Contour plot of stress component σ_{yy} above a symmetrically loaded crack tip at the origin.

5.11 THE KIRSCH PROBLEM: STRESS CONCENTRATION AROUND A CIRCULAR HOLE

This is the name often used to refer to the problem of remote loading of a plate containing a circular hole. We will consider the particular loading case of remote tension along the x-axis. The boundary conditions for this problem are given by

$$\sigma_{rr} = 0, \qquad r = a, \tag{5.71}$$

$$\sigma_{r\theta} = 0, \qquad r = a, \tag{5.72}$$

$$\sigma_{xy} = 0, \qquad r \to \infty, \tag{5.73}$$

$$\sigma_{yy} = 0, \qquad r \to \infty, \tag{5.74}$$

$$\sigma_{xx} = S, \qquad r \to \infty. \tag{5.75}$$

We seek the Airy stress function solution in the form of the following combination of terms from the Michell solution:

$$A_0(r,\theta) = c_1 r^2 + c_2 r^2 \cos 2\theta + c_3 \log r + c_4 \cos 2\theta + c_5 r^{-2} \cos 2\theta. \tag{5.76}$$

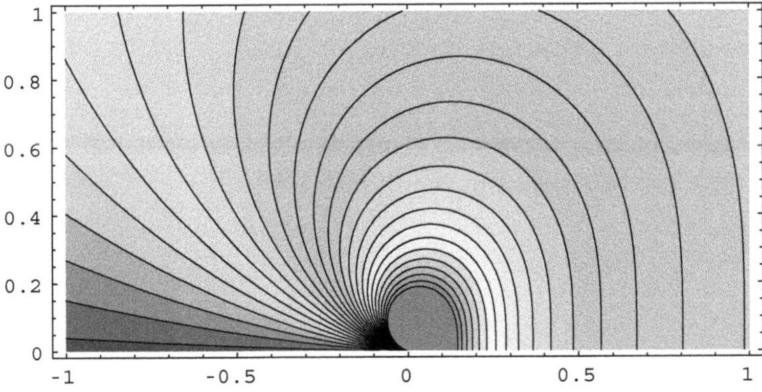

Figure 5.8. Contour plot of von Mises stress above a symmetrically loaded crack tip at the origin.

We use the algorithmic power of MATHEMATICA to solve the problem by direct application of the available tools.

The stress solution in polar coordinates is obtained by satisfying each of the boundary conditions. Note that the operator **Solve** can also deal with overdefined systems, provided they are consistent, so that a solution can be found.

```
cvec = Table["c" <> ToString[n], {n, 1, 5}]
fvec = {(r/a^2, (r/a)^2 Cos[2t], Log[r/a],
                Cos[2t], Cos[2t]/(r/a)^2}
A0[r_, a_, t_] = cvec . fvec
(Stress = Inc[{{0, 0, 0}, {0, 0, 0}, {0, 0, A0 [r, a, t]}}])
          // MatrixForm
StressCart = FieldToCartesian[Stress];
(FarStress = Limit[Expand[StressCart], r -> Infinity])
            // MatrixForm
(NearStress = Stress /. r -> a) // MatrixForm

eq1 = FarStress[[1, 1]]
eq2 = FarStress[[2, 2]]
eq3 = Coefficient[NearStress[[1, 2]], Sin[2t], 1]
eq4 = Coefficient[NearStress[[1, 1]], Cos[2t], 0]
eq5 = Coefficient[NearStress[[1, 1]], Cos[2t], 1]

csol =
 Solve[Thread[ {eq1,eq2,eq3,eq4,eq5} == {1,0,0,0,0}], cvec]
(KirschStress = Stress /. csol[[1]]) // MatrixForm
```

Now the stresses can be transformed to cartesian coordinates and plotted as contour maps.

The resulting plot is shown in Figure 5.9.

```
KirschStressCart = FieldToCartesian[KirschStress];
arule = a -> 1;
backrule =
 Thread[{r, t, z} -> CoordinatesFromCartesian[{x, y, z}]];
myrule =
 Thread[{Cos[t],Sin[t],Cos[2t],Sin[2t],r} ->
        {x/Sqrt[x^2+y^2], y/Sqrt[x^2+y^2],
         (x^2-y^2)/(x^2+y^2), xy/(x^2+y^2),
         Sqrt[x^2+y^2]
        }
        ];
(* Plot stress *)
S11 = KirschStressCart[[1, 1]]
       /. arule /. backrule /. myrule;
```

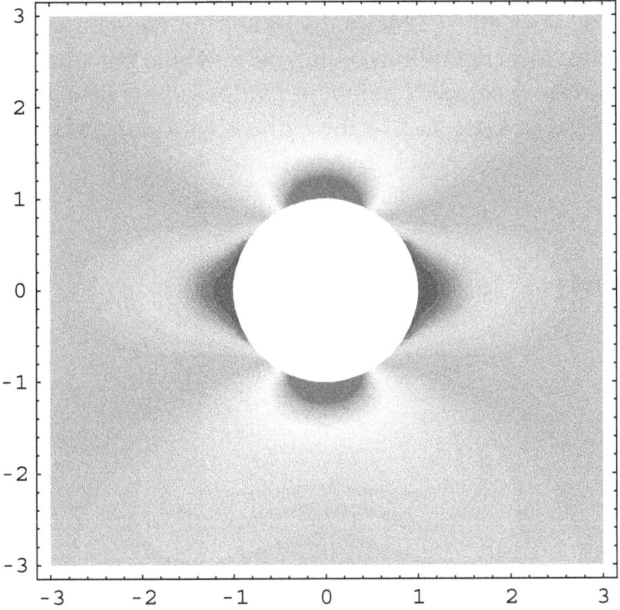

Figure 5.9. Contour plot of the horizontal stress component around a hole in an infinite plate (Kirsch problem) loaded by horizontal remote tension at infinity.

```
d = 3;
g1 = ContourPlot[N[S11], {x, -d, d}, {y, -d, d},
    PlotPoints -> 50, Contours -> 50, Compiled -> True,
    ContourShading -> True, ContourLines -> False,
    ColorFunction -> (Hue[2/3(1 - #)] &),
    DisplayFunction -> Identity]
g2 = Show[
       Graphics[{RGBColor[0, 0, 0], Disk[{0, 0}, 1]},
       AspectRatio -> Automatic, Axes -> Automatic],
      DisplayFunction -> Identity];
Show[{g1, g2}, DisplayFunction -> $DisplayFunction]
```

5.12 THE INGLIS PROBLEM: STRESS CONCENTRATION AROUND AN ELLIPTICAL HOLE

Inglis (1913) obtained the solution for the stress around an elliptical hole that paved the way for the development of the fundamentals of linear elastic fracture mechanics by Griffith (1921). The Inglis solution is most conveniently developed in elliptic cylindrical coordinates. We use this example to demonstrate MATHEMATICA's powers of manipulating tensor fields and differential operators in general curvilinear orthogonal coordinate systems.

The most convenient way to form the Airy stress function is to use the Goursat construction. For this purpose we introduce local definitions of the real and imaginary parts of a complex number. We further introduce complex numbers ζ in the (u, v) plane and Z in the (x, y) plane. Application of **ComplexExpand** to the expression $a \cosh \zeta$ shows that it indeed provides the transformation from ζ back to the z plane.

```
RePart[z_] := ComplexExpand[Re[z]];
```

```
ImPart[z_] := ComplexExpand[Im[z]];
```

```
zeta = u + I v;
```

```
Z = a Cosh[zeta];
```

```
ComplexExpand[Z]
```

Now define two complex functions $\phi(\zeta)$ and $\chi(\zeta)$ in the ζ plane and introduce a function Ψ by the Goursat construction.

```
phi = a(a1 Cosh[zeta] + (b1 + I b2) Sinh[zeta]);
chi = a^2((c1 + I c2) zeta + (d1 + I d2) Cosh[2 zeta]
        + (e1 + I e2) Sinh[2 zeta] );
Psi = RePart[Conjugate[Z] phi + chi];
```

Now group the scaling parameters (coefficients) appearing in Ψ and collect the terms containing each of the parameters in the expression for this function.

It is apparent that certain combinations of functions of coordinates u and v appear in the expressions. Using the operator **Biharmonic**, we can readily verify that all such combinations are indeed biharmonic. Thus the general form of the Airy stress function containing nine unknown coefficients is established.

```
par = {a1, b1, b2, c1, c2, d1, d2, e1, e2};
```

```
Collect[Psi, par]
```

```
Biharmonic[Cosh[u] Sinh[u]]
```

```
Biharmonic[Cos[v]^2 Cosh[u]^2 + Sin[v]^2 Sinh[u]^2]
```

Now the stress tensor is constructed using the Beltrami form of stress potential and the application of the operator **Inc**.

The rule **csrule** is introduced here to simplify some manipulations that follow.

Next a command **AtInf** is defined for the purpose of determining the behaviour of stresses at infinity. The command involves finding the limit of individual expressions accompanying each of the parameters, and then reassembling the formula for the stress component. The application of this command to the stress tensor together with **csrule** shows that the stress tensor approaches the limit that only depends on the elliptic angle **v**.

Note that the limiting stress tensor at infinity is referred to the local elliptic cylindrical coordinate system that makes an angle **v** with the cartesian system *x–y*.

```
Stress = Simplify[Inc[{{0,0,0},{0,0,0},{0,0,Psi}}]];

csrule = {Cos[2 v] -> cv, Sin[2v] -> sv,
          Cos[V] -> cv, Sin[V] -> sv};

(* Find stresses at the limit of infinity *)
AtInf[x_] :=
 Dot[
  Limit[Expand[Coefficient[Simplify[x],par]],u->Infinity],
  par]

LimitStress = Map[AtInf, Stress, {2}] /. csrule;

LimitStress // MatrixForm
```

To define an arbitrary remote stress state, we assume that the principal axes are associated with a frame *x′–y′*. This frame must be rotated by the angle **b** to return to the original frame *x–y*. In the principal axes frame *x′–y′* the stress state consists of a stress of magnitude unity acting in the direction *y′* and a stress of magnitude **L** acting in the direction *x′*.

Rotation matrices **Rv** and **Rb** are defined, and rotations are applied in the following order. Rotation by angle **b** allows the remote stress state to be expressed in the *x–y* system. Rotation by angle **v** allows the stress state to be expressed in the local coordinate system associated with the elliptic cylindrical coordinates. For convenience these rotations are applied in reverse order: first **Rv**, then **Rb**.

```
Rb = RotationMatrix3D[b, 0, 0];

Rv = RotationMatrix3D[v, 0, 0];

RemoteStress = {{1, 0, 0}, {0, L, 0}, {0, 0, 0}};

StressV =
 TrigReduce[Dot[Dot[Transpose[Rv], RemoteStress], Rv]]
   /. csrule;
(* csrule above used immediately to hold cv and sv *)

StressVB =
 TrigReduce[Dot[Dot[Transpose[Rb], StressV], Rb] ];
```

```
(* Use TrigReduce to simplify terms containing b *)

StressVB // MatrixForm
```

We are now ready to compare the remote stress state computed using the Airy stress function with the remote stress state prescribed by the principal stresses of unity and `L` and the angle `b`. Expression `S11` provides for such a comparison.

The command `EqMake` is introduced to generate linear equations for the unknown parameters from the requirement that an expression such as `S11` must be equal to zero. The application of this command to `S11` produces three independent equations denoted by `eq123`.

```
S11 = Collect[Take[Flatten[
        LimitStress - StressVB
                            ], 1], {cv, sv}]

EqMake[x_] :=
  Map[ # == 0 &,
     Drop[ Flatten[CoefficientList[x,{cv,sv}]],-1]]

eq123 = EqMake[S11]
```

The boundary of the elliptical hole is defined by a particular value of coordinate `u = u0`. Expecting `u` appearing in the stress expression to be replaced with `u0`, further equations are obtained by requiring that normal (`eq456`) and shear (`eq789`) tractions vanish on the hole boundary.

A solution for the unknown parameters is obtained by taking the `Union` of nine equations and using `Solve` to discover that a unique solution exisits for all parameters.

The rule `srule` ensures that the parameters are expressed in terms of the hole edge coordinate `u0`. This completes the solution of the Inglis problem in MATHEMATICA.

```
eq456 = EqMake[ Numerator[Stress[[1, 1]]] /. csrule ]

eq789 = EqMake[ Numerator[ Stress[[1, 2]] ] /. csrule ]

soln = Solve[ Union[ eq123 , eq456, eq789] , par];

srule = Simplify[soln] /. u -> u0
```

To illustrate the application of the solution obtained in this way, we select particular numerical values of problem parameters, as shown by the rule `nrule`. The parameter `edge` gives the numerical value of coordinate `u0` on the edge of the hole. Selecting the

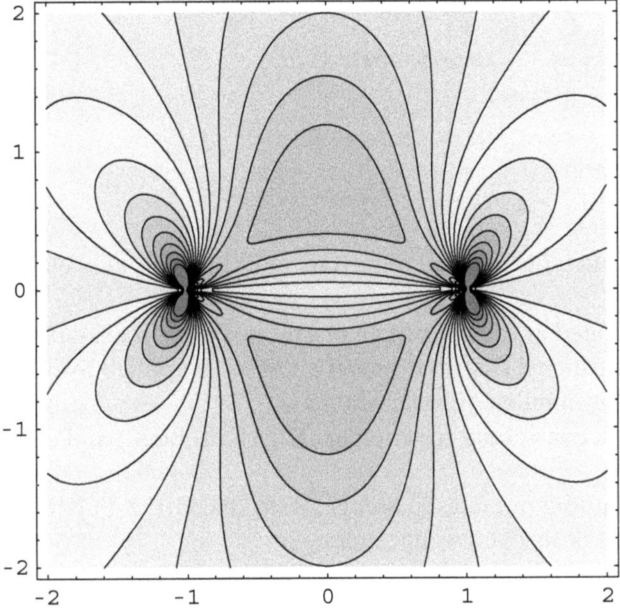

Figure 5.10. Contour plot of the vertical stress component around a horizontal crack in an infinite plate loaded by vertical remote tension at infinity; a particular case of the Inglis solution.

value of 0.01 for this parameter, together with $a = 1$, ensures that a slender elliptical hole of half-length unity along the x-axis is considered.

Solution for stress in the elliptic cylindrical system is found by the application of **srule** and **nrule**. The rotation matrix **R** is found using **RotateFromCartesian[]** command. It is then used to perform the rotation of the stress matrix in the opposite sense, that is, back from elliptic cylindrical coordinates to cartesian.

After algebraic substitutions described by rule **uv2xy**, the cartesian stress component σ_{22} is found.

```
edge = 0.01;
nrule = {a -> 1, u0 -> edge, L -> 0, b -> 0}

Str = Stress /. (srule /. nrule)[[1]];
R = RotateFromCartesian[];
uv2xy = {u -> N[Re[ArcCosh[(x +I y)/a]] /. a -> 1],
     v -> N[Im[ArcCosh[(x + I y)/a]] /. a -> 1]}
Scart = Dot[R, Dot[Str, Transpose[R]]];
S22 = RePart[Simplify[Scart[[2, 2]]]] /. uv2xy
```

The contours of the vertical stress component σ_{22} around a very slender elliptical hole are plotted in Figure 5.10. This is the picture of stress intensification by a crack that is familiar from fracture mechanics textbooks.

```
ContourPlot[N[S22], {x, -2, 2}, {y, -2, 2},
  PlotPoints -> 200, ColorFunction -> (Hue[0.42#] &),
  Contours -> 20, ContourLines -> True]
```

SUMMARY

Analyses presented in this chapter introduced some important solutions of plane elastic problems.

The solution for a point load applied at the apex of an elastic wedge is an example of a fundamental singular solution in elasticity. Distributions of point loads applied over the edge of a plate can be used to develop simple of elastic contacts.

Stress concentration at holes was considered for circular (Kirsch problem) and elliptical (Inglis problem) hole geometries.

Fundamental eigenfunction solutions for elastic wedges introduced the important concept of stress intensification at crack tips and wedge apices.

EXERCISES

1. Displacement field around a disclination

Demonstrate that the Airy stress function of the form

$$A(r, \theta) = Dr^2 \log r$$

corresponds to a *disclination*, that is, the stress and strain fields that arise if a wedge of material is inserted along the positive half of the x-axis.

A set of MATHEMATICA tools can be used:

- Stress tensor evaluation from a given Airy stress function:

  ```
  AiryStress[Airy_]:=Inc[ {{0,0,0},{0,0,0},{0,0,Airy}}]
  ```

- Fourth rank isotropic compliance tensor in terms of the Kolosov constant κ, denoted `K`:

  ```
  IC=IsotropicComplianceK[K]
  ```

  (For convenience this module assumes that $\mu = 1/2$ if the shear modulus is not given explicitly.)

- The application of `IntegrateStrain` generally leads to the appearance of rigid rotation and translation terms. These can be excluded by the application of the following rule:

  ```
  rulefun=Rule[#1,#2]& ;
  norigid=Thread[
  rulefun[Flatten[{Table[R[i], {i,3}],Table[T[i], {i,3}]}],Table[0,
                                  {i,6}]]].
  ```

  Obtain the stress tensor by applying `AiryStress[ ]` to the chosen $A(r, \theta)$,

  ```
  SIG=AiryStress[D r^2 Log[r]].
  ```

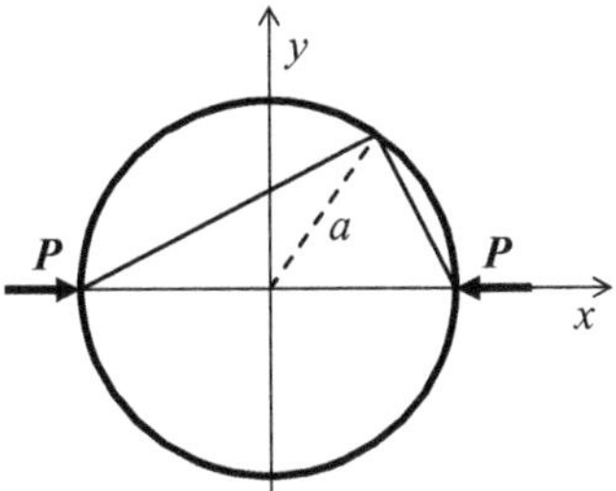

Figure 5.11. Cylindrical roller loaded by diametrically opposed concentrated forces.

Obtain the strain tensor by performing double convolution (double dot product) between the compliance tensor and the stress tensor,

```
DDot[T4_, t2_] := GTr[GDot[T4, t2, 1, 1], 1, 4]
EPS=DDot[IsotropicComplianceK[K], SIG].
```

Checking the compatibility of the strain tensor by applying operator `Inc` reveals that it is, in fact, incompatible. However, forcing the out-of-plane strain `EPS[[3,3]]` to be zero ensures compatibility.

Applying the `IntegrateStrain` procedure and excluding rigid body translation and rotation terms leads to the result of equation (5.39).

Hint: See notebook `C05_disclination_displacements.nb`.

2. Derivation of the dislocation and dislocation dipole solutions from the disclination solution

Consider the Airy stress function expression for a disclination located at point (ξ, η) in cartesian coordinates,

$$A(x-\xi, y-\eta) = \frac{1}{2}((x-\xi)^2 + (y-\eta)^2)\log((x-\xi)^2 + (y-\eta)^2).$$

Solutions for dislocations and dislocation dipoles can be obtained by differentiation with respect to the source position variables ξ and η. Note that trivial Airy stress function terms (constant and linear in cartesian coordinates) ought to be discarded.

Hint: See notebook `C05_disclination_family.nb`.

3. Kelvin solution

Obtain the Kelvin solution for a concentrated force applied at the origin of an infinite elastic plane by linear combination of two solutions: an axial force applied at the apex of a 2π wedge, and a dislocation. Determine the unknown coefficients from the conditions of static equilibrium and displacement continuity. For the latter, obtain displacements due to Airy stress fuction terms for the wedge apex force and dislocation using `IntegrateStrain`.

Hint: See notebook `C05_kelvin_displacements.nb`.

4. Diametrical compression of a cylinder by equal and opposite concentrated forces

Consider a cylindrical roller of radius a subjected to two diametrically opposing forces P per unit length (Figure 5.11). The deformation arising within the roller can be represented by the superposition of several terms. Initially consider the superposition of the following two terms:

- the Airy stress function for the Flamant problem for concentrated force P applied in the positive x direction at point $(-a, 0)$;

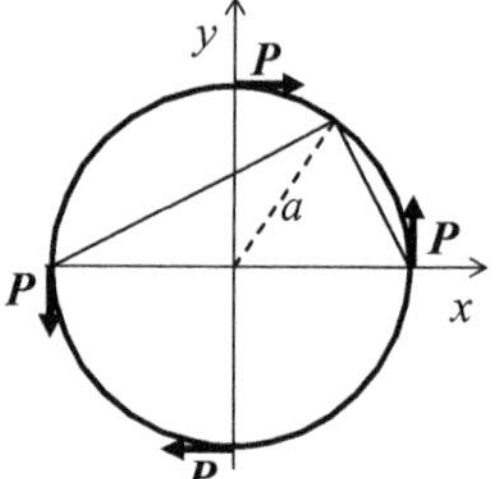

Figure 5.12. Surface loading of a cylinder by two self-equilibrated force couples.

- the Airy stress function for the Flamant problem for concentrated force P applied in the negative x direction at point $(a, 0)$.

Use MATHEMATICA tools to compute the stress field resulting from the superposition of these two solution in cartesian coordinates. Hence compute the stress state at the circumference of the cylinder.

What is the nature of the stress state? What Airy stress function term would give rise to this kind of stress state?

Hence identify the third Airy stress function term that must be superimposed to render the entire circumference of the cylinder traction-free.

Hint: See notebook `C05_nutcracker.nb`.

5. Surface loading of a cylinder by two equilibrated force couples

Consider a cylindrical roller of radius a subjected to the loading due to two force couples, as illustrated in Figure 5.12. It is convenient to assume $P = 1$.

The deformation arising within the roller can be represented by the superposition of four Airy stress function terms representing the boundary loading of a half-plane by a tangential concentrated force acting in appropriate directions and placed at positions $(a, 0)$, $(0, a)$, $(-a, 0)$, and $(0, -a)$.

Using MATHEMATICA tools, the resultant stress field may be readily computed and evaluated at the circumference in cylindrical coordinates associated with the roller axis.

What is the nature of the stress state? What are the unequilibrated surface tractions?

Consider the corrective Airy stress function term $A(r, \theta) = -2Pa\theta/\pi$ and demonstrate that its application renders the cylinder surface traction-free.

Hint: See notebook `C05_fourforces.nb`.

6. Concentrated and distributed loading at the surface of a half-plane

Consider the following Airy stress function:

$$A(x, y, \xi) = \frac{1}{2\pi}[(x-\xi)^2 + y^2)\arctan(y/(x-\xi)].$$

Show that the derivative of this function with respect to parameter ξ leads to the Flamant solution.

Hence show that the solution for boundary loading of a half-plane surface by uniformly distributed normal pressure on the segment $-a < x < a$ is given by the Airy stress function

$$A(x, y, a) - A(x, y, -a).$$

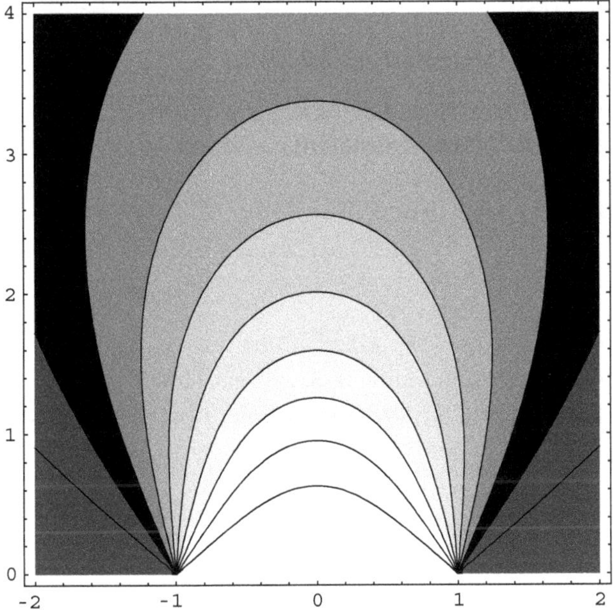

Figure 5.13. Stress distribution in a half-plane loaded by surface pressure on the segment $-1 < x/a < 1$.

Compute a contour plot of the stress component σ_{yy} in the region $-2a < x < 2a, 0 < y < 4a$ (Figure 5.13). Discuss the steps that need to be taken to develop a boundary element formulation for frictionless contact problems on this basis.

Hint: See notebook `C05_pressure.nb`.

7. Curved beam under shearing force at one end (Barber, 2002)

Consider a curved beam of finite thickness $a < r < b$ subjected to bending by a shearing force F applied at one end, with the other end built in (Figure 5.14). Seek an approximate

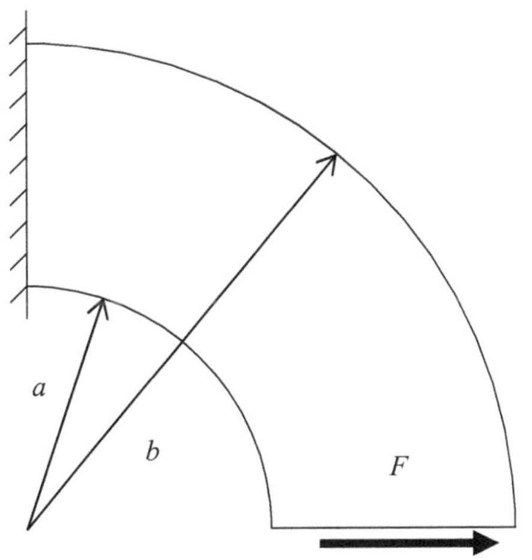

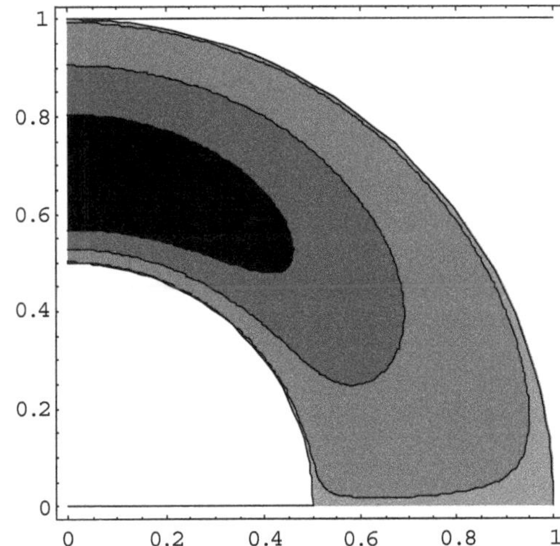

Figure 5.14. Curved beam under end shear loading.

elastic solution in the form of the Airy stress function in cylindrical polar coordinates:

$$A(r, \theta) = (Ar^3 + B/r + Cr \log r) \sin\theta + Dr\theta \cos\theta.$$

The unknown constants A, B, C, D must be found by satisfying the traction-free boundary conditions at $r = a, b$ and enforcing force and moment resultant conditions at $\theta = 0$:

$$\int_a^b \sigma_{\theta\theta} dr = \int_a^b \sigma_{\theta\theta} r \, dr = 0,$$

$$\int_a^b \sigma_{r\theta} dr = -F.$$

Using MATHEMATICA, construct and solve the system of linear equations. Hence determine the stress distribution in the curved beam, and produce a contour plot of the stress component σ_{rr} shown in Figure 5.14.

6 Displacement potentials

OUTLINE

In this chapter we consider the application of the methods of displacement potential and demonstrate the implementation of these methods in MATHEMATICA.

The fundamental expression for the Papkovich–Neuber representation of the elastic displacement fields is introduced first. Papkovich representations of the simple strain states are next considered, followed by the fundamental singular solutions for the centres of dilatation and rotation and the Kelvin solution for the concentrated force in an infinite solid. From the Kelvin solution the momentless force doublet and the force doublet with moment are derived by differentiation. The combination of three mutually perpendicular momentless force doublets is considered and is shown to be equivalent to the centre of dilatation. This example is used to demonstrate the nonuniqueness of the Papkovich description of elastic solutions. The combination of the centre of dilatation with a force doublet is also shown to correspond to a point eigenstrain solution. The point shear eigenstrain is compared with the combination of two force doublets.

Boussinesq and Cerruti solutions for the concentrated force applied at the boundary of a semi-infinite elastic solid are presented next. Solutions for concentrated forces applied at the vertex of an infinite cone are derived using the same principles from superpositions of known solutions for concentrated forces and lines of centres of rotation and dilatation. The formulation of the elastic problem in spherical coordinates is treated using spherical harmonics.

The Galerkin vector, and Love's strain function as its particular case, are introduced next. Their relationship with the Papkovich potentials is established, thus allowing any results available in the form of Galerkin vector or Love strain function to be reexamined in terms of the Papkovich potentials using the methods developed here.

At the beginning of executing each problem in MATHEMATICA the necessary packages must be loaded containing definitions of operators.

```
<< Tensor2Analysis.m
```

```
<< Displacement.m
```

Package **Displacement.m** contains standard definitions of strains in terms of displacements and of stresses in terms of strains. The latter requires the definition of the stiffness tensor for isotropic material, which is also provided. For simplicity and conciseness we assume the value of Young's modulus to be unity, $E = 1$, but the solutions for stresses can be multiplied by this factor to obtain correct dimensions.

```
IsotropicStiffness[nu_] :=
Array[ nu / (1 + nu)/(1 - 2 nu )
 KroneckerDelta[#1, #2] KroneckerDelta[#3, #4]
 + 1/ (1 + nu)/2(
KroneckerDelta[#1, #3] KroneckerDelta[#2, #4] +
KroneckerDelta[#1, #4]KroneckerDelta[#2, #3]) &, {3,3,3,3}]

Strain[u_] := (Grad[u] + Transpose[Grad[u]])/2

Stress[eps_] := DDot[ IsotropicStiffness[nu] , eps ]
```

6.1 PAPKOVICH–NEUBER POTENTIALS

The Navier equation of elastostatics can be written in the form

$$\Delta \boldsymbol{u} + \frac{1}{1-2\nu}\text{grad div } \boldsymbol{u} + \frac{2(1+\nu)}{E}\boldsymbol{b} = 0, \tag{6.1}$$

where $\boldsymbol{b}$ is the body force.

We note the following vector identity:

$$\triangle \boldsymbol{a} = \text{grad div } \boldsymbol{a} - \text{curl curl } \boldsymbol{a}. \tag{6.2}$$

To find a representation of an arbitrary vector $\boldsymbol{u}$ that vanishes at infinity, assume that

$$\boldsymbol{u} = \triangle \boldsymbol{a},$$

and hence

$$\boldsymbol{u} = \text{grad div } \boldsymbol{a} - \text{curl curl } \boldsymbol{a} = \boldsymbol{u}_I + \boldsymbol{u}_S, \tag{6.3}$$

where $\boldsymbol{u}_I = \text{grad div } \boldsymbol{a}$ is irrotational, because

$$\text{curl } \boldsymbol{u}_I = \text{curl grad div } \boldsymbol{a} = \boldsymbol{0},$$

and $\boldsymbol{u}_S = -\text{curl curl } \boldsymbol{a}$ is solenoidal, because

$$\text{div } \boldsymbol{u}_S = -\text{div curl curl } \boldsymbol{a} = 0.$$

The above statement is the Helmholtz theorem about decomposition of a vector into solenoidal and irrotational parts. Applied to the vector $(\boldsymbol{u} - \boldsymbol{u}_0)$, this theorem leads to

$$\boldsymbol{u} - \boldsymbol{u}_0 = \boldsymbol{u}_I + \boldsymbol{u}_S, \tag{6.4}$$

where vector $\boldsymbol{u}_0$ is chosen as the particular solution satisfying the equation

$$\Delta \boldsymbol{u}_0 + \frac{1}{1-2\nu}\text{grad div } \boldsymbol{u}_0 = -\frac{2(1+\nu)}{E}\boldsymbol{b}, \tag{6.5}$$

and for vectors $\boldsymbol{u}_S$ and $\boldsymbol{u}_I$ the equation becomes

$$2(1-\nu)\Delta(\boldsymbol{u}_I+\boldsymbol{u}_S)+\text{grad div }\boldsymbol{u}_I=0. \tag{6.6}$$

The vector identity (6.2) is next applied to each of the vectors $\boldsymbol{u}_I$ and $\boldsymbol{u}_S$ in the above equation to obtain

$$2(1-\nu)\,\text{grad div }\boldsymbol{u}_I=0, \tag{6.7}$$

$$\text{curl curl }\boldsymbol{u}_S=\boldsymbol{0}. \tag{6.8}$$

Substitution of these results back into equation (6.2) together with definition of $\boldsymbol{u}_I$ and $\boldsymbol{u}_S$ shows that both these vectors are harmonic, $\Delta\boldsymbol{u}_I=0,\ \ \Delta\boldsymbol{u}_S=0$.

The irrotational vector can be expressed as a gradient of a potential function,

$$\boldsymbol{u}_I=\text{grad }\chi, \tag{6.9}$$

whereas the solenoidal part can be thought to be an arbitrary harmonic vector,

$$\boldsymbol{u}_S=\boldsymbol{\psi},\quad \Delta\boldsymbol{\psi}=0,\quad \text{and} \tag{6.10}$$

$$\boldsymbol{u}=\boldsymbol{u}_0+\boldsymbol{\psi}+\text{grad }\chi. \tag{6.11}$$

Substitution of this expression into the Navier equation after simplification leads to the following relationship between vector $\boldsymbol{\psi}$ and scalar χ:

$$2(1-\nu)\Delta\,\chi+\text{div }\boldsymbol{\psi}=0. \tag{6.12}$$

In the Papkovich–Neuber formulation the relationship is taken in the form

$$\chi=-\frac{1}{1-4\nu}\,(\boldsymbol{r}\cdot\boldsymbol{\psi}+\phi)\,, \tag{6.13}$$

where $\boldsymbol{r}$ is the position vector and ϕ is an arbitrary harmonic function.

This is indeed a solution of equation (6.12), because for the harmonic vector $\boldsymbol{\psi}$

$$\Delta\,(\boldsymbol{r}\cdot\boldsymbol{\psi})=2\,\text{div }\boldsymbol{\psi}. \tag{6.14}$$

The Papkovich–Neuber representation for the displacement field is written in the form

$$\boldsymbol{u}=\boldsymbol{u}_0+\boldsymbol{\psi}-\frac{1}{(1-4\nu)}\text{grad }(\boldsymbol{r}\cdot\boldsymbol{\psi}+\phi)\,, \tag{6.15}$$

where $\boldsymbol{u}_0$ is the displacement field corresponding to the body force $\boldsymbol{b}$. The remaining terms depend on the harmonic vector potential $\boldsymbol{\psi}$ and harmonic scalar potential ϕ and represent the elastic displacement field in the absence of body forces.

Now that the Papkovich–Neuber form of the elastic solution has been derived on the basis of the Helmholtz decomposition, an important point needs to be made regarding the properties of the potentials. Although the vector potential $\boldsymbol{\psi}$ was assumed to be solenoidal ($\text{div}\,\boldsymbol{\psi}=0$) in the derivation, this requirement can now be relieved. Indeed, let $\boldsymbol{\psi}$ be an arbitrary *harmonic* vector potential and ϕ an arbitrary harmonic scalar potential. Then equation (6.12), which is the equivalent form of the Navier equation, is automatically satisfied by construction. However, it is no longer possible to assert that $\boldsymbol{u}_S=\boldsymbol{\psi}$; that is, vector potential $\boldsymbol{\psi}$ can no longer be thought to represent the solenoidal part of the Helmholtz decomposition.

Furthermore, a given displacement field $\boldsymbol{u}$ admits more than one representation in terms of the Papkovich potentials $\boldsymbol{\psi}$ and ϕ, as will be demonstrated below.

Uniform deformation and stress states

We begin consideration with the simple deformation and stress states that correspond to pure uniaxial strain or stress and to pure shear.

Uniform uniaxial strain state is described as follows:

Consider the following Papkovich potentials in cartesian coordinates: $\boldsymbol{\psi} = Cx\,\boldsymbol{e}_z$ and $\phi = 0$. The corresponding strain tensor is readily computed and found to correspond to *uniaxial strain* along the z-axis.

```
SetCoordinates[Cartesian[x, y, z]]
pos = {x, y, z}

psi = C z {0,0,1};
phi = 0;

u = Simplify[psi - 1/(4(1-nu)) Grad[Dot[pos, psi]+phi]]
eps=Strain[u]
sig=Stress[eps]
```

Uniform uniaxial stress state is described next.

In order to establish the Papkovich description of *uniaxial stress* state, consider a superposition of three components of the vector potential, as follows: $\psi = Cz\,\boldsymbol{e}_z + By\,\boldsymbol{e}_y + Bx\,\boldsymbol{e}_x$ and $\phi = 0$. We compute the strain and stress tensor and require that the transverse stress components σ_{yy} and σ_{xx} vanish and the axial stress component $\sigma_{zz} = \sigma_0$, so that both constants B and C are determined in terms of σ_0.

```
SetCoordinates[Cartesian[x, y, z]]
pos = {x, y, z}

psi = C z {0,0,1} + B y {0,1,0} + B x {1,0,0};
phi = 0;

u = Simplify[psi - 1/(4(1-nu)) Grad[Dot[pos, psi]+phi]]
eps = Strain[u]
sig = Simplify[Stress[eps]]

solB = Solve[sig[[2, 2]] == 0, B][[1, 1]]
sigg = Simplify[sig /. solB]

Solve[sigg[[1, 1]] == Sig0, {C}][[1, 1]]
```

Uniform pure shear deformation is described next:

To construct the Papkovich solution corresponding to the state of pure shear, a different approach is employed. Considering a particular vector function $\boldsymbol{a}$ and performing Helmholtz decomposition of the displacement field $\boldsymbol{u} = \Delta\boldsymbol{a}$ into irrotational part $\boldsymbol{u}_i$ and solenoidal part $\boldsymbol{u}_s$, it is found that $\boldsymbol{u}_i = \boldsymbol{0}$. We therefore set $\boldsymbol{\psi} = \boldsymbol{u}_s$ and select $\phi = -\boldsymbol{r}\cdot\boldsymbol{\psi}$, noting that this results in $\Delta\phi = 0$. Therefore the Papkovich expression for the displacement gives $\boldsymbol{u} = \boldsymbol{\psi} = y\boldsymbol{e}_x + x\boldsymbol{e}_y$, that is, the dispacement field for pure shear. This is confirmed by computing the strain and stress tensors.

```
SetCoordinates[Cartesian[x, y, z]]
pos={x,y,z}

a = {y^3/6, x^3/6, 0}

ui = Grad[Div[a]]
us = -Curl[Curl[a]]

psi = - Curl[Curl[a]]
phi = -pos.psi

u = psi - 1/(4(1 - nu))Grad[pos.psi + phi]
eps = Strain[u]
sig = Stress[eps]
```

Concentrated force in infinite space – the Kelvin solution

Probably the most important fundamental singular solution in the theory of elasticity corresponds to the concentrated force at the origin within an infinitely extended elastic isotropic material and is due to Kelvin. The Papkovich–Neuber potentials for the concentrated force applied at the origin and given by the vector $\boldsymbol{P}$ in cartesian coordinates are

$$\boldsymbol{\psi} = \boldsymbol{P}\frac{A}{R}, \qquad \phi = 0,$$

where $R = \sqrt{x^2 + y^2 + z^2}$ denotes the distance from the origin.

The concentrated force vector $\boldsymbol{P}$ is specified by its cartesian components P_x, P_y, P_z. However, because the problem is going to be solved in spherical coordinates, the Papkovich potential $\boldsymbol{\psi}$ must be defined as a vector field in spherical coordinates. The transformation of a vector field from cartesian to polar coordinates is a standard operation performed using the command **FieldFromCartesian**.

```
P = {Px, Py, Pz};

SetCoordinates[Spherical[R, p, t]]
pos = {R, 0, 0}
```

```
psi = FieldFromCartesian[P] A/R
phi = 0;
```

The Papkovich–Neuber potentials introduced in this way are harmonic, as can easily be verified. The corresponding deformation and stress state is determined as follows.

The dispacement field is computed according to the Papkovich–Neuber definition.

```
u = Simplify[psi-1/(4(1-nu))Grad[Dot[pos,psi]+phi]]
```

Following the definition of strain and stress, the divergence-free property of the stress field is confirmed by application of the `Div` operator.

```
eps = Strain[u]
```

```
sig = Simplify[Stress[eps]]
```

```
Simplify[Div[sig]]
```

Force equilibrium for any spherical volume centred around the origin is now enforced. The traction vector on this surface is determined by the dot product between the stress tensor and the normal. Resultant force determination requires integration of traction as a field referred to the fixed cartesian system, taking account of surface element area ($R^2 \sin\varphi$), found as the product of scale factors corresponding to coordinates $\varphi \to$ `p` and $\vartheta \to$ `t`. The unknown constant A is now determined by requiring the resultant force to be equal and opposite to $\boldsymbol{P}$.

```
eR = {1, 0, 0};
```

```
traction = Dot[sig, eR]
```

```
ScaleFactors[]
```

```
resforce = Simplify[Integrate[
      FieldToCartesian[traction] R^2 Sin[p],
{p, 0, Pi}, {t, 0, 2 Pi}]]
```

```
solA = Simplify[Solve[resforce == -P, A][[1, 1]]]
```

The final displacement solution is obtained by back substitution of the constant A.

```
usol = u /. solA
Simplify[usol]
```

The displacement field can now be transformed back to the cartesian system.

```
FieldToCartesian[usol] /.
  Thread[{R,p,t}->CoordinatesFromCartesian[{x,y,z}]]
```

A sequence of simplification operations is required to obtain the clear final form of the solution. In particular, the derivation involves the use of the exponential form of trigonometric functions.

```
TrigToExp[%];
Simplify[%];
ComplexExpand[%];
PowerExpand[%];
Simplify[%]
```

The solution for displacements in the spherical system of coordinates can be written as follows:

$$\boldsymbol{u} = \frac{(1+\nu)}{8\pi(1-\nu)E}\left[(3-4\nu)\frac{\boldsymbol{P}}{R} + \frac{(\boldsymbol{R}\cdot\boldsymbol{P})\boldsymbol{R}}{R^3}\right]. \tag{6.16}$$

Alternatively, using the vector $\boldsymbol{e}_R = \boldsymbol{R}/R$, the above equation can be rewritten as

$$\boldsymbol{u} = \frac{(1+\nu)}{8\pi(1-\nu)E}\frac{1}{R}\left[(3-4\nu)\boldsymbol{P} + (\boldsymbol{e}_R\otimes\boldsymbol{e}_R)\cdot\boldsymbol{P}\right]. \tag{6.17}$$

We recall here that $(\boldsymbol{e}_R\cdot\boldsymbol{P})\boldsymbol{e}_R = (\boldsymbol{e}_R\otimes\boldsymbol{e}_R)\cdot\boldsymbol{P}$.

Let now a distribution of forces $\boldsymbol{P}(\boldsymbol{\xi})$ be considered within some region Ω. The form of equation (6.16) remains unchanged if the point of application of the concentrated force is moved away from the origin, provided that the vector $\boldsymbol{R}'$ is replaced with the vector $\boldsymbol{R}'$ computed with respect to the new position of the force, $\boldsymbol{\xi}$; that is, $\boldsymbol{R}' = \boldsymbol{R} - \boldsymbol{\xi}$. Therefore, the displacement resulting from a force distribution over Ω can be obtained from equation (6.16) by superposition in the form

$$\boldsymbol{u} = \frac{(1+\nu)}{8\pi(1-\nu)E}\int_\Omega\left[(3-4\nu)\frac{\boldsymbol{P}(\boldsymbol{\xi})}{R} + \frac{(\boldsymbol{R}\cdot\boldsymbol{P}(\boldsymbol{\xi}))\boldsymbol{R}}{R^3}\right]d\boldsymbol{\xi}. \tag{6.18}$$

This is the original Kelvin solution of 1848 that gives a particular integral of the Navier equation.

Momentless force dipoles

A number of further singular solutions can be obtained from the Kelvin solution by superposition, differentiation, and integration. In this section we are concerned with pairs of concentrated forces that are equal in magnitude and opposite in sign and separated by an infinitesimal distance, that is, force dipoles.

First consider the particular case of a concentrated force $\boldsymbol{P} = P_z \boldsymbol{e}_z$ with the Papkovich–Neuber representation in cartesian coordinates

$$\boldsymbol{\psi} = \frac{1+\nu}{2\pi} \frac{P_z}{R} \boldsymbol{e}_z, \qquad \phi = 0. \tag{6.19}$$

We first obtain the solution for a force dipole without moment by considering two opposite concentrated forces of equal magnitude separated by a small distance a_z along the line of their action. We follow a limiting process whereby the distance a_z is allowed to diminish while P_z increases proportionately, so that the dipole strength D given by the product $D_{zz} = P_z a_z$ remains constant. The process is analogous to the definition of a derivative, and therefore the corresponding form of the Papkovich potentials can be obtained by differentiation of the above expression along the direction of the segment a_z, that is, with respect to the z coordinate. This results in

$$\boldsymbol{\psi}_{[zz]} = -\frac{1+\nu}{2\pi} \frac{z D_{zz}}{R^3} \boldsymbol{e}_z, \qquad \phi = 0. \tag{6.20}$$

Similar expressions can be obtained for the momentless force dipole along the x coordinate,

$$\boldsymbol{\psi}_{[xx]} = -\frac{1+\nu}{2\pi} \frac{x D_{xx}}{R^3} \boldsymbol{e}_x, \qquad \phi = 0,$$

and for the momentless force dipole along the y coordinate,

$$\boldsymbol{\psi}_{[yy]} = -\frac{1+\nu}{2\pi} \frac{y D_{yy}}{R^3} \boldsymbol{e}_y, \qquad \phi = 0.$$

It is now possible to consider a special case of superposition of three mutually orthogonal momentless force dipoles of identical strength $D = D_{xx} = D_{yy} = D_{zz}$, that is,

$$\boldsymbol{\psi} = \boldsymbol{\psi}_{[xx]} + \boldsymbol{\psi}_{[yy]} + \boldsymbol{\psi}_{[zz]} = -\frac{1+\nu}{2\pi} \frac{D}{R^3} (x \boldsymbol{e}_x + y \boldsymbol{e}_y + z \boldsymbol{e}_z) = -\frac{1+\nu}{2\pi} \frac{D}{R^2} \boldsymbol{e}_R, \qquad \phi = 0.$$

The resulting combination of Papkovich potentials is

$$\boldsymbol{\psi} = -\frac{1+\nu}{2\pi} \frac{D}{R^2} \boldsymbol{e}_R, \qquad \phi = 0. \tag{6.21}$$

It is interesting to note here that the above potentials give rise to a displacement field that is *identical* to the field arising from the potentials

$$\boldsymbol{\psi} = \boldsymbol{0}, \qquad \phi = \frac{A}{R}, \tag{6.22}$$

where A is another constant. This case demonstrates that the Papkovich–Neuber description of elastic fields is *nonunique*: the same elastic solution can be represented by two entirely different sets of potentials.

The classical combination of potentials introduced above, known as the *centre of dilatation*, is considered in the next section.

Centre of dilatation in the infinite space – the Lamé solution

The Papkovich–Neuber potential representation considered in this section corresponds to the singular solution of elasticity first studied by Lamé. Assume that the displacement

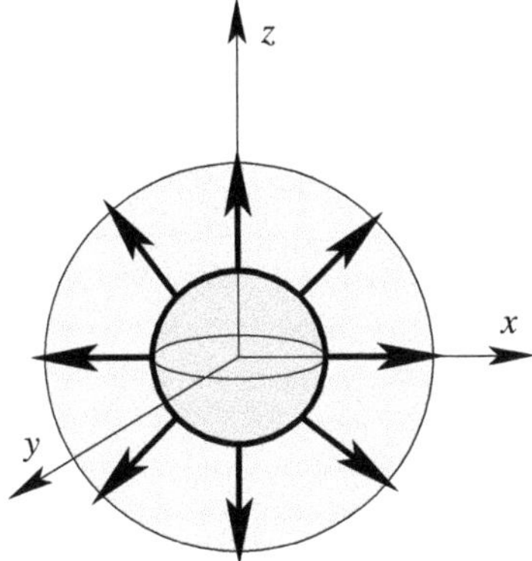

Figure 6.1. An illustration of a centre of dilatation.

field is simply given by the gradient of a scalar harmonic function, ϕ, or, in the form of equation (6.15)

$$\phi = \frac{A}{R}, \quad \boldsymbol{\psi} = \mathbf{0}.$$

To find the displacement, strain, and stress fields, the problem is solved in spherical coordinates. Vector potential $\boldsymbol{\psi}$ is defined as the zero vector, and scalar potential as $\phi = A/R$.

```
SetCoordinates[Spherical[R, p, t]]
pos = {R, 0, 0}
psi = {0,0,0};
phi = A/R;
```

The displacement field is constructed using equation (6.15), and the strain field is found. The volumetric strain can be computed as the trace of the strain tensor. It is interesting to note that it vanishes everywhere away from the origin.

```
u = Simplify[psi - 1/(4(1-nu)) Grad[Dot[pos, psi]+phi]]

eps = Strain[u]

veps = Tr[eps]
```

The stress tensor is found from strains and is diagonal everywhere in this coordinate system. Stress components $\sigma_{\vartheta\vartheta}$ and $\sigma_{\varphi\varphi}$ have half the magnitude of the radial stress component, σ_{RR}, and are opposite to it in sign. It is readily verified that the stress field is divergence-free.

```
sig = Simplify[Stress[eps]]

(sig) // MatrixForm

Simplify[Div[sig]]
```

To introduce a measure of intensity of the dilatation centre, we consider the traction on a spherical surface of radius R. The traction is purely radial and uniform, so that the solid lying outside the sphere of radius R is under uniform internal pressure. Denote by $\boldsymbol{n}$ the outward unit normal to the elastic solid, $\boldsymbol{n} = -\boldsymbol{e}_R$. Equating the traction to $-p\,\boldsymbol{n}$ at radius $R = a$ leads to the result $A = 2pa^3(1 - \nu^2)$.

```
eR = {1, 0, 0};
traction = Simplify[ Dot[sig, eR] ]

Solve[(traction /. R -> a) == {-p, 0, 0}, A]
```

The singular solution for the centre of dilatation at the origin can be obtained by considering a sequence of problems with pressure loading p applied within spherical cavities of diminishing radius a, so that $pa^3 = \text{const}$. The limit of the solutions is the dilatation centre with the strength $A = 2pa^3(1 - \nu^2)$. The displacement field is given by

$$\boldsymbol{u} = \frac{A}{4(1-\nu)r^2}\boldsymbol{e}_R \tag{6.23}$$

in the spherical coordinates.

Force dipoles with moment

We again begin the analysis with the particular case of concentrated force $\boldsymbol{P} = P_y\boldsymbol{e}_y$ with the Papkovich–Neuber representation in cartesian coordinates

$$\boldsymbol{\psi} = \frac{1+\nu}{2\pi}\frac{P_y}{R}\boldsymbol{e}_y, \qquad \phi = 0. \tag{6.24}$$

We now obtain the solution for a force dipole with moment by considering two opposite concentrated forces of equal magnitude separated by a small distance a_x perpendicular to the line of their action. We follow a limiting process whereby the distance a_x is allowed to diminish while P_y increases proportionately, so that the dipole strength D given by the product $D_{xy} = a_xP_y$ remains constant. The process is analogous to the definition of a derivative, and therefore the corresponding form of the Papkovich potentials can be obtained by differentiation of the above expression along the direction of the segment a_x, that is, with respect to the x coordinate. This results in

$$\boldsymbol{\psi}_{[xy]} = -\frac{1+\nu}{2\pi}\frac{D_{xy}\,x}{R^3}\boldsymbol{e}_y, \qquad \phi = 0. \tag{6.25}$$

A similar expression can be obtained for the force dipole with moment denoted by D_{yx}:

$$\boldsymbol{\psi}_{[yx]} = -\frac{1+\nu}{2\pi}\frac{D_{yx}\,y}{R^3}\boldsymbol{e}_x, \qquad \phi = 0.$$

The general expression for the Papkovich–Neuber potentials for the force dipole with moment D_{ij} is

$$\boldsymbol{\psi}_{[ij]} = -\frac{1+\nu}{2\pi}\frac{D_{ij}\,x_i}{R^3}\boldsymbol{e}_j, \qquad \phi = 0.$$

Centre of rotation in an infinite elastic solid

The moments associated with the force dipoles introduced in the previous section give rise to rotational displacements concentrated at the origin. Consider two force doublets with moment of equal intensity $D = D_{xy} = D_{yx}$ that form a superposition

$$\boldsymbol{\psi} = \boldsymbol{\psi}_{[xy]} - \boldsymbol{\psi}_{[yx]} = -\frac{1+\nu}{2\pi}\frac{D}{R^3}(x\boldsymbol{e}_y - y\boldsymbol{e}_x), \qquad \phi = 0. \tag{6.26}$$

Let us demonstrate that this deformation state corresponds to the centre of rotation arising from a concentrated moment M_0 at the origin.

Solution begins with defining the Papkovich potentials in cartesian coordinates according to the above formulas and verifying their harmonicity using the **Laplacian** operator. The displacement field is then computed using the standard formula.

```
SetCoordinates[Cartesian[x, y, z]]
pos = {x, y, z};
R = Sqrt[pos.pos];

psi = D{-y, x, 0} 1/R^3
phi = 0;
Simplify[Laplacian[psi]]
u = Simplify[psi - 1/(4(1 - nu)) Grad[pos.psi + phi]]
```

To obtain a simple representation of the displacement field, transformation into cylindrical polar coordinates is performed. For this purpose the command **SetCoordinates** is first used, the transformation rule **car2cyl** is defined, and expressions for potential and displacement vectors in cylindrical polar coordinates are obtained.

```
SetCoordinates[Cylindrical[r, t, z]]

car2cyl=Thread[{x,y,z} -> CoordinatesToCartesian[{r,t,z}]]

psicyl = Simplify[FieldFromCartesian[psi] /. car2cyl]

ucyl = Simplify[FieldFromCartesian[u] /. car2cyl]
```

The displacement field in cylindrical polar coordinates is given by the simple expression

$$\boldsymbol{u} = \frac{D\,r}{(r^2+z^2)^{3/2}}\,\boldsymbol{e}_\theta. \tag{6.27}$$

The displacement field is solenoidal, the only nonzero component being parallel to the basis vector $\boldsymbol{e}_\theta$, with a magnitude that is independent of θ, thus corresponding to torsional symmetry.

The strain and stress fields are computed using standard commands.

```
eps = Strain[ucyl]
sig = Stress[eps]
```

Nonzero strain components are $\varepsilon_{r\theta}$ and $\varepsilon_{z\theta}$ and give rise to the stresses

$$\sigma_{r\theta} = -\frac{3Dr^2}{2(1+\nu)(r^2+z^2)^{5/2}}, \tag{6.28}$$

$$\sigma_{z\theta} = -\frac{3Drz}{2(1+\nu)(r^2+z^2)^{5/2}}. \tag{6.29}$$

To determine the unknown constant D, the moment balance condition must be enforced.

The traction on the cylindrical surface is computed as a scalar product of the stress tensor with the surface normal vector. The resultant is computed by the integration of moments on elemental areas given by $r\, t_\theta\, r\, \mathrm{d}\theta\, \mathrm{d}z$. Noting that the result is independent of the cylinder radius, the moment balance condition is enforced by equating the integral to $-M_0$. Finally, the solution for the constant is found to be $D = M_0(1+\nu)/(4\pi)$, which can be substituted back into the expressions for displacements and the Papkovich vector potential $\boldsymbol{\psi}$.

```
t = sig.{1, 0, 0}

mom = Integrate[t[[2]] r^2,
    {t,0,2Pi},{z,-Infinity,Infinity}]

solD = Solve[mom0 == -M0, {D}][[1, 1]]

psicyl /. solD

ucyl /. solD
```

The same solution can be obtained using a slightly different approach by noting from the start that the vector potential $\boldsymbol{\psi}$ can be assumed to represent the solenoidal part of the displacement vector decomposition, with the irrotational part being equal to zero. The vector potential may therefore be written in the form

$$\boldsymbol{\psi} = \operatorname{grad}\left(\frac{1}{R}\right) \times \boldsymbol{e}_z. \tag{6.30}$$

This definition leads to identical results.

Strain nuclei in an infinite elastic solid

The term 'strain nuclei' is used here to refer to deformation states associated with force dipoles or strain states concentrated at a point. The elastic fields for these solutions can be obtained from the force dipole solutions described above.

Consider Papkovich–Neuber potentials for the superposition of two equal magnitude force dipoles $D_{xy} = D_{yx} = D$, as follows:

$$\boldsymbol{\psi} = \frac{1}{2}(\boldsymbol{\psi}_{[xy]} + \boldsymbol{\psi}_{[yx]}) = -\frac{1+\nu}{4\pi}\frac{D}{R^3}(y\boldsymbol{e}_x + x\boldsymbol{e}_y), \qquad \phi = 0. \tag{6.31}$$

This is equivalent to a point shear strain.

Solutions for direct strains concentrated at a point can be similarly developed. Their relationship to the force dipoles is given by a form of Hooke's law.

Concentrated load normal to the surface of a half space – The Boussinesq solution

This problem is solved in cylindrical coordinates. The solution is constructed by linear superposition of known solutions.

```
SetCoordinates[Cylindrical[r, t, z]]
R = Sqrt[r^2 + z^2];
pos = {r, 0, z};
```

The first term is the Kelvin solution with the force vector $\boldsymbol{P}$ having only one nonzero component in the z direction, and strength given by an unknown constant A.

```
psi = A {0, 0, 1/R}
```

The second term is the solution for a line of centres of dilatation along the negative z axis, for which the solution consists of only the scalar potential $\phi = B\log(z + R)$, where $R = \sqrt{r^2 + z^2}$, and B is a scaling constant.

```
phi = B Log[z + R]
```

The displacement field is constructed using the Papkovich–Neuber formula, and strain and stress fields are computed using commands defined previously.

```
u = Simplify[psi - 1/(4(1 - nu))Grad[Dot[pos, psi] + phi]]
```

```
eps = Strain[u]
sig = Simplify[Stress[eps]]
```

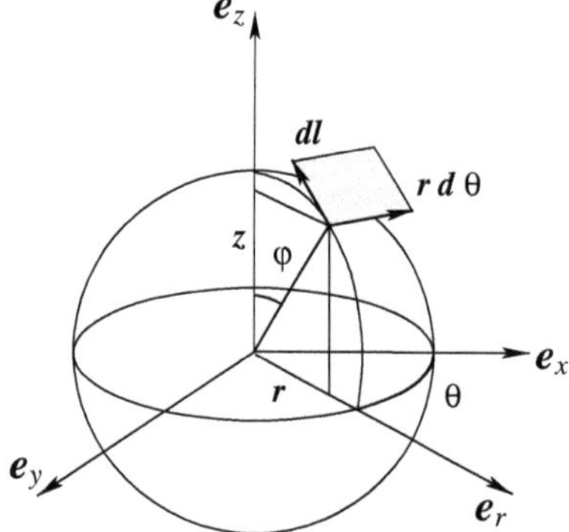

Figure 6.2. An illustration of spherical surface element in cylindrical polar coordinates.

Provided the functions for Papkovich potentials were chosen correctly, the stress field should be divergence-free. This is verified using the command **Div**.

```
Simplify[Div[sig]]
```

Next the relationship between constants A and B can be found by enforcing traction-free boundary conditions, $\boldsymbol{\sigma} \cdot \boldsymbol{n} = \mathbf{0}$, on the surface $z = 0$. This leads to the result $B = A(1 - 2\nu)$.

```
ez = {0, 0, 1};
surftraction = Simplify[ Dot[sig, ez] /. z -> 0]

solB = Solve[ %[[1]] == 0 , B][[1, 1]]
```

Next the force equilibrium on a semispherical volume of radius a around the origin is enforced. This requires first the determination of tractions on the hemispherical surface of radius a, using $\boldsymbol{t} = \boldsymbol{\sigma} \cdot \boldsymbol{e}_R$, where $\boldsymbol{e}_R$ is the radial vector field. Next the resultant force must be determined by integration. Figure 6.2 illustrates the procedure adopted for integration. Given that the spherical surface is described in cylindrical coordinates by $z = \sqrt{a^2 - r^2}$, the element of surface is computed as $\dfrac{a r \mathrm{d} r \mathrm{d}\theta}{\sqrt{a^2 - r^2}}$.

```
eR = {r, 0, z} / Sqrt[r^2 + z^2]
traction = Simplify[ Dot[ sig . eR ] ]
```

The resultant of internal tractions is computed with respect to the fixed system of coordinates, and the relationship between B and A is substituted. The result is independent of the radius a and has only one nonzero component in the z direction: $R_z = -\dfrac{A\pi}{2(1 - \nu^2)}$.

Assuming the external force P is applied in the positive z direction, the resultant internal force must be equal and opposite.

The solution is $A = -\frac{2(1-\nu^2)P}{\pi}$.

The solution for u is obtained by back substitution, and can be expressed in cartesian coordinates using the same set of commands given for the Kelvin solution.

```
resforce = Simplify[
  Integrate[
    (FieldToCartesian[traction] r a/Sqrt[a^2-r^2]) /.
        z -> Sqrt[ a^2 - r^2] ,
  {r, 0, a}, {t, 0, 2 Pi}]]

resforce = Simplify[resforce /. solB]

solA = Simplify[Solve[resforce == {0, 0, -P}, A][[1, 1]]]

usol = Simplify[ u /. solB  /. solA ]

FieldToCartesian[usol] /.
 Thread[ {r,t,z}->CoordinatesFromCartesian[{x,y,z}]]

TrigToExp[%];Simplify[%];ComplexExpand[%];PowerExpand[%];
Simplify[%]
```

The solution for displacements in the cylindrical system of coordinates is

$$u = u_r e_r + u_z e_z = \frac{P}{2\pi}\frac{(1+\nu)}{E}\left[\left(\frac{rz}{R^3} - \frac{(1-2\nu)}{R(R+z)}\right)e_r + \left(\frac{z^2}{R^3} + \frac{2(1-\nu)}{R}\right)e_z\right], \tag{6.32}$$

where $R = \sqrt{r^2+z^2}$.

Concentrated load parallel to the surface of a half space – The Cerruti solution

The solution to this problem begins in cartesian coordinates and is constructed by linear superposition of several known solutions.

```
SetCoordinates[Cartesian[x, y, z]]
R = Sqrt[x^2 + y^2 + z^2];
pos = {x, y, z};
P = {Px, 0, 0};
```

The first term is the Kelvin solution, with the force vector $\boldsymbol{P}$ having only one nonzero component in the x direction.

```
psi1 = {1, 0, 0}/R;
phi1 = 0;
```

The second term corresponds to a line of centres of rotation along the negative z-axis.

```
psi2 = -Simplify[PowerExpand[Curl[Log[R + z]{0, 1, 0}]]];
phi2 = -x/(R + z);
```

The third term corresponds to a double line of centres of dilatation, also along the negative z-axis.

```
psi3 = {0, 0, 0};
phi3 = D[z Log[z + R] - R, x];
```

The full Papkovich potentials are now assembled out of the three parts, with unknown coefficients A, B, and C.

```
psi = A psi1 + B psi2 + C psi3;
phi = A phi1 + B phi2 + C phi3;
```

The displacement field is constructed using the Papkovich–Neuber formula, and strain and stress fields are computed using commands defined previously.

```
u = Simplify[psi - 1/(4(1 - nu))Grad[Dot[pos, psi] + phi]]
eps = Strain[u]
sig = Simplify[Stress[eps]]
```

The stress field is verified to be divergence-free using the command `Div`.

```
Simplify[PowerExpand[Div[sig]]]
```

Now the surface $z = 0$ should be cleared of tractions everywhere (with the exception of a single point of load application at the origin). To this end we define the normal vector to the surface, $\boldsymbol{e}_z$, and compute the traction vector using $\boldsymbol{\sigma} \cdot \boldsymbol{e}_z$. Next the expressions are found for the constants A and B in terms of constant C by enforcing traction-free boundary conditions, $\boldsymbol{\sigma} \cdot \boldsymbol{n} = \boldsymbol{0}$, on the surface $z = 0$. This leads to the results $A = C/(1 - 2\nu)$, $B = -C/2(1 - \nu)$.

```
ez = {0, 0, 1};

surftraction = Simplify[ Dot[sig, ez] /. z -> 0]

solAB = Solve[ Thread[ surftraction == {0,0,0}]{A,B}][[1]]
```

To find the relationship between constant C and the applied force P_x, the force equilibrium on a semispherical volume of radius a around the origin is enforced. This calculation must be performed in fixed cartesian coordinates. The traction vector on the hemispherical surface of radius a is found using $\boldsymbol{t} = \boldsymbol{\sigma} \cdot \boldsymbol{e}_R$, where $\boldsymbol{e}_R$ is the radial vector field, and converted using the appropriate command. Next the resultant force is determined by integration, in a way entirely similar to that used for the Kelvin solution in an earlier section.

```
er = {x, y, z} /R

traction = Simplify[ Dot[ (sig /. solAB). er ] ];

traction = Simplify[traction /. Thread[{x,y,z} ->
     CoordinatesToCartesian[{a,t,p},Spherical]]]

resforce = PowerExpand[Simplify[
     Integrate[traction a^2 Sin[t], {p,0,2Pi}, {t,0,Pi/2}]]]
```

Note that the traction resultant is computed with respect to the fixed system of coordinates, and the dependence of A and B on C is substituted. The result is independent of the radius a and has only one nonzero component in the x direction: $R_x = -\dfrac{C\pi}{6(1-2\nu)(1-\nu^2)}$.

Because the external force P_x is applied in the positive x direction, the resultant internal force must be equal and opposite.

The solution is $C = -\dfrac{6(1-\nu^2)(1-2\nu)P_x}{\pi}$.

The solution for u is obtained by back substitution and can be expressed in cartesian coordinates using the same set of commands given for the Kelvin solution.

```
solC = Simplify[Solve[resforce == -P, C][[1, 1]]]

usol = Simplify[ PowerExpand[u /. solAB  /. solC ]]

SetCoordinates[Cylindrical[r, t, z]]

FieldFromCartesian[usol] /.
  Thread[ {x, y, z} -> CoordinatesToCartesian[{r, t, z}]]

TrigToExp[%];Simplify[%];ComplexExpand[%];PowerExpand[%];
Simplify[%]
```

The final form of the solution for displacements in the cartesian coordinate system can be written

$$\boldsymbol{u} = \frac{P}{2\pi}\frac{(1+\nu)}{E}\left[\frac{1}{R}\boldsymbol{e}_x + \frac{x\boldsymbol{R}}{R^3} + (1-2\nu)\frac{\partial}{\partial x}\left(\frac{x\boldsymbol{e}_x + y\boldsymbol{e}_y}{R+z} + \boldsymbol{e}_z \log(R+z)\right)\right], \tag{6.33}$$

where $R = \sqrt{r^2+z^2}$.

Concentrated loads at the tip of an elastic cone

In this section we consider both normal and transverse concentrated loads applied at the tip of an elastic cone of opening angle α with the axis coincident with the positive half of the z-axis. The solutions are constructed similarly to the Kelvin solution, for the case of axially applied load, and similarly to the Cerruti solution, for the case of load applied transversely to the cone axis. Spherical coordinates are used in both cases to simplify the satisfaction of the boundary conditions on the surface of the cone.

We first consider axial loading and use the sum of the Kelvin solution and the line of centres of dilatation along the negative z-axis.

The Kelvin solution for a force acting in the positive z-direction is written directly in terms of the Papkovich potentials found previously, with a coefficient A as yet unknown.

```
SetCoordinates[Spherical[R, p, t]]
pos={R,0,0};

(* Kelvin *)
psi1 = FieldFromCartesian[{0,0,1}] A(1 + nu)/(2 Pi R);
phi1 = 0;
```

Next the line of centres of dilatation along the negative z-axis is installed, with unknown intensity B to be determined.

```
(* Line of centres of dilatation *)
psi2 = {0, 0, 0};
phi2 = B Log[R(1 + Cos[p])];
```

The Papkovich potentials are next assembled, the solution for the displacement field found, the strains and stresses computed, and the divergence-free property of the stress field verified.

```
psi = psi1 + psi2;
phi = phi1 + phi2;

u = Simplify[psi - 1/(4(1 - nu))Grad[Dot[pos, psi] + phi]]
eps = Strain[u]
sig = Simplify[Stress[eps]]
Simplify[Div[sig]]
```

Now the traction vector is found on the cone surface, and the unknown constant $B[i]$ found by requiring that tractions vanish.

```
traction = sig.{0, 1, 0} /. t -> alpha
solB = Simplify[Solve[Thread[traction == {0,0,0}],{B}][[1,1]]]
```

The relationship between constants B and A that has been found at this stage can now be written in the form $B = A(1+\nu)(1-2\nu)(1+\cos\alpha)/(2\pi)$.

Now the resultant of the internal tractions is computed on a spherical cap and balanced against the externally applied force P_z.

```
traction = (sig /. solB).{1,0,0}
tracz = Simplify[FieldToCartesian[traction].{0, 0, 1}]
solA = Solve[
 Integrate[R^2 Sin[p] tracz,{p,0,alpha},{t,0,2 Pi}] + Pz == 0,
    {A}][[1, 1]]
```

The final solution for coefficients A and B in terms of the applied force P_z has the form

$$A = \frac{P_z(1-\nu)}{(1-\cos^3\alpha)) - (1-2\nu)\cos\alpha(1-\cos\alpha)},$$

$$B = \frac{4P_z(1-\nu^2)(1-2\nu)(1+\cos\alpha)}{\pi(3+4\nu\cos\alpha+\cos 2\alpha)(1-\cos\alpha)}.$$

The complete solution for the displacement, strain, and stress fields is now readily available through substitution of A and B into the expressions.

A similar procedure is now followed to derive the solution for the concentrated load applied transversely to the cone axis.

The three terms for Papkovich potentials that appeared in the Cerruti problem derivation are once again introduced with unknown coefficients in order to develop the solution for a concentrated load acting in the positive x-direction. Note that it is convenient to introduce several of the Papkovich potential terms in cartesian coordinates (referred using names with suffix '`car`').

```
SetCoordinates[Cartesian[x,y,z]];

psi2car = -Simplify[PowerExpand[
    Curl[Log[z + Sqrt[x^2+y^2+z^2]]{0,1,0}]]];

phi3car = Simplify[
    D[z Log[z + Sqrt[x^2+y^2+z^2]]-Sqrt[x^2+y^2+z^2],x]]
```

The spherical coordinate system is introduced now, and Papkovich potential expressions for the three solutions in question are written out, or obtained by conversion from cartesian fields. Rule `ruleR` is used to effect some additional simplifications.

```
SetCoordinates[Spherical[R, p, t]]
pos={R,0,0};
```

```
(* Kelvin *)
psi1 = FieldFromCartesian[{1, 0, 0}] /R
phi1 = 0;
(* Line of centres of dilatation *)
psi2 = Simplify[FieldFromCartesian[(psi2car /.
 Thread[{x,y,z}->CoordinatesToCartesian[{R,p,t}]] )]]/.ruleR
phi2 = - Cos[t] Sin[p]/(1 + Cos[p])
(* Double line of centres of dilatation *)
psi3 = {0, 0, 0};
phi3 = Simplify[phi3car/.
 Thread[{x,y,z}->CoordinatesToCartesian[{R,p,t}]] )]]/.ruleR
```

The Papkovich potentials are next assembled, the solution for displacement field is found, strains and stresses are computed, and the divergence-free property of the stress field is verified.

```
phi = A phi1 + B phi2 + C phi3;
psi = A psi1 + B psi2 + C psi3;

u = Simplify[psi - 1/(4(1 - nu))Grad[Dot[pos, psi] + phi]]
eps = Strain[u]
sig = Simplify[Stress[eps]]
Simplify[Div[sig]]
```

Now the traction vector is found on the cone surface and the unknown constants B and C are found by requiring that tractions vanish.

```
traction = sig.{0, 1, 0} /. p -> alpha
solBC=Simplify[Solve[Thread[traction=={0,0,0}],{B,C}] [[1]]]
```

The relationship between constants B, C, and A has now been found and can be written in the form

$$B = -A(1-2\nu)(1+\cos\alpha)/(2(1-\nu)),\, C = 2A(1-2\nu)(1+\cos\alpha).$$

Now the resultant of the internal tractions is computed on a spherical cap and balanced against the externally applied force P_x.

```
traction = (sig /. solBC).{1,0,0}

tracx = Simplify[FieldToCartesian[traction].{1,0,0}]

solA = Solve[
 Integrate[R^2 Sin[p] tracx,{p,0,alpha},{t,0,2 Pi}]+Px==0,
  {A}][[1, 1]]
```

The final solutions for coefficients A, B, and C in terms of the applied force P_x have the form

$$A = \frac{2P_x(1-\nu^2)}{\pi(1+\nu\cos\alpha)(1-\cos\alpha)^2},$$
$$B = \frac{-2P_x(1+\nu)(1-2\nu)(1+\cos\alpha)}{\pi(1+\nu\cos\alpha)(1-\cos\alpha)^3},$$
$$C = \frac{2P_x(1-\nu^2)(1-2\nu)(1+\cos\alpha)^2}{\pi(1+\nu\cos\alpha)(1-\cos\alpha)^2}.$$

The complete solution for the displacement, strain, and stress fields is now readily available through substitution of A, B, and C into the previously derived expressions.

Spherical harmonics

The deformation in axisymmetrically loaded bodies in spherical coordinates R, φ, ϑ is fully described by the Papkovich potentials having the form (Soutas-Little, 1973)

$$\begin{aligned}
\phi(R,\varphi) &= -\textstyle\sum_n R^n B_n P_n(\zeta),\\
\psi_R(R,\varphi) &= \textstyle\sum_n R^{n+1}\left[-A_n(n+1)P_n(\zeta) + C_n\zeta P_{n+1}(\zeta)\right],\\
\psi_\varphi(R,\varphi) &= -\textstyle\sum_n R^{n+1}\sin\varphi\left[A_n P_n'(\zeta) + C_n P_{n+1}(\zeta)\right],
\end{aligned} \tag{6.34}$$

where $\zeta = \cos\varphi$, $P_n(\zeta)$ denotes the Legendre polynomial of the first kind of degree n, and the prime denotes differentiation.

We begin by developing the tools to demonstrate that both the scalar potential ϕ and the vector potential $\boldsymbol{\psi} = (\psi_R, \psi_\varphi, 0)$ are in fact harmonic.

Spherical coordinates are introduced, and **CoordinatesToCartesian** provides a convenient method of checking the coordinate definition. Here we use a convention that always refers to the azimuthal angle as ϑ, and the elevation is expressed by the angle φ measured from the north pole (Figure 6.2). The Papkovich potential components are defined according to the above formulae.

```
SetCoordinates[Spherical[R, p, t]]
CoordinatesToCartesian[{R, p, t}]


pos = {R, 0, 0};
zeta = Cos[p];
psiR = R^(n + 1)(-A[n](n + 1)LegendreP[n, zeta] +
        C[n]zeta LegendreP[n + 1, zeta])
psip = -R^(n + 1)Sin[p]
   (A[n]D[LegendreP[n,zeta],zeta] + C[n]LegendreP[n+1,zeta])
phin = -B[n]R^n LegendreP[n, zeta]
```

Legendre polynomials of the first kind obey the recursion relation

$$nP_n(\zeta) = (2n-1)\zeta P_{n-1}(\zeta) - (n-1)P_{n-2}(\zeta), \tag{6.35}$$

which also can be written in the form

$$(n+1)P_{n+1}(\zeta) = (2n+1)\zeta P_n(\zeta) - nP_{n-1}(\zeta). \tag{6.36}$$

We define a series of simplification rules based on the recursion relation for several successive values of n.

```
rul := LegendreP[-2+n,x_]->
  ((x-2 n x) LegendreP[-1+n,x]+n LegendreP[n,x])/(1-n)
rulP := rul /. n -> n - 1
rulM := rul /. n -> n + 1
```

It is now possible to apply the laplacian operator to the scalar and vector potentials and to demonstrate that both vanish.

```
psi[n_] = {psiR, psip, 0}
phi[n_] = phin
Simplify[Laplacian[psi[n]] /. rulP /. rul /. rulM]
Simplify[Laplacian[phi[n]] /. rul]
```

We now have access to the series representation of the axially symmetric Papkovich solution in spherical coordinates.

Define the displacement vector. The particular case of $C_n = 0$ is considered, for which the strains and stresses are computed, and it is demonstrated that the stress field is divergence-free.

```
u[n_] = Simplify[psi[n]-1/(4(1-nu))Grad[pos.psi[n]+phi[n]]]
eps[n_] = Simplify[Strain[u[n]] /. C[n]->0 ];
sig[n_] = Simplify[Stress[eps]] /. rul;
Simplify[Div[sig[n]] /. rul /. rulP]
```

The tractions arising from the solution have a zero resultant. The tractions on a spherical surface are first found as the dot product of the stress tensor with the outward normal on a spherical surface. The traction vector field is then converted to fixed cartesian coordinates.

```
tracR[n_] = Simplify[sig[n].{1, 0, 0}]

tracRcar = Simplify[FieldToCartesian[tracR]]
```

The tractions must now be integrated appropriately over ϑ and φ after multiplying by the element of surface area given by the product of scale factors for coordinates φ and

ϑ, which amounts to $R^2 \sin\varphi$. For purposes of analysis it is convenient to perform the change of variable to $\zeta = \cos\varphi$, $d\zeta = \sin\varphi d\varphi$, which leads to the limits of integration being set at $-1 < \zeta < 1$.

```
Rez[n_] = R^2 Simplify[
    Integrate[
      (Simplify[tracRcar/.p->ArcCos[zet]])
          ,{t,0,2 Pi},{zet,-1,1}
          ]
                  ]
```

MATHEMATICA output shows that the ϑ and φ components of the resultant **Rez[n]** evaluate explicitly to zero. The z-component, on the other hand, appears in integral form, containing expressions of the type

$$\int_{-1}^{1} \zeta P_n(\zeta) d\zeta, \int_{-1}^{1} P_{n-1}(\zeta) d\zeta.$$

From the orthogonality property of Legendre polynomials it follows immediately that these integrals evaluate to zero for $n > 1$, due to the orthogonality of $P_n(\zeta)$ and $P_1(\zeta) = \zeta$, and $P_{n-1}(\zeta)$ and $P_0(\zeta) = 1$. For $n = 0$ and $n = 1$ it can be explicitly verified in MATHEMATICA that **Rez[0]** and **Rez[1]** evaluate to zero.

```
Rez[0]
```

```
Rez[1]
```

It has been shown therefore that equations (6.34) with $C_n = 0$ provide solutions for spheres loaded by axisymmetric distributions of surface tractions with zero resultants. It is now possible to consider boundary value problems with either tractions or displacement prescribed on the spherical surface. Assuming these may be expanded into series of Legendre polynomials, the solution may then be sought in the form of equations (6.34).

To develop a demonstration of this method we wish to consider the internal stresses within a gravitating rotating sphere of radius R_0. However, in order to tackle this problem we must first develop the procedure for finding particular solutions to elastic problems with prescribed body forces.

We shall assume that the body force $\boldsymbol{b}(\boldsymbol{x})$ is conservative and can be expressed in terms of potential Γ as

$$\boldsymbol{b} = -\operatorname{grad}\Gamma. \tag{6.37}$$

We seek a particular solution of the Navier equation

$$\Delta\boldsymbol{u} + \frac{1}{1-2\nu}\operatorname{grad}\operatorname{div}\boldsymbol{u} = \frac{1}{G}\operatorname{grad}\Gamma, \tag{6.38}$$

where $G = E/2(1+\nu)$ is the shear modulus. Assuming that this particular solution is irrotational and can be represented by its potential,

$$\boldsymbol{u} = \operatorname{grad}\chi,$$

the Navier equation reduces to

$$\Delta\chi = \frac{1-2\nu}{(1-\nu)G}\Gamma. \tag{6.39}$$

This equation can be readily solved for χ in two cases:

1. If function Γ depends only on R, then equation (6.39) assumes the form

$$\frac{1}{R^2}\frac{\partial}{\partial R}\left(R^2\frac{\partial\chi}{\partial R}\right) = \frac{1-2\nu}{(1-\nu)G}\Gamma(R),$$

   and can be readily integrated.
2. If function Γ is harmonic, for example,

$$\Gamma = c_n R^n P_n(\zeta),$$

   then χ is biharmonic, and can be written as R^2 times a harmonic function, that is, in the form

$$d_n R^{n+2} P_n(\zeta).$$

   The relationship between constants is then readily established as

$$d_n = \frac{1-2\nu}{(1-\nu)G}\frac{c_n}{4(2n+3)},$$

   so that the displacement field becomes

$$\boldsymbol{u} = \frac{1-2\nu}{(1-\nu)G}\frac{c_n}{4(2n+3)}R^{n+1}\left[(n+2)P_n(\zeta)\boldsymbol{e}_R + \frac{dP_n(\zeta)}{d\varphi}\boldsymbol{e}_\varphi\right].$$

The procedure for solving the problem about a rotating gravitating sphere then consists of the following steps:

- Identify potential functions Γ for the graviation force and inertial forces due to rotation.
- Solve the Navier equation for this case and obtain the particular solution denoted $\boldsymbol{u}_0$.
- Determine the surface tractions due to the solution $\boldsymbol{u}_0$ and impose equal and opposite tractions as boundary conditions for the problem about unknown displacement $\boldsymbol{u}$.
- Superimpose $\boldsymbol{u}_0$ and $\boldsymbol{u}$ to obtain the final solution and calculate strains and stresses.

Denoting the sphere radius by R_0 and the specific gravity on its surface by g, the body force field can written in the form

$$\boldsymbol{b}_g = -g\boldsymbol{R}/R_0$$

and the potential in the form

$$\Gamma_g = \frac{g}{R_0}\frac{R^2}{2}.$$

This corresponds to case (i) above. Integration leads to the results

$$\chi_g = \frac{(1-2\nu)}{80G(1-\nu)}\frac{gR^4}{R_0}$$

and

$$\boldsymbol{u}_g = \frac{(1-2\nu)}{20G(1-\nu)}\frac{gR^3}{R_0}\boldsymbol{e}_R.$$

The body force due to the rotation of the sphere with the angular velocity ω produces the inertial body force

$$\boldsymbol{b}_\omega = \rho\omega^2 r\boldsymbol{r},$$

where ρ denotes the mass density, and $\boldsymbol{r}$ is the radius vector in cylindrical coordinates, which may be expressed as

$$\boldsymbol{r} = \boldsymbol{e}_R \sin\varphi + \boldsymbol{e}_\varphi \cos\varphi.$$

This force field can be represented by the potential function

$$\Gamma_\omega = -\frac{\rho\omega^2 R^2}{2}\sin^2\varphi = -\frac{\rho\omega^2 R^2}{3} + \frac{\rho\omega^2 R^2}{3}P_2(\zeta);$$

that is, it is given by the sum of two potentials conforming to cases (i) and (ii) above, respectively. Integration of this field to obtain the displacements leads to

$$\boldsymbol{u}_\omega = \frac{\rho\omega^2 R^3}{3}\frac{(1-2\nu)}{G(1-\nu)}\left[\frac{13}{140}\boldsymbol{e}_R + \frac{1}{28}\frac{dP_n(\zeta)}{d\varphi}\boldsymbol{e}_\varphi\right].$$

Adding the particular solutions $\boldsymbol{u}_g$ and $\boldsymbol{u}_\omega$, one obtains the general form of the particular displacement solution for the rotating gravitating sphere as

$$\boldsymbol{u}_0 = aR^3P_2(\zeta)\boldsymbol{e}_R + bR^3\frac{dP_2(\zeta)}{d\varphi}\boldsymbol{e}_\varphi = aR^3\frac{3\cos^2\varphi - 1}{2}\boldsymbol{e}_R - 3bR^3\cos\varphi\sin\varphi\,\boldsymbol{e}_\varphi, \tag{6.40}$$

where coefficients a and b can be found explicitly from the above results.

We can now readily determine the surface tractions $\boldsymbol{t}_0$ due to the particular solution $\boldsymbol{u}_0$ and require that

$$\boldsymbol{t}_0 + \boldsymbol{t} = \boldsymbol{0} \qquad \text{on } R = R_0.$$

This task is readily accomplished in MATHEMATICA.

Define the displacement field of the particular solution $\boldsymbol{u}_0$ and compute strains, stresses, and tractions on a spherical surface of radius R_0.

```
u0 = R^3{a  LegendreP[2,Cos[p]], b D[LegendreP[2,Cos[p]],p],0}

eps0 = Simplify[(Grad[u0] + Transpose[Grad[u0]])/2]
sig0 = Simplify[Stress[eps0]]
tracR0 = Simplify[sig0.{1, 0, 0}] /. R -> R0
```

Taking the general solution for axisymmetric surface loading of an elastic sphere, consider the form of surface tractions on the sphere of radius R_0. Note that the constant C_2 can be set to zero, because the remaining two constants A_2 and B_2 provide sufficient flexibility to satisfy the two traction-free boundary conditions written in terms of stress components σ_{RR} and $\sigma_{R\varphi}$.

```
Simplify[tracR[2] /. C[2] -> 0 /. R -> R0]
```

The unknown coefficients A_2 and B_2 can now be found using the MATHEMATICA **Solve** command.

```
Simplify[
  Solve[(tracR[2] /. C[2]->0 /. R->R0) + tracR0 == 0,
        {A[2], B[2]}]]
```

The final solution is the superposition of the field determined by the displacements $\boldsymbol{u}_0$ of equation (6.40) and the field defined by the Papkovich–Neuber potentials in the form of equation (6.34) with the only two nonzero coefficients given by

$$A_2 = \frac{(1-\nu)}{(1-2\nu)(7+5\nu)}\left[a(2+\nu) - 2b(1+\nu)\right],$$

$$B_2 = \frac{(1-\nu)(3+2\nu)}{(1-2\nu)(7+5\nu)}\left[12b\nu - a(7-4\nu)\right] R_0^2.$$

Solutions for other important axisymmetric geometries, notably cylinders, can be developed similarly by observing that the requirements of harmonicity of Papkovich potentials allow the general form of a variable separable solution to be established and a series formulation to be developed.

6.2 GALERKIN VECTOR

Galerkin (1930) introduced an alternative general description of the elastic displacement field of the general form

$$\boldsymbol{u} = \Delta \boldsymbol{g} - \frac{1}{2(1-\nu)} \text{grad div } \boldsymbol{g}, \tag{6.41}$$

where $\boldsymbol{g}$ is a vector field that satisfies the equation

$$\Delta\Delta \boldsymbol{g} = -\frac{\boldsymbol{b}}{1-\nu}, \tag{6.42}$$

where $\boldsymbol{b}$ is the field of body forces. Then $\boldsymbol{u}$ satisfies the Navier equation for elastic displacements in the presence of body forces.

If the displacement field is written in the form

$$\boldsymbol{u} = \boldsymbol{u}_0 + \Delta \boldsymbol{g} - \frac{1}{2(1-\nu)} \text{grad div } \boldsymbol{g}, \tag{6.43}$$

where $\boldsymbol{u}_0$ is a particular solution of the Navier equation with body forces, then the Galerkin vector $\boldsymbol{g}$ must be *biharmonic*; that is,

$$\triangle\triangle \boldsymbol{g} = \boldsymbol{0}. \tag{6.44}$$

In cartesian coordinates this is equivalent to the requirement that each component g_i, $i =$ 1–3 of the Galerkin vector is a biharmonic scalar function. In general curvilinear coordinates the requirement of biharmonicity of a vector is more complex. However, for a given candidate vector this can be readily verified using the command **`Biharmonic[]`** provided as part of the package **`Tensor2Analysis.m`**.

By inspection of equations (6.43) and (6.15) the relationship between the biharmonic Galerkin vector and the Papkovich–Neuber harmonic potentials can be established in the form

$$\phi = 2\text{div}\,\boldsymbol{g} - \boldsymbol{r}\cdot\triangle\boldsymbol{g}, \tag{6.45}$$

$$\boldsymbol{\psi} = \triangle\boldsymbol{g}. \tag{6.46}$$

Similarly to the Papkovich–Neuber potential description, the Galerkin vector description of elastic displacement fields is nonunique, that is, the same displacement field may be described by two different Galerkin vectors that differ by a nonzero term (Westergaard, 1952).

Equations (6.46) provide the conversion from Papkovich–Neuber potential description to Galerkin vector. Therefore, solution techniques previously developed and applied in the Papkovich description can also be used if the Galerkin description is preferred. In practice there is no particular reason for such preference to be made, because both descriptions ultimately lead to expressions for displacements, from which strains, stresses, and tractions can be readily computed with the help of MATHEMATICA.

One special case where Galerkin vector description does offer particular advantages is when the three-dimensional elasticity problems considered possess axial symmetry. The formulation used in this case, known as the Love strain function, is discussed in the next section.

6.3 LOVE STRAIN FUNCTION

Consider a special case when the Galerkin vector $\boldsymbol{g}$ possesses only one nonzero component,

$$\boldsymbol{g} = Z\boldsymbol{e}_z, \quad \triangle\triangle Z = 0.$$

It is apparent that the requirement of biharmonicity of the Galerkin vector in this case is reduced to the scalar biharmonic equation for function Z. This formulation was originally introduced by Love in 1906 (Love, 1944).

Although in principle Z may depend on three spatial coordinates, of particular interest is the use of this formulation for axisymmetric problems described in cylindrical polar coordinates. In this case dependence solely on coordinates r and z is assumed: $Z = Z(r, z)$.

Expressions for the displacements and stresses can be readily written down for this case as follows:

$$u_r = -\frac{1}{2(1-\nu)}Z_{,rz}, \quad u_z = \Delta Z - \frac{1}{2(1-\nu)}Z_{,zz}, \tag{6.47}$$

$$\sigma_{rr} = \frac{E}{2(1-\nu^2)}\left(\nu\Delta Z_{,z} - Z_{,rrz}\right), \quad \sigma_{rz} = \frac{E}{2(1-\nu^2)}\left((1-\nu)\Delta Z_{,r} - Z_{,rzz}\right), \tag{6.48}$$

$$\sigma_{zz} = \frac{E}{2(1-\nu^2)}\left((2-\nu)\Delta Z_{,z} - Z_{,zzz}\right), \quad \sigma_{\theta\theta} = \frac{E}{2(1-\nu^2)}\left(\nu\Delta Z_{,z} - \frac{1}{r}Z_{,rz}\right). \tag{6.49}$$

The above formulae give elastic displacement and stress fields for any biharmonic function $Z(r, z)$. The search for the general elastic solution of the form given by Love's strain function is therefore once again reduced to the search for biharmonic functions.

Various techniques can be utilised for this purpose. For example, let $\Psi(r, z)$ be a harmonic function. Then it is readily verified that

$$R^2\Psi(r, z) = (r^2 + z^2)\Psi(r, z)$$

is biharmonic, thus allowing a series of solutions to be generated.

To identify *harmonic* functions that have the form of homogeneous polynomials in r and z of order n,

$$\Psi_n(r, z) = \sum_{i=0}^{n} a_i r^i z^{n-i},$$

the Mathematica implementation shown below can be used.

Commands are defined as a function of arbitrary order n to generate the set of powers and coefficients, and hence homogeneous polynomials, by using the dot product.

```
SetCoordinates[Cylindrical[r, t, z]]

mypowers[n_] := Table[r^i z^(n - i), {i, 0, n}]
mycoef[n_] := Table[ToExpression["a"<>ToString[i]],{i,0,n}]
mypoly[n_] := mycoef[n].mypowers[n]
```

Now for an arbitrary chosen order **nn** the laplacian of the polynomial is found to contain homogeneous terms of order $n - 2$ and an additional term of the form

$$a_1 z^{n-1}/r.$$

Hence, restricting our attention to nontrivial cases $n > 2$, according to **mysub** $a_1 = 0$, and $a_0 = 1$ is chosen without loss of generality. Coefficients of different power terms in **lap** are assembled in **cof**, and the solution **mysol** is obtained. Backsubstitution generates the polynomial **harpoly**, which is shown to be harmonic.

```
nn = 10;

Collect[(lap = Laplacian[mypoly[nn]]), mypowers[nn]]

mysub = {a0 -> 1, a1 -> 0};
cof = Coefficient[lap, mypowers[nn - 2]];
zer = cof - cof;
```

```
mysol = Solve[(Thread[cof == zer] /. mysub),
        Drop[mycoef[nn], 2]][[1]];

harpoly = (mycoef[nn]/.Join[mysub,mysol]).mypowers[nn]

Simplify[Laplacian[harpoly]]
```

Soutas-Little (1973) demonstrates how an equivalent system of harmonic polynomials can be generated using spherical harmonics in the form of Legendre polynomials.

Biharmonic functions given by polynomials of order n can therefore be written as

$$Z(r,t) = \Psi_n(r,z) + (r^2 + z^2)\Psi_{n-2}(r,z). \tag{6.50}$$

Using the homogeneous polynomial form of biharmonic functions given above, Timoshenko and Goodier (1951) build Love strain function solutions for the concentrated force in an infinite solid (Kelvin problem), and also for a momentless force doublet and for the centre of dilatation. The same approach can also be used (Soutas-Little, 1973) to obtain solutions for simple stress states of uniaxial (σ_{zz}) or equibiaxial (σ_{rr}) states of tension or compression, as well as bending solutions for thick plates.

Also included in consideration must be biharmonic functions that are not expressible as polynomials, and in fact may be singular due to the presence of logarithmic terms or negative powers, such as

$$(r^2 + z^2)^{1/2},\ (r^2 + z^2)^{-1/2},\ \ldots,\ \log r,\ z\log r,\ z^2 \log r\, r^2 \log r,\ \ldots.$$

The corresponding elastic fields are singular at the origin, but can be useful in solving problems involving concentrated loads or elastic solids with spherical cavities, because they generate constant and linearly varying traction distributions on spherical surfaces (Barber, 2002).

For example, the solution of the Kelvin problem about a concentrated force P acting in the z-direction and applied at the origin can be expressed by the Love strain function

$$Z = B(r^2 + z^2)^{1/2}. \tag{6.51}$$

The MATHEMATICA tools developed above allow displacements, strains, and stresses to be computed readily, so that traction boundary conditions (or equilibrium with the externally applied load) can be satisfied to find the value of the unknown constant B.

Integral transform methods

An alternative form of biharmonic function of r and z is given by

$$Z(r,z) = \left[AJ_0(\alpha r) + BY_0(\alpha r) + r\frac{\partial}{\partial r}(CJ_0(\alpha r) + DY_0(\alpha r)) + E\alpha z J_0(\alpha r) + F\alpha z Y_0(\alpha r)\right]\exp(\alpha z), \tag{6.52}$$

where α can be imaginary or real, positive or negative, and J_0 and Y_0 are Bessel functions of order zero of the first and second kind, which must be replaced with the modified Bessel functions I_0 and K_0 if α is imaginary. This form of $Z(r, z)$ is particularly useful if boundary conditions are specified on cylindrical surfaces, for example, axisymmetric tractions on the side surfaces $r = r_0$ of a rod of radius r_0, or loading of plates containing circular holes.

A generalisation of the above formulation leads to the use of Hankel transform methods. The Hankel transform of order zero of the Love strain function is defined as

$$\bar{Z}^0(\xi, z) = \int_0^\infty rZ(r, z)J_0(\xi r)dr. \tag{6.53}$$

Applying the Hankel transform to the biharmonic equation

$$\Delta\Delta Z(r, z) = 0$$

and using the differentiation properties of Bessel function J_0 leads to the ordinary differential equation in z for the unknown function

$$\left(\frac{d^2}{dz^2} - \xi^2\right) \bar{Z}^0(\xi, z) = 0. \tag{6.54}$$

The solution of this equation is given by

$$\bar{Z}^0(\xi, z) = (A(\xi) + B(\xi)\xi z) \exp(\pm\xi z). \tag{6.55}$$

The Hankel transform must also now be applied to displacement and/or traction boundary conditions of the form (6.47), (6.48) or (6.49) in order to find the unknown functions $A(\xi)$ and $B(\xi)$.

Soutas-Little (1973) gives an example of using this approach to obtain the Boussinesq solution for the concentrated load P acting normal to the surface of a half space. The Hankel transform of Love strain function can be sought in the form

$$\bar{Z}^0(\xi, z) = \frac{CP}{2\pi\xi^2} [2\nu + \xi z] \exp(-\xi z), \tag{6.56}$$

where the constant C is found by satisfying equlibrium between the externally applied force P and internal stresses.

Axisymmetric contact problems involving elastic half-spaces are a class of so-called mixed boundary value problems, since on part of the surface the boundary condition is prescribed in terms of displacements, and elsewhere traction boundary conditions apply. The integral transform formulation of Love strain solution provides an effective way of addressing such problems, because it allows the condition on some part of the boundary (e.g., the traction-free requirement) to be satisfied by construction, thus leading to a single integral equation formulation.

SUMMARY

The displacement potential approach to the solution of elastic problems is presented. Harmonic Papkovich–Neuber potentials are first used as the basis for the analysis, and various fundamental solutions are considered (Kelvin, Boussinesq, and Cerruti). Biharmonic potentials (Galerkin vector and Love strain function) are introduced next, and their application to the solution of special problems is discussed, for example, problems involving cylindrical or spherical symmetry.

EXERCISES

1. Simple stress and strain states in terms of the Love strain function

Using the procedures for identifying biharmonic homogeneous polynomials described in the chapter, generate the following third-order polynomial Love strain function solutions:

$$Az(r^2 + z^2) \quad \text{and} \quad B(2z^3 - 3r^2 z).$$

Compute the displacements arising due to the superposition of these solutions. Derive the strain and stress states, and show that they are uniform. Determine the relationship between constants A and B required to obtain a uniaxial strain state and uniaxial and hydrostatic stress states.

Hint: See notebook `C06_Love_3-poly.nb`

2. Kelvin solution using Love strain function

Demonstrate that the Love strain function of the form

$$Z(r, z) = B\sqrt{r^2 + z^2}$$

provides the solution to the Kelvin problem about the concentrated force P applied at the origin and acting in the positive z direction. Determine the unknown constant B.

Hint: See notebook `C06_Love_Kelvin.nb` and follow the procedures used for deriving this solution in terms of the Papkovich–Neuber potentials.

3. Momentless force dipole

Using the Kelvin solution for the concentrated force at the origin as the starting point, derive the Love strain function solution for the momentless force doublet at the origin. Verify that the stress field is divergence-free and self-equilibrated.

Hint: See notebook `C06_Love_momentless_force_dipole.nb`

4. Stresses around a spherical cavity

Papkovich–Neuber potentials may be used to construct solutions of significant practical interest, for example, for stresses around notches and defects.

Consider a spherical cavity of radius a within an infinitely extended elastic solid subjected to remote tension. Build the solution using the superposition of the following terms:

- Papkovich–Neuber potentials for a uniform uniaxial tensile stress field $\sigma_{zz} = S$.
- Papkovich–Neuber potentials for a centre of dilatation.
- Papkovich–Neuber potentials for a momentless force dipole along the direction of tensile loading.
- Papkovich–Neuber potential solution of the form $\psi = 0, \phi = C(r^2 - 2z^2)(r^2 + z^2)^{-5/2}$.

Compute tractions on the spherical surface of radius a (with respect to spherical coordinates). Requiring that this surface be traction-free, obtain expressions for constants A, B, and C in terms of the remote tension S.

Hint: It is more convenient to carry out manipulations with potentials in cartesian coordinates and transform to spherical coordinates for the last step of relieving the spherical surface of tractions. *See notebook* `C06_PN_sph_cavity_car.nb`

5. Torsion of a cylindrical shaft (Barber, 2002)

Using cylindrical polar coordinates, consider the Papkovich–Neuber solution of the form

$$\boldsymbol{\psi} = \operatorname{curl}\,(Bz(r^2 + z^2)\boldsymbol{e}_z), \quad \phi = -\boldsymbol{r} \cdot \boldsymbol{\psi}.$$

Compute the displacement, strain, and stress fields. Find tractions in the section $z = $ const. and calculate the torque. Hence find the torsional stiffness of a circular shaft, that is, the ratio of applied torque to angle of twist per unit length.

Hint: See notebook **`C06_PN_shaft_torsion.nb`**

6. Torsion of a conical shaft (Barber, 2002)

Using cartesian coordinates, consider the Papkovich–Neuber solution for a centre of rotation,

$$\boldsymbol{\psi} = \operatorname{curl}\,(\boldsymbol{e}_z/R), \quad \phi = -\boldsymbol{r} \cdot \boldsymbol{\psi},$$

where $R = \sqrt{x^2 + y^2 + z^2}$.

Transform the stress state into spherical coordinates, and show that any cone with the apex at the origin is free from surface tractions.

Transform the same stress state into cylindrical polar coordinates, and show that the tractions in any section $z = $ const. reduce to just the shear component $\tau_{r\theta}$. Compute the torque transmitted across the section.

Hint: See notebook **`C06_PN_cone_torsion.nb`**

7. Torsion of a cylindrical shaft with a spherical hole (Barber, 2002)

Using cartesian coordinates, construct the Papkovich–Neuber solution for a shaft or radius b with a spherical hole of radius $a < b$ using the form

$$\boldsymbol{\psi} = \operatorname{curl}\,((A/R + BzR^2)\boldsymbol{e}_z), \quad \phi = -\boldsymbol{r} \cdot \boldsymbol{\psi},$$

where $R = \sqrt{x^2 + y^2 + z^2}$.

Transform the stress state into spherical coordinates and show that the surface $r = a$ can be rendered traction-free with a suitable choice of the relationship between constants A and B.

Hint: See notebook **`C06_PN_shaft_hole_torsion.nb`**.

7 Energy principles and variational formulations

OUTLINE

In this chapter the classical variational principles of elasticity are introduced and applied. The solution of the elastic problem is shown to furnish an extremum value of a scalar potential, either the minimum of the strain energy potential or, respectively and simultaneously, the maximum of the complementary energy potential. These principles provide a complete characterisation of solutions of the regular (well-posed) elasticity problem defined in the preceeding chapters. Moreover, they allow the introduction of classes of approximate solutions through the optimisation of potentials over a subspace of the space of admissible fields. Equations of elastostatics are considered first, followed by a discussion of free vibrations.

7.1 STRAIN ENERGY AND COMPLEMENTARY ENERGY

Strain energy

Suppose that the elastic body Ω is characterised by a positive definite symmetric tensor of elastic moduli $\boldsymbol{C}$.

Define the space of *kinematically admissible* displacement fields as displacement fields that obey displacement boundary conditions prescribed on a part $\partial\Omega^d$ of the boundary $\partial\Omega$:

$$\mathcal{K}(\boldsymbol{u}^D, \partial\Omega^d) = \{ \quad \boldsymbol{v} | \quad \boldsymbol{v} = \boldsymbol{u}^D \quad \text{on } \partial\Omega^d\}. \tag{7.1}$$

Here $\partial\Omega^d \subset \partial\Omega$ denotes that part of the boundary $\partial\Omega$ where displacements $\boldsymbol{u}^D$ are prescribed. We recall briefly that in a regular elasticity problem the boundary $\partial\Omega$ is subject to complementary partition into $\partial\Omega_i^d$ with prescribed displacements u_i^d and $\partial\Omega^t$ with prescribed tractions t_i^t in each direction x_i, as discussed previously in Chapter 4.

Strain energy is equal to the actual mechanical work done by internal and external forces between the initial and the actual state. The initial state is defined by zero displacement field, whereas the actual state is characterised by a kinematically admissible displacement field $\boldsymbol{v}$.

The strain energy of an elastic body is defined over an arbitrary kinematically admissible displacement field $\boldsymbol{v}$ as

$$\mathcal{U}_i(\boldsymbol{v}) = \int_\Omega \left(\boldsymbol{\sigma}_0 : \boldsymbol{\varepsilon}[\boldsymbol{v}] + \frac{1}{2}\boldsymbol{\varepsilon}[\boldsymbol{v}] : \boldsymbol{C} : \boldsymbol{\varepsilon}[\boldsymbol{v}] - \boldsymbol{\varepsilon}[\boldsymbol{v}] : \boldsymbol{C} : \boldsymbol{A}\theta^D\right) dv \tag{7.2}$$

with $\boldsymbol{v} \in \mathcal{K}(\boldsymbol{u}^D, \partial\Omega^d)$, and $\boldsymbol{\sigma}_0$ denoting the tensor of initial stresses, $\boldsymbol{C}$ the tensor of elastic moduli, and $\boldsymbol{A}$ the tensor of thermal expansions coefficients.

It is important to note that the strain energy so defined is equal to the actual internal energy of an elastic body in the thermodynamic sense only if the initial (residual) stresses $\boldsymbol{\sigma}_0$ and the prescribed temperature change field θ^D vanish. The strain energy introduced above should be understood merely as a scalar potential of internal forces.

The work of *given* external forces $\boldsymbol{b}^D$ and $\boldsymbol{t}^D$ is defined over a kinematically admissible displacement field $\boldsymbol{v}$ as

$$\mathcal{U}_e(\boldsymbol{v}) = -\int_\Omega \boldsymbol{v} \cdot \boldsymbol{b}^D \, dv - \int_{\partial\Omega^t} \boldsymbol{v} \cdot \boldsymbol{t}^D \, ds. \tag{7.3}$$

The strain energy potential is the sum of the strain energy and the work done by external forces over an arbitrary kinematically admissible displacement field $\boldsymbol{v}$:

$$\mathcal{U}_p(\boldsymbol{v}) = \mathcal{U}_e(\boldsymbol{v}) + \mathcal{U}_i(\boldsymbol{v}) \tag{7.4}$$

$$= \int_\Omega \left(\boldsymbol{\sigma}_0 : \boldsymbol{\varepsilon}[\boldsymbol{v}] + \frac{1}{2}\boldsymbol{\varepsilon}[\boldsymbol{v}] : \boldsymbol{C} : \boldsymbol{\varepsilon}[\boldsymbol{v}] - \boldsymbol{\varepsilon}[\boldsymbol{v}] : \boldsymbol{C} : \boldsymbol{A}\theta^D\right) dv - \int_\Omega \boldsymbol{v} \cdot \boldsymbol{b}^D \, dv - \int_{\partial\Omega^t} \boldsymbol{v} \cdot \boldsymbol{t}^D \, ds. \tag{7.5}$$

Subscript p in $\mathcal{U}_p$ is used to indicate that the strain energy potential is being considered.

Complementary energy

The space of *statically admissible* stress fields is defined as those stress fields that are in equilibrium with the presecribed external body forces $\boldsymbol{b}^D$ and surface tractions $\boldsymbol{t}^D$,

$$\mathcal{S}(\boldsymbol{b}^D, \boldsymbol{t}^D, \partial\Omega^t) = \{ \ \boldsymbol{s} | \quad \operatorname{div} \boldsymbol{s} + \boldsymbol{b}^D = 0 \quad \text{in } \Omega, \quad \boldsymbol{s} \cdot \boldsymbol{n} = \boldsymbol{t}^D \quad \text{on } \partial\Omega^t \}, \tag{7.6}$$

where $\partial\Omega^t \subset \partial\Omega$ denotes the part of the boundary $\partial\Omega$ where given boundary tractions $\boldsymbol{t}^D$ are prescribed, and $\boldsymbol{b}^D$ are the given body forces.

The complementary energy is defined as the work of a statically admissible stress field over internal strains $\boldsymbol{s}$,

$$\mathcal{U}_i^*(\boldsymbol{s}) = -\frac{1}{2}\int_\Omega (\boldsymbol{s} - \boldsymbol{\sigma}_0 + \boldsymbol{C} : \boldsymbol{A}\theta^D) : \boldsymbol{C}^{-1} : (\boldsymbol{s} - \boldsymbol{\sigma}_0 + \boldsymbol{C} : \boldsymbol{A}\theta^D) \, dv, \tag{7.7}$$

with $\boldsymbol{s} \in \mathcal{S}(\boldsymbol{b}^D, \boldsymbol{t}^D, S^t)$. The complementary energy equals the actual internal energy of the body only if the initial (residual) stresses $\boldsymbol{\sigma}_0$ and the temperature change field θ^D vanish, and must be understood merely as a scalar potential function for the strains.

The work of tractions due to a statically admissible stress field $\boldsymbol{s}$ over *given* external displacements $\boldsymbol{u}^D$ is given by

$$\mathcal{U}_e^*(\boldsymbol{s}) = \int_{\partial\Omega} \boldsymbol{u}^D \cdot (\boldsymbol{s} \cdot \boldsymbol{n}) \, ds. \tag{7.8}$$

The complementary energy potential is the sum of the complementary energy and the work done over the prescribed displacements $\boldsymbol{u}^D$:

$$\mathcal{U}_p^*(\boldsymbol{s}) = \mathcal{U}_e^*(\boldsymbol{s}) + \mathcal{U}_i^*(\boldsymbol{s}) \tag{7.9}$$

$$= -\frac{1}{2}\int_\Omega (\boldsymbol{s} - \boldsymbol{\sigma}_0 + \boldsymbol{C} : \boldsymbol{A}\theta^D) : \boldsymbol{C}^{-1} : (\boldsymbol{s} - \boldsymbol{\sigma}_0 + \boldsymbol{C} : \boldsymbol{A}\theta^D)\, dv + \int_{\partial\Omega} \boldsymbol{u}^D \cdot (\boldsymbol{s} \cdot \boldsymbol{n})\, ds. \tag{7.10}$$

Equality of potentials

The particular physical significance of the strain energy potential and the complementary energy potential becomes particularly clear in the case where initial (residual) stresses $\boldsymbol{\sigma}_0$ and temperature change fields θ vanish. In this case the following result can be be proven:

Theorem: Equality of potentials If $[\boldsymbol{u}, \boldsymbol{\varepsilon}, \boldsymbol{\sigma}]$ is the solution of a regular elastic problem with vanishing residual stresses and temperature change field,

$$\boldsymbol{\sigma}_0 = 0, \qquad \theta^D = 0,$$

then the following equality between the strain energy potential and the complementary energy potential holds:

$$\mathcal{U}_p(\boldsymbol{u}) = \mathcal{U}_p^*(\boldsymbol{\sigma}).$$

The proof of this theorem is straightforward. Using expressions for the strain energy potential and the complementary energy potential for the case of vanishing residual stress and temperature change field, consider

$$\mathcal{U}_p(\boldsymbol{u}) - \mathcal{U}_p^*(\boldsymbol{\sigma}) = \frac{1}{2}\int_\Omega \boldsymbol{\varepsilon} : \boldsymbol{C} : \boldsymbol{\varepsilon}\, dv + \frac{1}{2}\int_\Omega \boldsymbol{\sigma} : \boldsymbol{C}^{-1} : \boldsymbol{\sigma}\, dv \tag{7.11}$$

$$- \int_\Omega \boldsymbol{u} \cdot \boldsymbol{b}^D\, dv - \int_{\partial\Omega^t} \boldsymbol{u} \cdot \boldsymbol{t}^D\, ds - \int_{\partial\Omega^d} \boldsymbol{u}^d \cdot (\boldsymbol{\sigma} \cdot \boldsymbol{n})\, ds. \tag{7.12}$$

Note that the elastic constitutive law in this case reduces to

$$\boldsymbol{\sigma} = \boldsymbol{C} : \boldsymbol{\varepsilon}.$$

Now take into account that the solution consists of both kinematically and statically admissible fields:

$$\boldsymbol{u} \in \mathcal{K}(\boldsymbol{u}^D, \partial\Omega^d) \quad \text{and} \quad \boldsymbol{\sigma} \in \mathcal{S}(\boldsymbol{b}^D, \boldsymbol{t}^D, \partial\Omega^t).$$

Because displacement and traction boundary conditions are defined on a complementary partition of the boundary $\partial\Omega$, one obtains

$$\mathcal{U}_p(\boldsymbol{u}) - \mathcal{U}_p^*(\boldsymbol{\sigma}) = \int_\Omega \boldsymbol{\varepsilon} : \boldsymbol{C} : \boldsymbol{\varepsilon} - \int_\Omega \boldsymbol{u} \cdot \boldsymbol{b}^D\, dv - \int_{\partial\Omega} \boldsymbol{u} \cdot (\boldsymbol{\sigma} \cdot \boldsymbol{n})\, ds = 0. \tag{7.13}$$

The last equality follows from the Stokes theorem, an argument already used in deriving the virtual work Theorem.

7.2 EXTREMUM THEOREMS

The purpose of this section is to demonstrate that elastic solutions deliver extremum values of the potentials defined in the previous section.

The displacement field of the elastic solution minimises the strain energy potential over all kinematically admissible displacement fields. In other words, the minimum principle of the strain energy potential asserts that the sum of the strain energy and the work of prescribed given external forces is smaller for the displacement field of the elastic solution than for any other kinematically admissible displacement field.

The complementary energy potential stands in a dual relationship to the strain energy potential. That is, the stress field of the elastic solution maximises the complementary energy potential over all statically admissible stress fields. In other words, this maximum principle asserts that the sum of the complementary energy and the external work over prescribed boundary displacements is larger for the stress field of the elastic solution than for any other statically admissible stress field.

The combination of the two extremum principles provides a means of complete characterisation of the elastic solution fields, in the sense that they furnish a formulation that is mathematically equivalent to the system of PDEs described in previous chapters.

Before giving rigorous statements of the extremum principles and their proof, it is worth mentioning that the key underlying property that ensures their validity is the positive definiteness of the tensor of elastic moduli $\boldsymbol{C}$, which ensures that the strain energy potential and the complementary energy potential are a convex and a concave functional, respectively.

Theorem: The minimum property of the strain energy potential The displacement field $\boldsymbol{u}$ of the solution $[\boldsymbol{u}, \boldsymbol{\varepsilon}, \boldsymbol{\sigma}]$ of a regular thermoelatic problem *minimises* the strain energy potential over all kinematically admissible displacement fields:

$$\mathcal{U}_p(\boldsymbol{v}) \geq \mathcal{U}_p(\boldsymbol{u}) \qquad \forall \boldsymbol{v} \in \mathcal{K}(\boldsymbol{u}^D, \partial\Omega^d). \tag{7.14}$$

Consider a kinematically admissible displacement field $\boldsymbol{v} \in \mathcal{K}(\boldsymbol{u}^D, S^u)$ and the displacement field $\boldsymbol{u}$ of the elastic solution. The proof will proceed by first showing that

$$\mathcal{U}_p(\boldsymbol{v}) - \mathcal{U}_p(\boldsymbol{u}) = \mathcal{U}_i(\boldsymbol{v}) - \mathcal{U}_i(\boldsymbol{u}) - \mathrm{D}\mathcal{U}_i(\boldsymbol{u})(\boldsymbol{v} - \boldsymbol{u}), \tag{7.15}$$

where $\mathrm{D}\mathcal{U}_i(\boldsymbol{u})$ is the differential of the strain energy.

Second, the inequality:

$$\mathcal{U}_i(\boldsymbol{v}) - \mathcal{U}_i(\boldsymbol{u}) - D\mathcal{U}_i(\boldsymbol{u})(\boldsymbol{v} - \boldsymbol{u}) \geq 0 \quad \forall \boldsymbol{v} \in \mathcal{K}$$

will be established based on the convexity of the strain energy $\mathcal{U}_i$.

From the definition of work of prescribed external forces we obtain

$$\mathcal{U}_e(\boldsymbol{v}) - \mathcal{U}_e(\boldsymbol{u}) = -\int_\Omega (\boldsymbol{v} - \boldsymbol{u}) \cdot \boldsymbol{b}^D \, dv - \int_{\partial\Omega^t} (\boldsymbol{v} - \boldsymbol{u}) \cdot \boldsymbol{t}^D \, ds \tag{7.16}$$

$$= -\int_\Omega (\boldsymbol{v} - \boldsymbol{u}) \cdot \boldsymbol{b}^D \, dv - \int_{\partial\Omega} (\boldsymbol{v} - \boldsymbol{u}) \cdot \boldsymbol{\sigma}[\boldsymbol{u}] \cdot \boldsymbol{n} \, ds. \tag{7.17}$$

The last equality is ensured by the traction boundary condition $\boldsymbol{\sigma}[\boldsymbol{u}] = \boldsymbol{t}^D$ on $\partial\Omega^t$, with the observation that $\boldsymbol{v} - \boldsymbol{u} = 0$ on $\partial\Omega^d$, and from the complementarity of $\partial\Omega^d$ and $\partial\Omega^t$.

An application of the Stokes theorem shows that

$$\mathcal{U}_e(\boldsymbol{v}) - \mathcal{U}_e(\boldsymbol{u}) = -\int_\Omega \boldsymbol{\sigma}[\boldsymbol{u}] : (\boldsymbol{\varepsilon}[\boldsymbol{v}] - \boldsymbol{\varepsilon}[\boldsymbol{u}])\, dv. \tag{7.18}$$

To demonstrate the validity of equation (7.15) we now need to show that the last expression in the above equation is in fact the negative of the differential of strain energy $-\mathrm{D}\mathcal{U}_i[\boldsymbol{u}]$ $(\boldsymbol{v} - \boldsymbol{u})$.

The differential of a functional can be expressed through the Taylor series as

$$\mathcal{U}_i(\boldsymbol{u} + \eta\boldsymbol{w}) = \mathcal{U}_i(\boldsymbol{u}) + \eta \mathrm{D}\mathcal{U}_i[\boldsymbol{u}](\boldsymbol{w}) + o(\eta^2),$$

with η a small parameter. The strain energy definition used here requires that the displacement field be kinematically admissible; that is,

$$\boldsymbol{u} + \eta\boldsymbol{w} = \boldsymbol{u}^D \quad \text{on} \quad \partial\Omega^d.$$

This is indeed the case, provided that

$$\boldsymbol{w} = 0 \quad \text{on} \quad \partial\Omega^d \quad \text{or, equivalently,} \quad \boldsymbol{w} \in \mathcal{K}(0, \partial\Omega^d).$$

From the following series of equalities,

$$\boldsymbol{\sigma}_0 : \boldsymbol{\varepsilon}[\boldsymbol{u} + \eta\boldsymbol{w}] = \boldsymbol{\sigma}_0 : \boldsymbol{\varepsilon}[\boldsymbol{u}] + \eta\boldsymbol{\sigma}_0 : \boldsymbol{\varepsilon}[\boldsymbol{w}] \tag{7.19}$$

$$\boldsymbol{\varepsilon}[\boldsymbol{u} + \eta\boldsymbol{w}] : \boldsymbol{C} : \boldsymbol{\varepsilon}[\boldsymbol{u} + \eta\boldsymbol{w}] = \boldsymbol{\varepsilon}[\boldsymbol{u}] : \boldsymbol{C} : \boldsymbol{\varepsilon}[\boldsymbol{u}] + 2\eta\boldsymbol{\varepsilon}[\boldsymbol{u}] : \boldsymbol{C} : \boldsymbol{\varepsilon}[\boldsymbol{v}] + o(\eta^2) \tag{7.20}$$

$$\boldsymbol{A}\theta^D : \boldsymbol{\varepsilon}[\boldsymbol{u} + \eta\boldsymbol{w}] = \boldsymbol{A}\theta^D : \boldsymbol{\varepsilon}[\boldsymbol{u}] + \eta\boldsymbol{A}\theta^D : \boldsymbol{\varepsilon}[\boldsymbol{w}], \tag{7.21}$$

one deduces that

$$\begin{aligned}\mathrm{D}\mathcal{U}_i[\boldsymbol{u}](\boldsymbol{w}) &= \int_\Omega (\boldsymbol{\sigma}_0 : \boldsymbol{\varepsilon}[\boldsymbol{w}] + \boldsymbol{\varepsilon}[\boldsymbol{u}] : \boldsymbol{C} : \boldsymbol{\varepsilon}[\boldsymbol{w}] - \boldsymbol{\varepsilon}[\boldsymbol{u}] : \boldsymbol{A}\theta^d : \boldsymbol{\varepsilon}[\boldsymbol{w}])\, dv = \int_\Omega \boldsymbol{\sigma}[\boldsymbol{u}] : \boldsymbol{\varepsilon}[\boldsymbol{w}]\, dv,\\ &= \int_\Omega \boldsymbol{\sigma}[\boldsymbol{u}] : (\boldsymbol{\varepsilon}[\boldsymbol{v}] - \boldsymbol{\varepsilon}[\boldsymbol{u}])\, dv,\end{aligned}$$

which finally demonstrates (7.15).

The last step required is the proof of the inequality

$$\mathcal{U}_p(\boldsymbol{v}) - \mathcal{U}_p(\boldsymbol{u}) = \mathcal{U}_i(\boldsymbol{v}) - \mathcal{U}_i(\boldsymbol{u}) - \mathrm{D}\mathcal{U}_i(\boldsymbol{u})(\boldsymbol{v} - \boldsymbol{u}) \geq 0. \tag{7.22}$$

This inequality is valid due to the convexity of $\mathcal{U}_i$, which can be defined by one of the equivalent statements (Figure 7.1)

$$\begin{aligned}&\mathcal{U}_i(t\boldsymbol{u} + (1-t)\boldsymbol{v}) \leq t\mathcal{U}_i(\boldsymbol{u}) + (1-t)\mathcal{U}_i(\boldsymbol{v}) \qquad \forall t \in [0,1] \quad \forall \boldsymbol{u}, \boldsymbol{v} \in \mathbb{V}\\ &\mathcal{U}_i(\boldsymbol{v}) - \mathcal{U}_i(\boldsymbol{u}) - \mathrm{D}\mathcal{U}_i(\boldsymbol{u})(\boldsymbol{v} - \boldsymbol{u}) \geq 0\\ &D^2\mathcal{U}_i(\boldsymbol{u}, \boldsymbol{w}) \geq 0 \quad \forall \boldsymbol{u}, \boldsymbol{w} \in \mathbb{V},\end{aligned}$$

where $\mathbb{V}$ is the space on which $\mathcal{U}_i$ is defined.

Strain energy is indeed a convex functional, for example, according to the last definition above, because it is a second-order functional of strains based on a positive definite tensor $\boldsymbol{C}$.

This completes the proof.

Theorem: The maximum property of the complementary energy potential The stress field $\boldsymbol{\sigma}$ of the elastic solution $[\boldsymbol{u}, \boldsymbol{\varepsilon}, \boldsymbol{\sigma}]$ of a regular thermoelatic problem *maximises* the

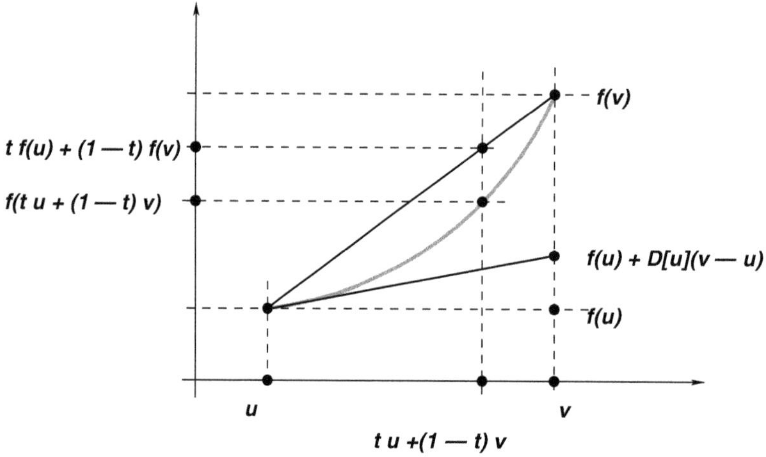

Figure 7.1. A schematic illustration of a convex function and its properties.

complementary energy potential over all statically admissible stress fields:

$$\mathcal{U}_p^*(\boldsymbol{\sigma}) \geq \mathcal{U}_p^*(\boldsymbol{s}) \qquad \forall \boldsymbol{s} \in \mathcal{S}(\boldsymbol{b}^D, \boldsymbol{t}^D, \partial\Omega^t). \tag{7.23}$$

The proof follows a line of argument similar to that used for the minimal principle of the strain energy potential. We therefore only provide a sketch of the proof here.

As the first step, one shows by the definition of the complementary energy potential, the Stokes theorem, and the application of the boundary conditions that

$$\mathcal{U}_p^*(\boldsymbol{s}) - \mathcal{U}_p^*(\boldsymbol{\sigma}) = \mathcal{U}_i^*(\boldsymbol{s}) - \mathcal{U}_i^*(\boldsymbol{\sigma}) - D\mathcal{U}_i^*[\boldsymbol{\sigma}](\boldsymbol{\sigma} - \boldsymbol{s}). \tag{7.24}$$

The inequality

$$\mathcal{U}_i^*(\boldsymbol{s}) - \mathcal{U}_i^*(\boldsymbol{\sigma}) - D\mathcal{U}_i^*[\boldsymbol{\sigma}](\boldsymbol{\sigma} - \boldsymbol{s}) \leq 0 \tag{7.25}$$

is verified through the use of concavity of the complementary energy potential, ensured by the negative sign of the complementary energy and the positive definiteness of $\boldsymbol{C}^{-1}$.

Based on the extremum theorems, a series of inequalities and the uniqueness theorem are established.

Theorem: The inequality of strain energy potential and complementary energy potential Let $[\boldsymbol{u}, \boldsymbol{\varepsilon}, \boldsymbol{\sigma}]$ be the solution of a regular elastic problem with vanishing residual stresses and temperature change field:

$$\boldsymbol{\sigma}_0 = 0 \qquad \theta^D = 0.$$

Then the following inequalities between the strain energy potential and the complementary energy potentials hold,

$$\mathcal{U}_p(\boldsymbol{v}) \geq \mathcal{U}_p(\boldsymbol{u}) = \mathcal{U}_p^*(\boldsymbol{\sigma}) \geq \mathcal{U}_p^*(\boldsymbol{s}),$$

where $\boldsymbol{v}$ and $\boldsymbol{s}$ are respectively the kinematically admissible displacement and statically admissible stress fields (see Figure 7.2 for illustration),

$$\boldsymbol{v} \in \mathcal{K}(\boldsymbol{u}^D, \partial\Omega^d) \qquad \boldsymbol{s} \in \mathcal{S}(\boldsymbol{b}^D, \boldsymbol{t}^D, \partial\Omega^t).$$

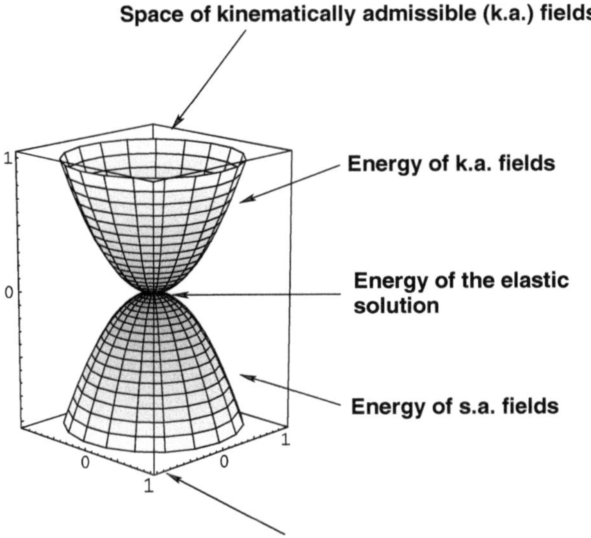

Figure 7.2. A schematic representation of the strain energy potential and the complementary energy potential over admissible fields. The image serves merely as an illustration and should not be interpreted as if a direct relationship between kinematically and statically admissible fields were implied.

Theorem: Uniqueness of the elastic solution Solution $[\boldsymbol{u}, \boldsymbol{\varepsilon}, \boldsymbol{\sigma}]$ of a regular elastic problem is unique to within a rigid body displacement field.

This result is a direct consequence of convexity of the strain energy potential $\mathcal{U}_i$ over kinematically admissible fields. Considering two putative solutions of a given regular thermoelastic problem $[\boldsymbol{u}_i, \boldsymbol{\varepsilon}_i, \boldsymbol{\sigma}_i]$ $(i = 1, 2)$, we immediately observe that

$$[\boldsymbol{u}_1 - \boldsymbol{u}_2, \boldsymbol{\varepsilon}_1 - \boldsymbol{\varepsilon}_2, \boldsymbol{\sigma}_1 - \boldsymbol{\sigma}_2]$$

is also the solution of a problem with zero external loads and zero imposed displacements, and vanishing residual stresses, and temperature change field. Moreover,

$$\boldsymbol{u}_1 - \boldsymbol{u}_2 \in \mathcal{K}(\mathbf{0}, \partial\Omega^d) \qquad \boldsymbol{\sigma}_1 - \boldsymbol{\sigma}_2 \in \mathcal{S}(\mathbf{0}, \mathbf{0}, \partial\Omega^t).$$

Using previously obtained results for the energy potentials, it is easy to show that

$$\mathcal{U}_i(\boldsymbol{u}_1 - \boldsymbol{u}_2) = 0,$$

and therefore

$$\boldsymbol{\varepsilon}_1 = \boldsymbol{\varepsilon}_2.$$

The integration of strain fields to obtain displacements leads to the following conclusion about the equality of displacements to within a rigid body displacement:

$$\boldsymbol{u}_1 = \boldsymbol{u}_2 + \boldsymbol{a} + \boldsymbol{b} \times \boldsymbol{x} \qquad \boldsymbol{a}, \boldsymbol{b} \in \mathbb{R}^3.$$

If the part of the boundary $\partial\Omega^d$ on which displacement boundary condition are imposed is nonzero, then

$$\boldsymbol{u}_1(\boldsymbol{x}) = \boldsymbol{u}_2(\boldsymbol{x}) \qquad \boldsymbol{x} \in \partial\Omega^d,$$

and therefore $\boldsymbol{a} = 0$ and $\boldsymbol{b} = 0$, thus ensuring the uniqueness of the displacement solution.

The extrema of continuous differentiable functions are usually determined from the condition of vanishing derivative. This approach can also be applied to the strain energy potential and complementary energy potential of linear thermoelasticity, as is apparent from the arguments developed in this section. Equilibrium equations of linear thermoelasticity can be obtained as the Euler–Lagrange variational equations for the strain energy potential and complementary energy potential.

This property can be stated in the following form:

Theorem: Properties of solution at extremum Consider the solution $(\boldsymbol{u}, \boldsymbol{\varepsilon}, \boldsymbol{\sigma})$ of a regular thermoelastic problem. Then the equalities below hold for the derivatives of the energy potentials:

- Strain energy potential:

$$\mathrm{D}\mathcal{U}_p[\boldsymbol{u}](\boldsymbol{v}) = 0 \qquad \forall \boldsymbol{v} \in \mathcal{K}(\boldsymbol{0}, \partial\Omega^d). \tag{7.26}$$

  The above equality can also be expressed as an extended integral formula:

$$\int_\Omega \boldsymbol{\sigma}[\boldsymbol{u}] : \boldsymbol{\varepsilon}[\boldsymbol{v}]\, dv - \int_\Omega \boldsymbol{b}^D \cdot \boldsymbol{v}\, dv - \int_{\partial\Omega^t} \boldsymbol{t}^D \cdot \boldsymbol{v}\, ds = 0 \qquad \forall \boldsymbol{v} \in \mathcal{K}(\boldsymbol{0}, \partial\Omega^d). \tag{7.27}$$

- Complementary energy potential:

$$\mathrm{D}\mathcal{U}_p^*[\boldsymbol{\sigma}](\boldsymbol{s}) = 0 \qquad \forall \boldsymbol{s} \in \mathcal{S}(\boldsymbol{0}, \boldsymbol{0}, \partial\Omega^t). \tag{7.28}$$

  The above equality can also be expressed as an extended integral formula:

$$-\int_\Omega (\boldsymbol{\sigma}[\boldsymbol{u}] - \boldsymbol{\sigma}_0 - \boldsymbol{C} : \boldsymbol{A}\theta^D) : \boldsymbol{\varepsilon}[\boldsymbol{v}]\, dv + \int_{\partial\Omega^d} \boldsymbol{u}^D \cdot (\boldsymbol{s} \cdot \boldsymbol{n})\, ds = 0 \quad \forall \boldsymbol{s} \in \mathcal{S}(\boldsymbol{0}, \boldsymbol{0}, \partial\Omega^t). \tag{7.29}$$

7.3 APPROXIMATE SOLUTIONS FOR PROBLEMS OF ELASTICITY

Having established the extremal properties of the elastic solution as the miminum of the strain energy potential and the maximum of the complementary energy potential over respective admissible fields, we may now introduce *approximate solutions* over subsets or subspaces of admissible fields as solutions delivering the extremal value within the respective subset or subspace.

- The approximate displacement solution $\boldsymbol{u}^A$ within the subset of admissible fields $\mathcal{K}_A \subset \mathcal{K}(\boldsymbol{u}^D, \partial\Omega^d)$ is the solution that minimises the strain energy potential over the subset $\mathcal{K}_A$ of admissible displacement fields; that is,

$$\boldsymbol{u}^A = \arg\min_{v \in \mathcal{K}_A} \mathcal{U}_p[\boldsymbol{v}].$$

- The approximate stress solution $\boldsymbol{\sigma}^A$ within the subset of statically admissible stress fields $\mathcal{S}_A \subset \mathcal{S}(\boldsymbol{b}^D, \boldsymbol{t}^D, \partial\Omega^t)$ is the solution that maximises the complementary energy potential over the subset $\mathcal{S}_A$ of admissible stress fields; that is,

$$\boldsymbol{\sigma}^A = \arg\max_{s \in \mathcal{S}_A} \mathcal{U}_p^*[\boldsymbol{s}].$$

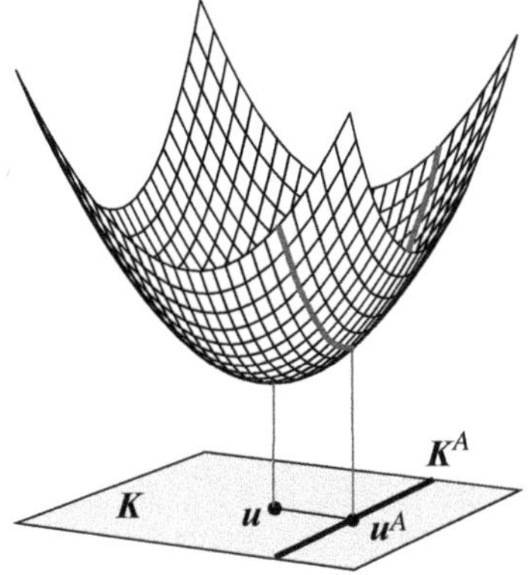

Figure 7.3. A schematic indication of location of the exact $\boldsymbol{u}$ and the approximate solution $\boldsymbol{u}^A$ on the strain energy potential surface and in the space of kinematically admissible fields.

It can be shown in a mathematically rigorous way that the approximate solutions defined in this way also minimise the norm of the difference between the exact solution $[\boldsymbol{u}, \boldsymbol{\varepsilon}, \boldsymbol{\sigma}]$ and the subspace of admissible fields $\mathcal{K}_A$ or respectively $\mathcal{S}_A$. A schematic illustration of this argument is shown in Figure 7.3.

The corresponding definition of the norm is related to the strain energy and is associated with the scalar product, defined as

$$\langle \boldsymbol{u}, \boldsymbol{v} \rangle := \int_{\Omega} \boldsymbol{\varepsilon}[\boldsymbol{u}] : \boldsymbol{C} : \boldsymbol{\varepsilon}[\boldsymbol{v}] \, dv, \tag{7.30}$$

with the norm given by

$$\|\boldsymbol{u}\| := \langle \boldsymbol{u}, \boldsymbol{u} \rangle^{\frac{1}{2}}. \tag{7.31}$$

7.4 THE RAYLEIGH–RITZ METHOD

A particular technique for the determination of approximate solutions based on the energy minimisation approach is the Rayleigh–Ritz method. This method occupies a place of particular importance in numerical analysis applied to continuum mechanics because it provides a rigorous basis for the construction of a number of other techniques. Notably this includes the finite element method, one of the most widely used techniques for computing continuum mechanics solutions for engineering structures.

To simplify the presentation, without restricting the generality of treatment, we assume that in the elastic problem under consideration, initial (residual) stresses and the temperature change field are zero:

$$\boldsymbol{\sigma}_0 = \boldsymbol{0}, \qquad \theta^D = 0.$$

Furthermore we shall assume that a zero displacement field $\boldsymbol{u}^D = 0$ is prescribed on $\partial\Omega^d$.

We seek an approximate solution in the subspace of admissible displacement fields $\mathcal{K}^A$ spanned by a finite number of modal displacement fields described by known functions $\boldsymbol{w}_m \in \mathcal{K}(\boldsymbol{0}, \partial\Omega^d)$, $m = 1, \ldots, M$.

The unknown displacement field that corresponds to the approximate solution in the sense of the previous section $\boldsymbol{v} \in \mathcal{K}^A$ is assumed to be given by a linear combination of modal displacement fields,

$$\boldsymbol{v}[\boldsymbol{\alpha}] = \sum_{m=1}^{M} \alpha_m \boldsymbol{w}_m, \tag{7.32}$$

where $\alpha_m \in \mathbb{R}$ are as yet unknown constant coefficients that are to be determined. The infinite-dimensional problem of finding the approximate solution is thus reduced to the problem of determining a finite-dimensional unknown vector of coefficients $[\boldsymbol{\alpha}] \in \mathbb{R}^M$ given by

$$[\boldsymbol{\alpha}] = [\alpha_1, \alpha_2, \ldots, \alpha_M]^T \in \mathbb{R}^M.$$

Coefficients $[\boldsymbol{\alpha}]$ that correspond to the approximate solution $\boldsymbol{u}^A$ deliver a minimum of the strain energy potential over all kinematically admissible fields of the form (7.32), that is,

$$[\boldsymbol{\alpha}] = \arg \min_{[\alpha] \in \mathbb{R}^M} \mathcal{U}_p(\boldsymbol{v}[\boldsymbol{\alpha}]).$$

A series of elementary manipulations lead to the strain energy potential in the form

$$\begin{aligned}
\mathcal{U}_i(\boldsymbol{v}[\boldsymbol{\alpha}]) &= \frac{1}{2} \int_\Omega \boldsymbol{\varepsilon}\left[\boldsymbol{v}[\boldsymbol{\alpha}]\right] : \boldsymbol{C} : \boldsymbol{\varepsilon}\left[\boldsymbol{v}[\boldsymbol{\alpha}]\right] \, dv \\
&= \frac{1}{2} \int_\Omega \left(\sum_{m=1}^{M} \alpha_m \boldsymbol{\varepsilon}[\boldsymbol{w}_m] \right) : \boldsymbol{C} : \left(\sum_{\ell=1}^{M} \alpha_\ell \boldsymbol{\varepsilon}[\boldsymbol{w}_\ell] \right) dv \\
&= \frac{1}{2} [\boldsymbol{\alpha}]^T \cdot [\boldsymbol{K}] \cdot [\boldsymbol{\alpha}],
\end{aligned}$$

where the components of the matrix $[\boldsymbol{K}] \in (\mathbb{R}^M \times \mathbb{R}^M)$ are given by

$$K_{m\ell} = \int_\Omega \boldsymbol{\varepsilon}(\boldsymbol{w}_m) : \boldsymbol{C} : \boldsymbol{\varepsilon}(\boldsymbol{w}_\ell) \, dv \qquad m, \ell = 1, \ldots, M.$$

Matrix $[\boldsymbol{K}]$ is symmetric by construction and positive definite as a consequence of the positive definiteness of the tensor of elastic moduli $\boldsymbol{C}$.

The expression for the mechanical work of external forces is written similarly as

$$\begin{aligned}
\mathcal{U}_e(\boldsymbol{v}[\boldsymbol{\alpha}]) &= \int_\Omega \boldsymbol{v}[\boldsymbol{\alpha}] \cdot \boldsymbol{b}^D \, dv + \int_{\partial\Omega} \boldsymbol{v}[\boldsymbol{\alpha}] \cdot \boldsymbol{t}^D \, ds \\
&= \int_\Omega \left(\sum_{m=1}^{M} \alpha_m \boldsymbol{w}_m \right) \cdot \boldsymbol{b}^D \, dv + \int_{\partial\Omega} \left(\sum_{m=1}^{M} \alpha_m \boldsymbol{w}_m \right) \cdot \boldsymbol{t}^D \, ds \\
&= -[\boldsymbol{\beta}]^T \cdot [\boldsymbol{\alpha}]
\end{aligned}$$

$$\beta_m = \int_\Omega \boldsymbol{w}_m \cdot \boldsymbol{b}^D \, dv + \int_{\partial\Omega} \boldsymbol{w}_m \cdot \boldsymbol{t}^D \, ds. \tag{7.33}$$

The complete approximate strain energy potential is then written in the following form:

$$\mathcal{U}_p(\boldsymbol{v}[\boldsymbol{\alpha}]) = \frac{1}{2} [\boldsymbol{\alpha}]^T \cdot [\boldsymbol{K}] \cdot [\boldsymbol{\alpha}] - [\boldsymbol{\beta}]^T \cdot [\boldsymbol{\alpha}].$$

The condition of attaining the minimum requires the derivative to vanish,

$$\mathrm{D}_{[\boldsymbol{\alpha}]}\mathcal{U}_p(\boldsymbol{v}[\boldsymbol{\alpha}]) = 0,$$

resulting in the following linear system of equations in the unknowns $[\boldsymbol{\alpha}]$:

$$[\boldsymbol{K}] \cdot [\boldsymbol{\alpha}] - [\boldsymbol{\beta}] = [0]. \tag{7.34}$$

An alternative method of reducing the solution of the elastic problem to the solution of a finite system of linear algebraic equations is the Galerkin method. This method also uses the representation of the unknown displacements in the form of a linear combination of trial functions, as in equation (7.32). However, the approximate solution is obtained by enforcing the Navier equation by ensuring that the residue is orthogonal (in the sense of the scalar product defined in the previous section) to the subspace spanned by the system of trial functions.

Unlike the Raleigh–Ritz method, the Galerkin method does not necessarily give rise to a linear system with symmetric positive definite matrix. This may potentially lead to additional difficulties in obtaining numerical solutions.

Example: compression of a cylinder between two fully adhered rigid platens

Consider a cylindrical slab occupying in the initial configuration the domain $\Omega(r, z) = [0, R] \times [-H, H]$, defined in terms of the cylindrical coordinate system (r, θ, z) with its axis oriented in direction z and its origin at its geometrical centre. The slab is compressed between two rigid platens that adhere perfectly to the plane faces of the cylinder defined by $z = \pm H$. The boundary displacement field is therefore required to be of the form

$$\boldsymbol{u}(r, \theta, \pm H) = \mp\delta \boldsymbol{e}_z, \tag{7.35}$$

and the cylindrical surface must be traction-free.

The Poisson effect in the slab will lead to widening of its middle sections, thus producing a 'barrelling' effect. It is apparent, however, that the amount of barrelling will not be uniform over the height of the slab, because adhesion in the extreme sections acts to prevent it.

We shall call the ratio between the deformed radius and the initial radius at a given height the barrelling function, and write

$$r = f(z)R \quad f : [-H, H] \longrightarrow \mathbb{R}.$$

This function cannot be estimated by a closed form solution as in the trivial case of frictionless platens. The purpose of this section is to obtain approximate forms of the barrelling function using the strain energy minimum principle.

We begin with a kinematically admissible field of the general form

$$\boldsymbol{u} = \frac{\delta}{H}\left(\nu \frac{r}{R} f(z)\boldsymbol{e}_r - \frac{z}{Z}\boldsymbol{e}_z\right).$$

Hereafter we shall assume different classes of functions $f(z)$ and identify the best choice of $f(z)$ in each class using the strain energy minimum principle. Although the solutions found in this way are guaranteed to be optimal within their own class, it is worth noting

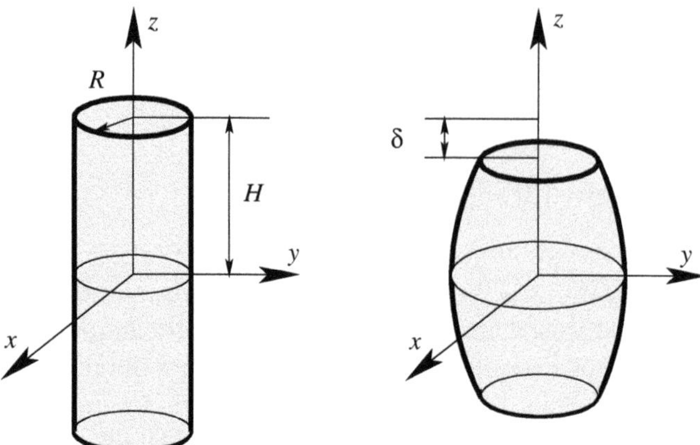

Figure 7.4. A cylindrical slab compressed between two rigid perfectly adhered platens.

that all of them are still likely to be approximate, since the chosen form of displacement field keeps transverse sections $z =$ const. transverse, that is, does not allow deplanation.

We start by loading the required packages. The **Displacement** package contains the formulas of linear isotropic elasticity that simplify the presentation.

The form of the kinematically admissible displacement is defined and the boundary conditions are verified.

We parameterise the problem by introducing the aspect ratio of the slab:

$$\mathbf{ee} = \frac{H}{R}.$$

```
<< Tensor2Analysis.m
<< Displacement.m

SetCoordinates[Cylindrical[r, t, z]]
CoordinatesToCartesian[{r, t, z}]

u := {nu delta / H r  ee/ H  f[z / H ], 0,  - delta / H z}

Thread[ Simplify[ u /. z -> H ] == {0, 0, -delta}]
```

The strain and stress tensors are computed. We can then check the static admissibility conditions for the stresses. One can remark that the stresses do not satisfy the traction-free condition at the external surface of the cylinder, so that one cannot expect to obtain the exact solution of the problem.

The quantity **w** denotes the strain energy density, which can be computed from the strain and stress tensors in several equivalent ways.

```
(eps = Strain[u] ) // MatrixForm
(sig = Stress[eps]) // MatrixForm
```

```
Simplify[Div[ sig ]]
sig . {1, 0, 0}


w = 1/ 2 Simplify[Tr[ DDot[ IsotropicStiffness[nu], eps]]]

w = 1 / 2  Simplify[Tr[ eps   . sig ]]
```

The potential energy is next computed by integration. The use of the `ScaleFactors` command gives a convenient way of evaluating the volume element required in the calculation.

In order to thread the `Integrate` operator over the expressions containing f, and to isolate integrals of f and its derivatives, we define some appropriate rules by the following operations:

$$\int_0^H ah(z/H)\,dz = a\,H\int_0^1 h(s)ds$$

$$\int_0^H a(h(z/H))^n\,dz = a\,H\int_0^1 (h(s))^2 ds$$

$$\int_0^H a(f'(z/H))^2\,dz = a\,H\int_0^1 (f'(s))^2 ds.$$

```
sf = ScaleFactors[]

wintr = 4 Pi Integrate[ w  r , { r, 0, H / ee}]

energy = Thread[ Integrate[ Expand[ wintr],{z,0,H}],Plus]

rule1 =  Integrate[ a___ h___[z/H], {z, 0, H}] ->
  If[ D [a, z] == 0, a H  Integrate[h[s], {s, 0, 1}]]

rule2 = Integrate[ a___ Power[h___[z/H], n_], {z, 0, H}] ->
  If[ D [a, z] == 0, a H Integrate[Power[h[s],n],{s,0,1}]]

rule3 = Integrate[
  Times[ a___, Power[Derivative[1][f][z/H], 2]], {z, 0, H}] ->
 If[ D [a, z] == 0,
  Times[a,H,Integrate[Power[Derivative[1][f][s],2],{s,0,1}]] ]

energy = energy  /. rule1 /. rule2 /. rule3

ww = Simplify[%]
```

Triangular solution

As the first approximation let us compute the best choice of barrelling function f in the class of triangular functions,

$$f(z) = q\left(1 - \frac{|z|}{H}\right) \qquad q \in \mathbb{R}_+,$$

where the parameter q that represents the maximal barrelling is found from the minimization of the strain energy potential.

We compute the strain energy potential by replacing the barrelling function f with its triangular form and integrating over the volume of the cylinder.

The extremum point of the strain energy potential,

$$\frac{\partial \mathcal{U}_p}{\partial q} = 0,$$

defines the optimal value of q,

$$q = \frac{12\mathbf{ee}\,H}{3 + 8\mathbf{ee}^2 - 6\nu}.$$

```
Wtriang = Collect[Simplify[ ww /. f -> (q (1 - #) &)], q]

qsol = Simplify[ Solve[ D[Wtriang, q] == 0, q]][[1]]

wsol = Simplify[Wtriang /. qsol]

utriang =  Simplify[ u /.   f -> (q (1 - #) &) /. qsol ]
```

Series solutions

As the next class of barrelling functions we can explore the series expressions

$$f(s) = \sum_{i=1}^{N} c_i g_i(s) \qquad c_i \in \mathbb{R}, \tag{7.36}$$

where c_i are the coefficients to be determined in the minimisation process.

The minimisation operations are algorithmic and can easily be packaged into a **`Module`** function defined for any given form of potential and test functions. The following steps need to be taken:

- evaluate the potential for the given test function f
- differentiate the potential obtained with respect to the list of unknown coefficients and solve the resulting system of equations
- find the optimal function f and the optimal value of the strain energy potential

```
MinEnergy[ testfunction_ , testcoef_List , potential_, f_] :=
  Module[{},
    testpotential = Evaluate[ potential /. f -> testfunction  ];
    eq = Map[ D[  testpotential ,   #] == 0 & , testcoef ];
    solcoef = Solve[ eq , testcoef ];
    optfunction = Simplify[ testfunction /. solcoef ][[1]];
    {optfunction, Simplify[ potential   /. f -> optfunction]}
    ]
```

We can now explore a polynomial series

$$f(s) = \sum_{i=1}^{N} c_i (1-s)^i (1+s)^i \qquad c_i \in \mathbb{R}.$$

The optimal barrelling functions can be plotted for different values of the aspect ratio **ee**.

Exploration of other forms of barrelling test functions is left to the reader as an exercise.

```
testcoef = Map[ ToExpression["c" <> ToString[#]] & , Range[1] ]

testfunction = Evaluate[Sum[
  testcoef[[i]] (1 - #)^i (1 + #)^i,
      {i, Length[testcoef]}] ] &

sol = MinEnergy[testfunction, testcoef, ww, f]

ff[ee_,H_,nu_] = (f[s]/.f -> sol[[1]] ) nu ee / H

optww[ ee_, H_, nu_ ] = sol[[2]]

Plot[ {ff[10, 1, -0.1], ff[1, 1, -0.1]} , {s, 0, 1} ]
```

Calculus of variations

The approximate solution can be improved further if the class of test functions is extended to include new terms. The largest admissible class is C^1, that is, all continuously differentiable functions. Functions f must satisfy the boundary condition

$$f(\pm H) = 0.$$

All calculations can be performed explicitly in Mathematica, but one can also use the package **VariationalMethods**.

Before beginning the computation, it is convenient to simplify the expression for strain energy potential and to get rid of a constant that does not affect the result.

The optimal value of an integral defines **EulerEquations** for its integrand.

```
<< Calculus`VariationalMethods`

wintr
cc = Pi delta^2 /
   (4 ee^2 H^2 (-1+nu+2 nu^2))
int = wintr/cc   /. z/H -> s

eueq = EulerEquations[ int , f[s], s ]

eusol = DSolve[ eueq  , f, s ][[1]]

eq = {f[s] == 0 /. eusol /. s -> -1,
      f[s] == 0 /. eusol /. s -> 1}
csol = Solve[ eq, {C[1], C[2] }]

fopt = Simplify[eusol /. csol][[1, 1]]

Evaluate[f[s] /. fopt /. nu -> 0. /. ee -> 1 /. H -> 1]
Plot[% , {s, 0, 1}, PlotRange -> All]
```

7.5 EXTREMAL PROPERTIES OF FREE VIBRATIONS

Well-posed problem of elastodynamics

The solution $[\boldsymbol{u}, \boldsymbol{\varepsilon}, \boldsymbol{\sigma}]$ of the elastodynamic problem satisfies the partial differential equation

$$\operatorname{div}(\boldsymbol{C} : \boldsymbol{\varepsilon}[\boldsymbol{u}]) + \boldsymbol{b}^D = \rho \ddot{\boldsymbol{u}} \qquad \text{on } \Omega \times [0, T], \tag{7.37}$$

where the strains and stresses are related to the displacements by

$$\boldsymbol{\varepsilon}[\boldsymbol{u}] = \frac{1}{2}(\nabla \boldsymbol{u} + \nabla^T \boldsymbol{u}), \qquad \boldsymbol{\sigma}[\boldsymbol{u}] = \boldsymbol{\sigma}_0 + \boldsymbol{C} : (\boldsymbol{\varepsilon}[\boldsymbol{u}] - \boldsymbol{A}\theta^D).$$

To obtain a well-posed problem in the sense of Hadamard, that is, to guarantee the existence of a unique solution continuous with respect to the prescribed conditions, we shall assume that the following data are provided:

- Material data: the elasticity tensor $\boldsymbol{C}$ and the tensor of thermal expansions $\boldsymbol{A}$ on Ω
- Initial (residual) stresses $\boldsymbol{\sigma}_0$ on Ω and the temperature change field θ^D on $\Omega \times [0, T]$
- Body forces $\boldsymbol{b}^D$ on $\Omega \times [0, T]$

- Boundary conditions on complementary parts of the boundary $\partial\Omega^d$ and $\partial\Omega^t$ in the form

$$\boldsymbol{u} = \boldsymbol{u}^D \qquad \text{on } \partial\Omega^d \times [0, T]$$
$$\boldsymbol{\sigma} \cdot \boldsymbol{n} = \boldsymbol{t}^D \qquad \text{on } \partial\Omega^t \times [0, T]$$

- Initial conditions at $t = 0$ for displacements and velocities in the form

$$\boldsymbol{u} = \dot{\boldsymbol{u}}_0, \qquad \dot{\boldsymbol{u}} = \boldsymbol{v}_0 \qquad \text{on } \Omega. \tag{7.38}$$

The spectrum of free vibrations

Hereafter we consider a particular class of solutions of the elastodynamic problem, namely the case of free vibrations. We make the following assumptions:

- the evolution is isothermal, $\theta^D = 0$
- the body is free of initial (residual) stress, $\boldsymbol{\sigma}_0 = 0$
- the body forces are absent, $\boldsymbol{b}^D = \boldsymbol{0}$ on $\Omega \times [0, T]$
- the boundary conditions consist of zero displacements and traction-free surfaces:

$$\boldsymbol{u} = \boldsymbol{0} \qquad \text{on } \partial\Omega^d \times [0, T] \tag{7.39}$$
$$\boldsymbol{\sigma} \cdot \boldsymbol{n} = \boldsymbol{0} \qquad \text{on } \partial\Omega^t \times [0, T]. \tag{7.40}$$

To obtain free vibrations the solutions of the elastodynamic equation (7.37) are sought under the hypothesis of separation of variables:

$$\boldsymbol{u}(\boldsymbol{x}, t) = h(t)\boldsymbol{w}(\boldsymbol{x}) \qquad (\boldsymbol{x}, t) \in \Omega \times [0, T]. \tag{7.41}$$

Due to zero displacement boundary conditions (7.39) $\boldsymbol{w}$ must satisfy the same condition, that is,

$$\boldsymbol{w} = \boldsymbol{0} \qquad \text{on } \partial\Omega^d. \tag{7.42}$$

The linearity of equations leads to

$$\boldsymbol{\varepsilon}[\boldsymbol{u}](\boldsymbol{x}, t) = h(t)\,\boldsymbol{\varepsilon}[\boldsymbol{w}](\boldsymbol{x}), \qquad \boldsymbol{\sigma}[\boldsymbol{u}](\boldsymbol{x}, t) = h(t)\,\boldsymbol{\sigma}[\boldsymbol{w}](\boldsymbol{x}), \tag{7.43}$$

and as a consequence the elastodynamic equation (7.37) becomes

$$h(t)\,\mathrm{div}\,\boldsymbol{\sigma}[\boldsymbol{w}] = \ddot{h}(t)\rho\boldsymbol{w}. \tag{7.44}$$

After multiplication of the last equation by $\boldsymbol{w}$, integration over Ω, and application of the Stokes theorem using boundary conditions (7.39) and (7.41), one obtains

$$-h(t)\int_\Omega \boldsymbol{\epsilon}[\boldsymbol{w}] : \boldsymbol{C} : \boldsymbol{\epsilon}[\boldsymbol{w}]\,dv = \ddot{h}(t)\int_\Omega \boldsymbol{w} \cdot \boldsymbol{w}\,dv, \tag{7.45}$$

or, equivalently,

$$-\frac{\ddot{h}(t)}{h(t)} = \frac{\displaystyle\int_\Omega \boldsymbol{\epsilon}[\boldsymbol{w}] : \boldsymbol{C} : \boldsymbol{\epsilon}[\boldsymbol{w}]\,dv}{\displaystyle\int_\Omega \rho\boldsymbol{w} \cdot \boldsymbol{w}\,dv}. \tag{7.46}$$

Because $\boldsymbol{w}$ depends only on the spatial variable $\boldsymbol{x}$, the right-hand side of the above equation is a constant. Because $\boldsymbol{C}$ is positive definite, this constant is nonnegative, and can be denoted by ω^2.

We distinguish two cases:

- $\omega^2 = 0$

  The positive definiteness of $\boldsymbol{C}$ implies that $\boldsymbol{\varepsilon}[\boldsymbol{w}] = 0$. By integration one obtains the result that h is a linear function of time and $\boldsymbol{w}$ is a rigid body displacement:

$$h(t) \quad = \alpha + \beta t \qquad \alpha, \beta \in \mathbb{R} \tag{7.47}$$

$$\boldsymbol{w}(\boldsymbol{x}) = \boldsymbol{a} + \boldsymbol{b} \times \boldsymbol{x} \qquad \boldsymbol{a}, \boldsymbol{b} \in \mathbb{R}^3. \tag{7.48}$$

  The complete displacement field is given by

$$\boldsymbol{u}(\boldsymbol{x}, t) = (\alpha + \beta t)(\boldsymbol{a} + \boldsymbol{b} \times \boldsymbol{x}). \tag{7.49}$$

  This solution, representing rigid body motion, may only exist if there are no encastre (built-in) displacement conditions; that is, $\partial\Omega^d = \emptyset$, meaning that the entire boundary of the body, $\partial\Omega$, is a traction-free surface.

- $\omega^2 > 0$

  The solution for the temporal part of the displacement function is given by

$$h(t) = \alpha \cos(\omega t + \beta),$$

  meaning that displacements of the body vary periodically with time, as expected for free vibrations.

  Pairs ω and $\boldsymbol{w}$ are obtained as the solution of the following eigenvalue problem :

$$\int_\Omega \boldsymbol{\epsilon}[\boldsymbol{w}] : \boldsymbol{C} : \boldsymbol{\epsilon}[\boldsymbol{w}]\, dv - \omega^2 \int_\Omega \rho \boldsymbol{w} \cdot \boldsymbol{w}\, dv = 0. \tag{7.50}$$

In the nontrivial case, $\omega^2 > 0$, ω is referred to as the *cyclic eigenfrequency* and $\boldsymbol{w}$ as the *eigenmode*. The pairs of cyclic eigenfrequencies and eigenmodes possess certain important properties:

1. The pairs of cyclic eigenfrequencies and eigenmodes form an infinite discrete set $\{(\omega_n, \boldsymbol{w}_n) | n \in \mathbb{N}\}$ called the *spectrum of free vibrations.*
2. The set of eigenmodes $\{\boldsymbol{w}_n | n \in \mathbb{N}\}$ provides an orthogonal vector basis of the space of admissible displacement fields with built-in boundary conditions $\mathcal{K}(\boldsymbol{0}, \partial\Omega^d)$. In other words, every $\boldsymbol{v} \in \mathcal{K}(\boldsymbol{0}, \partial\Omega^d))$ can be expressed as

$$\boldsymbol{v} = \sum_{n=1}^{\infty} \alpha_n \boldsymbol{w}_n.$$

   The orthogonality of eigenmodes is understood in terms of the scalar products

$$\langle \boldsymbol{w}, \boldsymbol{v} \rangle_{\mathcal{Q}} = \int_\Omega \boldsymbol{\epsilon}[\boldsymbol{w}] : \boldsymbol{C} : \boldsymbol{\epsilon}[\boldsymbol{v}]\, dv \tag{7.51}$$

$$\langle \boldsymbol{w}, \boldsymbol{v} \rangle_{\mathcal{M}} = \int_\Omega \rho \boldsymbol{w} \cdot \boldsymbol{v}\, dv \tag{7.52}$$

with two associated norms:

$$|\,\boldsymbol{v}\,|_{\mathcal{Q}}^2 = \langle \boldsymbol{v}, \boldsymbol{v} \rangle_{\mathcal{Q}} \qquad |\,\boldsymbol{v}\,|_{\mathcal{M}}^2 = \langle \boldsymbol{v}, \boldsymbol{v} \rangle_{\mathcal{M}}.$$

3. The functional defined by

$$\mathcal{R}[\boldsymbol{w}] = \frac{|\,\boldsymbol{v}\,|_{\mathcal{Q}}^2}{|\,\boldsymbol{v}\,|_{\mathcal{M}}^2} = \frac{\langle \boldsymbol{w}, \boldsymbol{w} \rangle_{\mathcal{Q}}}{\langle \boldsymbol{w}, \boldsymbol{w} \rangle_{\mathcal{M}}} = \frac{\displaystyle\int_\Omega \epsilon[\boldsymbol{w}] : \boldsymbol{C} : \epsilon[\boldsymbol{w}]\, dv}{\displaystyle\int_\Omega \rho \boldsymbol{w} \cdot \boldsymbol{v}\, dv}$$

takes extreme values over the set of kinematically admissible fields $\mathcal{K}(\mathbf{0}, \partial\Omega^d)$ for the eigenmodes $\{\boldsymbol{w}_n | n \in \mathbb{N}\}$. These extreme values are given by the squares of cyclic eigenfrequencies,

$$\mathcal{R}[\boldsymbol{w}_n] = \omega^2.$$

In other words, for every eigenmode $\boldsymbol{w}_n$, we have

$$\mathrm{D}_w \mathcal{R}[\boldsymbol{w}_n](\boldsymbol{v}) = 0 \qquad \forall \boldsymbol{v} \in \mathcal{C}(\mathbf{0}, \partial\Omega^d).$$

Statements (ii) and (iii) can be demonstrated using the properties of eigenfrequencies and eigenmodes described previously.

Because $\alpha \cos(\omega_n t + \beta)\boldsymbol{w}_n(\boldsymbol{x})$ is a solution of the elastodynamic equation for every $n \in \mathbb{N}$, it follows that

$$\operatorname{div} \boldsymbol{C} : \boldsymbol{\varepsilon}[\boldsymbol{w}_n] = \omega_n^2 \boldsymbol{w}_n. \tag{7.53}$$

Multiplying both sides of the above equation by a test function $\boldsymbol{w}_m$ and integrating over Ω leads to the expression

$$\int_\Omega \boldsymbol{\varepsilon}[\boldsymbol{w}_n] : \boldsymbol{C} : \boldsymbol{\varepsilon}[\boldsymbol{w}_m]\, dv - \omega_n^2 \int_\Omega \boldsymbol{w}_n \cdot \boldsymbol{w}_m\, dv = 0. \tag{7.54}$$

Exchanging indices n and m, shows that

$$\int_\Omega \boldsymbol{\varepsilon}[\boldsymbol{w}_n] : \boldsymbol{C} : \boldsymbol{\varepsilon}[\boldsymbol{w}_m]\, dv - \omega_m^2 \int_\Omega \boldsymbol{w}_n \cdot \boldsymbol{w}_m\, dv = 0. \tag{7.55}$$

Because eigenfrequencies are distinct, $w_n \neq w_m$, the two above equalities demonstrate that

$$\int_\Omega \boldsymbol{\varepsilon}[\boldsymbol{w}_n] : \boldsymbol{C} : \boldsymbol{\varepsilon}[\boldsymbol{w}_m]\, dv = 0, \qquad \int_\Omega \boldsymbol{w}_n \cdot \boldsymbol{w}_m\, dv = 0.$$

It is thus shown that eigenmodes $\boldsymbol{w}_n, \boldsymbol{w}_m$ are orthogonal in terms of both scalar products $\langle \cdot, \cdot \rangle_{\mathcal{Q}}$ and $\langle \cdot, \cdot \rangle_{\mathcal{M}}$.

To show that that the functional $\mathcal{R}$ takes extreme values at eigenmodes, we consider the first variation of this functional using the classical formula

$$\mathcal{R}[\boldsymbol{w}_n + \eta \boldsymbol{v}] = \mathcal{R}[\boldsymbol{u}_n] + \eta \mathrm{D}_w \mathcal{R}[\boldsymbol{w}_n](\boldsymbol{v}) + o(\eta^2)$$

and show that the first-order term in η is equal to zero.

Considering first-order expansions of the scalar products

$$\langle \boldsymbol{w}_n + \eta \boldsymbol{v}, \boldsymbol{w}_n + \eta \boldsymbol{v} \rangle_{\mathcal{Q}} = \langle \boldsymbol{w}_n, \boldsymbol{w}_n \rangle_{\mathcal{Q}} + 2\eta \langle \boldsymbol{w}_n, \boldsymbol{v} \rangle_{\mathcal{Q}} + o(\eta^2), \tag{7.56}$$

$$\langle \boldsymbol{w}_n + \eta \boldsymbol{v}, \boldsymbol{w}_n + \eta \boldsymbol{v} \rangle_{\mathcal{M}} = \langle \boldsymbol{w}_n, \boldsymbol{w}_n \rangle_{\mathcal{M}} + 2\eta \langle \boldsymbol{w}_n, \boldsymbol{v} \rangle_{\mathcal{M}} + o(\eta^2), \tag{7.57}$$

and applying Taylor series expansion in the form

$$\frac{1}{a + \eta b} = \frac{1}{a(a + \eta \frac{b}{a})} = \frac{1}{a}\left(1 - \eta \frac{b}{a}\right) + o(\eta^2), \tag{7.58}$$

one obtains the first-order term in the expansion of $\mathcal{R}$ as

$$\mathrm{D}_w \mathcal{R}[\boldsymbol{w}_n](\boldsymbol{v}) = \frac{2}{\displaystyle\int_\Omega \rho \boldsymbol{w}_n \cdot \boldsymbol{w}_n \, dv} \left[\int_\Omega \boldsymbol{\varepsilon}[\boldsymbol{w}_n] : \boldsymbol{C} : \boldsymbol{\varepsilon}[\boldsymbol{v}] - \omega_n^2 \int_\Omega \boldsymbol{w}_n \cdot \boldsymbol{v} \, dv \right] = 0. \tag{7.59}$$

The equality to zero is derived from equations (7.53) and (7.54).

Approximate spectra

Similarly to the approximate solutions in elastostatics, approximate spectra of free vibrations can be defined in terms of cyclic eigenfrequencies and eigenmodes $(\omega_A, \boldsymbol{w}_A)$ as the pairs delivering extreme values of the functional $\mathcal{R}$ over a subset of kinematically admissible displacement fields $\mathcal{K}_A \subset \mathcal{K}(\boldsymbol{u}^D, \partial\Omega^d)$,

$$\mathcal{R}[\boldsymbol{w}_A] = (\omega_A)^2, \qquad \mathrm{D}_w \mathcal{R}[\boldsymbol{w}^A](\boldsymbol{v}) = 0, \qquad \forall \boldsymbol{v} \in \mathcal{K}_A.$$

To explore the definition of approximate spectra, let us consider the case for which a subset of kinematically admissible fields is defined by a linear combination of basis modal shapes, similarly to the approach taken in the Raleigh–Ritz method (see Section 7.4).

We seek an approximate solution in the subspace of admissible displacement fields $\mathcal{K}^A$ spanned by a finite number of fixed displacement fields $\boldsymbol{v}_m \in \mathcal{C}(\boldsymbol{0}, S^u)$, $m = 1, \ldots, M$.

The displacement field $\boldsymbol{w} \in \mathcal{K}^A$ has the form

$$\boldsymbol{w}[\boldsymbol{\alpha}] = \sum_{m=1}^{M} \alpha_m \boldsymbol{v}_m, \tag{7.60}$$

where $\alpha_m \in \mathbb{R}$ are constant coefficients to be determined and the finite-dimensional vector $[\boldsymbol{\alpha}] \in \mathbb{R}^M$ is defined as

$$[\boldsymbol{\alpha}] = [\alpha_1, \alpha_2, \ldots, \alpha_M]^T \in \mathbb{R}^M.$$

Coefficients $[\boldsymbol{\alpha}]$ corresponding to the approximate displacement solution $\boldsymbol{u}^A$ minimise the strain energy potential over all kinematically admissible displacement fields of the form (7.60); that is,

$$[\boldsymbol{\alpha}] = \arg \min_{[\alpha] \in \mathbb{R}^M} \mathcal{U}_p(\boldsymbol{v}[\boldsymbol{\alpha}]).$$

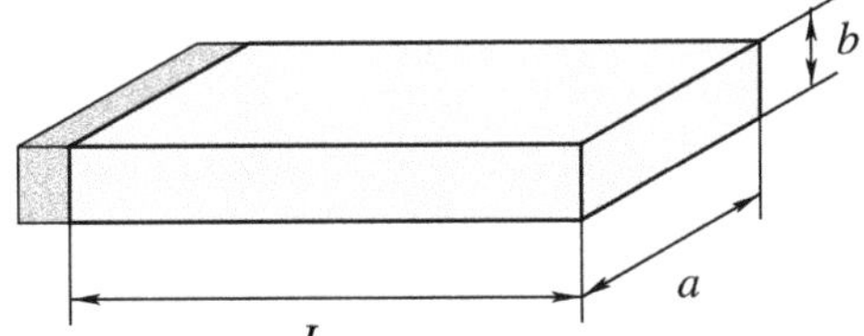

Figure 7.5. Encastre cantilever beam of rectangular cross section.

A series of elementary manipulations shows that

$$\mathcal{R}[\boldsymbol{w}[\alpha]] == \frac{\int_\Omega \epsilon[\boldsymbol{w}[\alpha]] : \boldsymbol{C} : \epsilon[\boldsymbol{w}[\alpha]]\, dv}{\int_\Omega \rho \boldsymbol{w}[\alpha] \cdot \boldsymbol{w}[\alpha]\, dv} = \frac{[\alpha][\boldsymbol{K}][\alpha]}{[\alpha][\boldsymbol{M}][\alpha]},$$

where the components of the matrices $[\boldsymbol{K}]$ and $[\boldsymbol{M}]$ are defined in terms of functions $\boldsymbol{v}_i$ by

$$K_{ij} = \int_\Omega \boldsymbol{\varepsilon}[\boldsymbol{v}_i] : \boldsymbol{C} : \boldsymbol{\varepsilon}[\boldsymbol{v}_j]\, dv,$$

$$M_{ij} = \int_\Omega \rho \boldsymbol{v}_i \cdot \boldsymbol{v}_j \, dv.$$

Matrices $[\boldsymbol{K}]$ and $[\boldsymbol{M}]$ are referred to as the stiffness and the mass matrix, respectively, and define the following approximate eigenvalue system for the extremal values of $\mathcal{R}$:

$$([\boldsymbol{K}] - \omega_A^2[\boldsymbol{M}])[\alpha] = 0. \tag{7.61}$$

Approximate modal shapes are obtained from the eigenvectors $\boldsymbol{\alpha}_A$ of the above equation.

Example: vibration of a cantelever beam

Consider as an example of application of the above technique the calculation of the natural frequencies of an encastre cantelever beam with rectangular section as shown in Figure 7.5. The beam occupies the domain $\Omega = [0, L] \times [-\frac{a}{2}, \frac{a}{2}] \times [-\frac{b}{2}, \frac{b}{2}]$ and is encastre at the end section $x = 0$. The remaining surface is traction-free. The material of the beam is isotropic and linear elastic.

Let us define the class of kinematically admissible fields,

$$\boldsymbol{u} = -y\frac{\partial w}{\partial y}\boldsymbol{e}_x + w(y)\boldsymbol{e}_y,$$

with

$$w = A\,(\cos(\pi x/L) - 1) + B\,(\cos(2\pi x/L) - 1)\,,$$

where $A, B \in \mathbb{R}$ are the coefficients to be determined.

The encastre condition $\boldsymbol{u} = 0$ on $x = 0$ is satisfied by the choice of trial displacement fields.

In the computation of strain and stresses, please note the explicit introduction of Young's modulus `EE`. This is necessary because the problem does not simply scale with respect to this variable, so that the solution depends on its value.

We verify the traction-free surface condition for the stress field and note from the series expansion of the traction vector that this condition is not satisfied unless the thickness of the beam is small compared to its length: $a/L \ll 1$.

```
<< Tensor2Analysis.m
<< Displacement.m


SetCoordinates[Cartesian[x, y, z]]


domain = {{x, 0, L}, {y, -a/2, a/2}, {z, -b/2, b/2}}
w = A ( Cos[Pi*x/L] - 1) + B ( Cos[2*Pi*x/L] - 1)
u = {-y D[w, x], w, 0}


u /. x -> 0


(eps = Strain[u]) // MatrixForm
(sig = EE Stress[eps]) // MatrixForm


normal = {0, -1, 0}


traction = sig . normal  /. y -> -a/2
Series[ traction /. a -> aoL L,{aoL ,0,1}] /. aoL -> a/L
```

We can now proceed to compute the mass matrix and the stiffness matrix. They can be obtained from the equalities

$$[\boldsymbol{\alpha}]^T[\boldsymbol{M}][\boldsymbol{\alpha}] = \int_\Omega \rho \boldsymbol{u}^2 \, dv$$

$$[\boldsymbol{\alpha}]^T[\boldsymbol{K}][\boldsymbol{\alpha}] = \int_\Omega \varepsilon[\boldsymbol{u}] : \boldsymbol{C} : \varepsilon[\boldsymbol{u}] \, dv,$$

where $[\boldsymbol{\alpha}] = [A\,B]^T$.

To compute the integrals, we **Apply** (**@@**) the **Integrate** command to the integrand **Prepend**'ed to the integration domain.

The results show that we can isolate the two constants in order to simplify the computations:

$$c_m = a\,b\,L\,\rho c_k = \frac{Pi^4}{24}\frac{a^3 b}{L^3}\frac{(-1+nu)E}{(1-2\nu)(1+\nu)}.$$

```
M = Integrate @@ Prepend[domain, rho u . u]
cf = CoefficientList[M, {A, B}]
MM = {{cf[[1,3]], cf[[2,2]]/2}, {cf[[2,2]]/2, cf[[3,1]]}}


K = Integrate @@ Prepend[domain, Flatten[eps] . Flatten[sig]]
cf = CoefficientList[M, {A, B}]
```

```
KK = {{cf[[1,3]], cf[[2,2]]/2}, {cf[[2,2]]/2, cf[[3,1]]}}
```

```
cm = a b L rho
Mc = Simplify[MM/cm]
```

```
ck = a^3 b EE (-1 + nu) Pi^4 / ( 24 L^3 (-1 + nu + 2 nu ^2) )
Kc = Simplify[KK/ck]
```

As $[\boldsymbol{K}] = c_k[\boldsymbol{K_c}]$ and $[\boldsymbol{M}] = c_m[\boldsymbol{M_c}]$, the cyclic eigenfrequency ω can be computed from the solution ℓ of the equation:

$$\det\left([\boldsymbol{K_c}] - \ell[\boldsymbol{M_c}]\right) = 0 \qquad \omega = \sqrt{\ell \frac{c_k}{c_m}}.$$

```
elsol = Solve[ Det[ Kc - el Mc] == 0, el]
```

In the case $a/L \ll 1$ we can obtain a simple expression for ℓ and implicitly for the cyclic eigenfrequency and also compute the eigenmode in this limiting case.

A series expansion of the matrices provides a linear system of equations that must be satisfied by the eigenmode.

The two approximate eigenvalues and eigenmodes in the case $a/L \ll 1$ are defined by

$$\ell_{\pm} = \frac{1}{5}(51 \pm \sqrt{2281})$$

and

$$A_{\pm} = -\frac{\pm 143 + 3\sqrt{2281}}{\pm 51\sqrt{2281}} B.$$

The proportionality between A and B indicates that the eigenvector defines only a direction in the space of kinematically admissible fields, without defining its amplitude.

```
ellim = Simplify[
   Map[(Series[#/.a->aoL,{aoL,0,1}] &), elsol], L>0 ]
```

```
N[ellim]
```

```
Mclim = Normal[ Series[ Mc /. a -> aoL , {aoL, 0, 1}] ]
Kclim = Kc
```

```
system = Thread[ (Kclim - el Mclim). {A, B} == 0  ] /. ellim
```

```
Map[ (Solve[ # , {A, B} ][[1, 1]] & ) , system ]
```

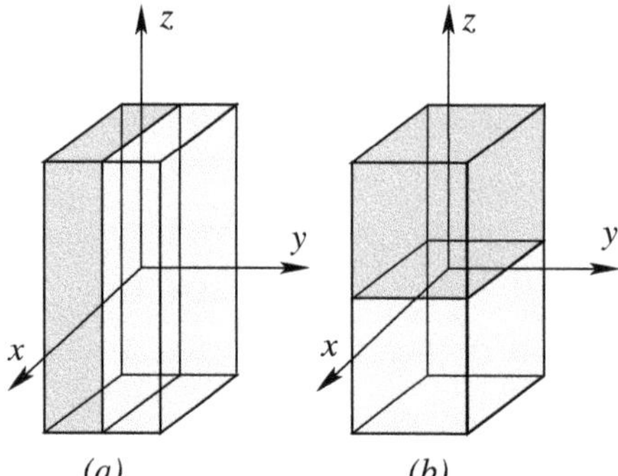

Figure 7.6. Two configurations of a bar composed of two different elastic materials.

SUMMARY

Variational principles in elastostatics and vibration analysis were introduced in this chapter, leading to the development of approximate methods of solution that are illustrated in the form of MATHEMATICA code.

EXERCISES

1. Extension of bimaterial rods

Consider two bars composed of two elastic isotropic materials as illustrated in Figure 7.6. The bars occupy the domain $\Omega = [-L, L] \times [-L, L] \times [-H, H]$ and the elastic moduli will be denoted by (E_i, ν_i) and (λ_i, μ_i) for $i = 1, 2$.

We shall subject both bars to tension along the axis Oz by applying uniform vertical displacement:

$$u_z = \pm\delta \qquad \text{at both ends} \qquad z = \pm H.$$

(a) Write down conditions of continuity for tractions and displacement vectors at each of the interfaces.
(b) Show that for bar (a) an exact solution can be constructed.

2. Bending of a plate

Consider a parallelepipedal plate Ω that in its initial configuration occupies the domain $[0, L] \times [-e, e] \times [-H, H]$. The plate is made of an isotropic homogenous elastic material defined by the constants (E, ν). The prescribed boundary displacements are

- $\boldsymbol{u}(0, y, z) = 0$, that is, the face $x = 0$ is encastre.
- The displacement component along $\boldsymbol{e}_y$ is given by $u_y(L, y, z) = d$ on the face $x = L$.
- On the faces $z = \pm H$, the displacement component along $\boldsymbol{e}_z$ is given by $u_y(x, y, \pm H) = 0$.
  (a) Under the assumption that all other necessary components of the boundary conditions are such that the surfaces are traction-free, complete the boundary conditions such that the problem is well-posed.

(b) Using problem symmetry and the given boundary conditions, justify that the problem is consistent with the plane strain hypothesis,

$$\boldsymbol{u}(x, y, z) = u_x(x, y)\boldsymbol{e}_x + u_y(x, y)\boldsymbol{e}_y,$$

and determine the consequences for the stress field.

(c) Determine the general expression for the stress field under the assumption that

$$\sigma_{xy}(x, y) = a^2 y^2 + b y + c \qquad a, b, c \in \mathbb{R}.$$

(d) Determine the unknown coefficients of the stress field from the condition of minimization of the complementary potential energy.

(e) Is the obtained solution the exact solution?

(f) Discuss the case of a thin plate, that is, $e \ll L$.

Hint: See notebook `C07_bending_plate.nb`.

3. Torsional stiffness of a cylindrical rod

Consider a cylindrical rod $\Omega = S \times [0, L]$ made from an isotropic elastic material. The rod is subject to torsional deformation, in the absence of body forces, defined by the following boundary conditions:

- On the end section $S_{z=0}$:

$$u_x = 0 \quad u_y = 0 \quad t_z = 0.$$

- On the end section $S_{z=L}$:

$$u_z = -\alpha y \quad u_y = +\alpha x \quad t_z = 0.$$

- Lateral surface $\partial S \times [0, L]$ is free from tractions:

$$\boldsymbol{t} = 0.$$

Above, u_i and t_j are components of the displacement and traction vector field on respective surfaces.

(a) Show that the field

$$\boldsymbol{u} = \frac{\alpha z}{L}\boldsymbol{e}_z \times (x\,\boldsymbol{e}_x + y\,\boldsymbol{e}_y) + \frac{\alpha z}{L}\varphi(x, y)\boldsymbol{e}_z$$

is a *kinematically admissible displacement field* for each harmonic function φ on S obeying the Neumann boundary condition

$$\frac{\partial \varphi}{\partial \boldsymbol{n}}(x, y) = \operatorname{grad}\varphi \cdot \boldsymbol{n} = y\,n_x - x\,n_y \qquad (x, y) \in \partial S,$$

where $\boldsymbol{n} = (n_x\,\boldsymbol{e}_x + n_y\,\boldsymbol{e}_y)$ is the unit normal to ∂S.

(b) Show that the following field is a *statically admissible stress field*:

$$\begin{aligned}\boldsymbol{\sigma} &= \alpha\mu\left(\text{curl} + \text{curl}^T\right)\left(\psi(x, y)\,\boldsymbol{e}_z \otimes \boldsymbol{e}_z\right)\\ &= \alpha\mu\frac{\partial \psi}{\partial y}(x, y)\,(\boldsymbol{e}_x \otimes \boldsymbol{e}_z + \boldsymbol{e}_z \otimes \boldsymbol{e}_x) - \alpha\mu\frac{\partial \psi}{\partial x}(x, y)\,(\boldsymbol{e}_y \otimes \boldsymbol{e}_z + \boldsymbol{e}_z \otimes \boldsymbol{e}_y).\end{aligned}$$

(c) Define the torsional stiffness K_T throught the linear relation

$$M_z = K_T\alpha,$$

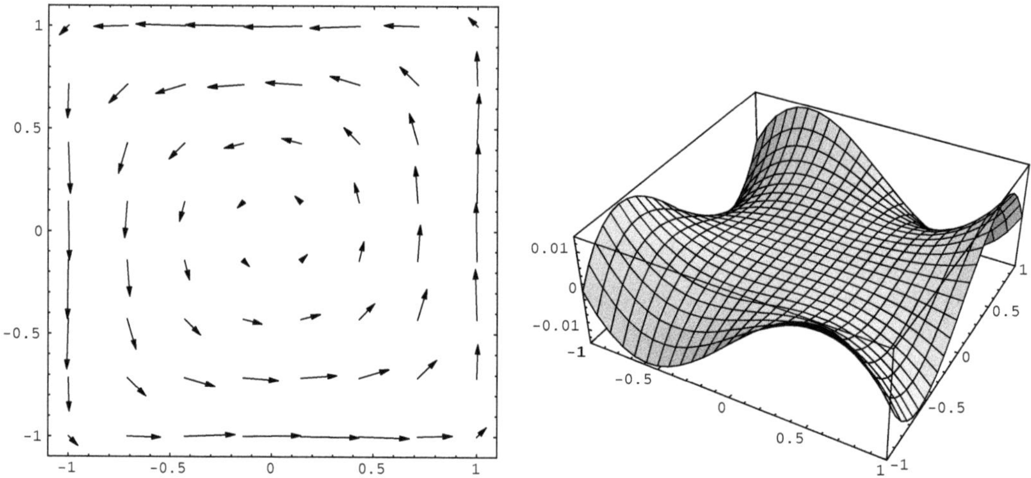

Figure 7.7. Approximate surface tractions and warping (axial displacement) of the end section of a bar with a square cross section subjected to torsion.

where $M_z \boldsymbol{e}_z$ is the applied angular moment on the end section $S_{z=L}$ and α is the angle between the end sections $S_{z=0}$ and $S_{z=L}$. Using previously defined kinematically and statically admissible fields and the variational principles, prove that the following inequalities hold:

$$\mathcal{U}_p^*(\psi) \leq \frac{1}{2} K_T \alpha^2 \leq \mathcal{U}_p(\varphi) \tag{7.62}$$

with

$$\mathcal{U}_p^*(\psi) = -L\alpha^2 \left(\int_S \frac{1}{2\mu} \mid \text{grad}\, \psi \mid^2 + (x\boldsymbol{e}_x + y\boldsymbol{e}_y) \cdot \text{grad}\, \psi \, ds \right)$$

$$\mathcal{U}_p(\psi) = \frac{\alpha^2 \mu}{2L} \int_S \left(\left(\frac{\partial \varphi}{\partial y} + x \right)^2 + \left(\frac{\partial \varphi}{\partial x} - x \right)^2 \right) ds.$$

(d) Consider a bar with a square cross section: $\Omega = [-1, 1] \times [-1, 1] \times [0, L]$. Show that the following choice of the admissible function,

$$\psi(x, y) = m(1 - x^2)(1 - y^2) + n(1 - x^4)(1 - y^4) \qquad m, n \in \mathbb{R}$$
$$\varphi(x, y) = xy(p + q(x^2 - y^2)),$$

leads to the following bounds on the effective torsional stiffness K_T of the bar:

$$2.24803\mu \frac{\alpha^2}{2L} \leq \frac{1}{2} K_T \alpha^2 \leq 2.25185\mu \frac{\alpha^2}{2L}.$$

(e) Plot the traction field and the warping function for the end sections using the **ParametricPlot3D** and **PlotVectorField** MATHEMATICA commands (see Figure 7.7).

(f) Plot the distribution of the Tresca equivalent stress f_T for the end section. Recall that

$$f_T(\boldsymbol{\sigma}) = 1/2 \max_{i,j} \mid \sigma_i - \sigma_j \mid,$$

where σ_i $(i = 1, 3)$ are the eigenvalues of the stress tensor.

Hint: See notebook **C07_torsion_triangle.nb**

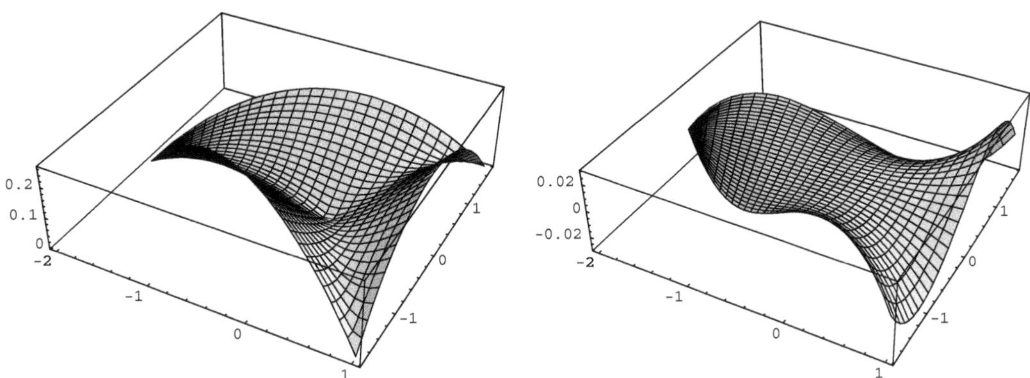

Figure 7.8. Plots of the Tresca equivalent stess and the warping function (axial displacement) for the end section of a bar with triangular cross section subjected to torsion.

4. Torsional stiffness of a cylindrical rod with triangluar cross section (Obala, 1997)

Consider a cylindrical rod $\Omega = S \times [0, L]$ with triangular cross section. S is an equilateral triangle with an edge of length $2a\sqrt{3}$ and comes that correspond in the cartesian coordinate system to the points $(-2a, 0)$, $(a, a\sqrt{3})$, $(a, -a\sqrt{3})$. The cylinder is made out of an isotropic elastic material and is subjected to torsion characterised by the angle α between its end sections.

(a) Show that the function

$$\psi(x, y) = c(a - x)(x - y\sqrt{3} + 2a)(x + y\sqrt{3} + 2a) \qquad c \in \mathbb{R}$$

defines a *statically admissible stress field* through the relation

$$\begin{aligned}\boldsymbol{\sigma} &= \alpha\mu \left(\text{curl } + \text{curl }^T\right) \left(\psi(x, y)\, \boldsymbol{e}_z \otimes \boldsymbol{e}_z\right) \\ &= \alpha\mu \frac{\partial \psi}{\partial y}(x, y)\, (\boldsymbol{e}_x \otimes \boldsymbol{e}_z + \boldsymbol{e}_z \otimes \boldsymbol{e}_x) - \alpha\mu \frac{\partial \psi}{\partial x}(x, y)\, (\boldsymbol{e}_y \otimes \boldsymbol{e}_z + \boldsymbol{e}_z \otimes \boldsymbol{e}_y).\end{aligned}$$

(b) Determine the parameter c in two ways:
- by determining the approximate solution in the space of statically admissible stresses;
- by showing that the corresponding strain is compatible and constructing the exact solution of the problem.

  Compare the two values.

(c) Plot the traction field and the warping function for the end sections using the `ParametricPlot3D` and `PlotVectorField` Mathematica commands (see Figure 7.8).

(d) Plot the distribution of the Tresca equivalent stress f_T for the end section. Recall that

$$f_T(\boldsymbol{\sigma}) = 1/2 \max_{i,j} \left| \sigma_i - \sigma_j \right|,$$

where σ_i $(i = 1, 3)$ are the eigenvalues of the stress tensor (see Figure 7.8).

Hint: See notebook `C07_torsion_triangle.nb`

5. A dam loaded by the water pressure and its own weight (Obala, 1997)

Consider a dam of height H with an apex angle α made out of an isotropic elastic material with the cross section illustrated in Figure 7.9. The dam is considered to extend over a

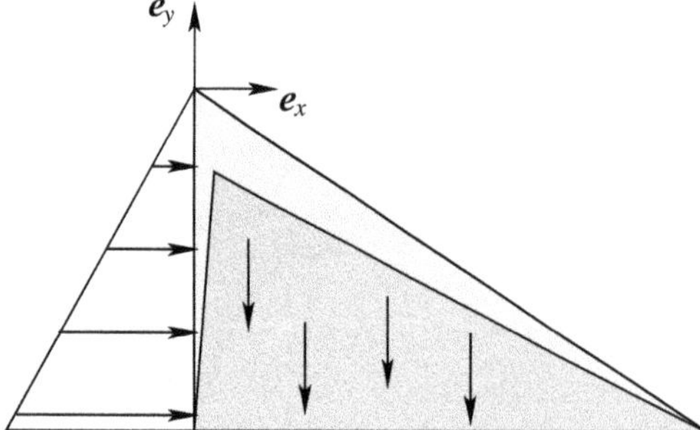

Figure 7.9. Deformation of a dam under water pressure and its own weight.

finite length in the Oz direction and to be locked at the end sections so that displacements along the Oz direction are zero everywhere.

The dam is subjected to the following loading:

- Encastre at the ground level, $y = -H$.
- Water pressure on the face $x = 0$, imposed through the weight of water with the density ρ_w.
- Body force due to gravity that corresponds to the density of concrete, ρ_c.

The gravitational acceleration is denoted by g.

(a) Compute an approximate solution within the space of displacement fields of the form

$$\boldsymbol{u}(x, y, z) = (a_x x + b_x y + c_x)\boldsymbol{e}_x + (a_y x + b_y y + c_y)\boldsymbol{e}_y.$$

(b) Is this solution exact ? Is the obtained solution physically valid ?
Hint: See notebook **C07_dam.nb**.

6. A heated disk glued to a planar rigid substrate

Consider a cylinder of radius R and height Z. The disk is made from an isotropic thermoelastic material with material parameters (λ, μ, α). The disk is glued to a rigid planar surface over its lower section $z = 0$ and uniformly heated to a temperature difference Θ.

(a) Compute an approximate solution starting from kinematically admissible fields:
   - $\boldsymbol{u}(r, \theta, z) = az\boldsymbol{e}_z$
   - $\boldsymbol{u}(r, \theta, z) = abz\boldsymbol{e}_r$

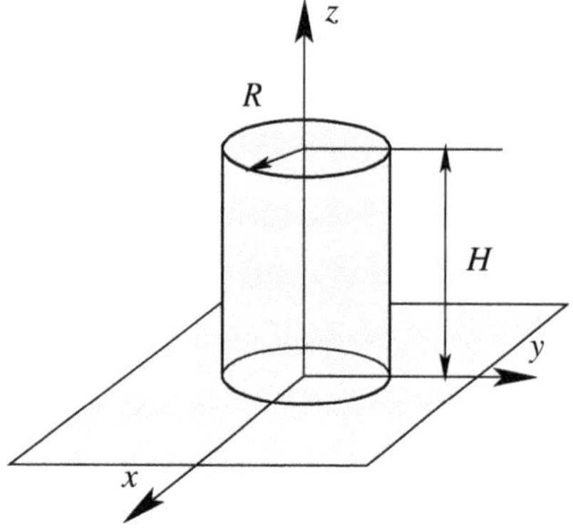

Figure 7.10. A heated disk glued to a rigid planar surface.

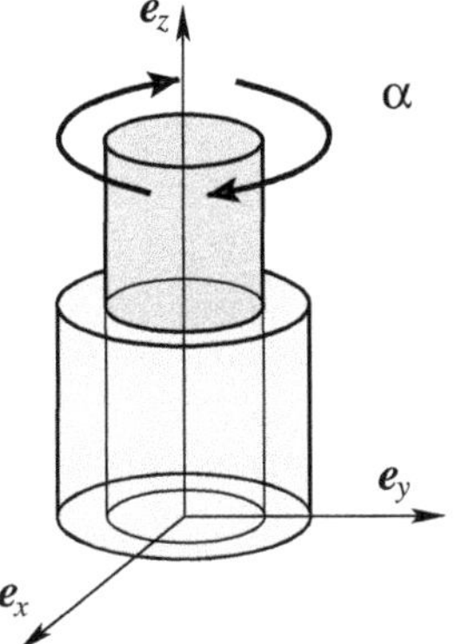

Figure 7.11. A cylindrical adhesive bond subjected to torsion.

Which one do you consider to be a better approximation? Does it have any physical justification?

(b) Construct a MATHEMATICA function using the **Module** operator for computing the solution and explore other admissible fields.

Hint: Use a function similar to the ones proposed for the compression test of this chapter and inspect the code contained in the notebook **C07_heating_disk.nb**.

7. Torsion applied to an adhesive bond (Dumontet et al., 1998)

Consider an adhesive bond, illustrated in Figure 7.11. The glue has the form of a cylindrical tube with internal and external radii denoted by R_i and R_e, respectively, made out of an isotropic elastic material. Its inner surface is perfectly attached to a rigid shaft that may rotate, while its outer surface is attached to a rigid hollow shaft that is stationary. A rotation angle α is imposed on the shaft and causes torsional deformation of the bond material. (The problem formulation simulates a glued joint, or may be thought of as a representation of a drilling procedure, in which case the bond represents the process zone where concentrated deformation occurs.)

The solution will be constructed in the cylindrical coordinate system (r, θ, z) oriented in a standard way with respect to the cartesian system of coordinates.

(a) Describe the well-posed problem for the loading of the joint.

(b) Determine the *statically admissible stress fields* of the form

$$\boldsymbol{\sigma}(r, \theta, z) = \sigma_{r\theta}(r)\,(\boldsymbol{e}_r \otimes \boldsymbol{e}_\theta + \boldsymbol{e}_\theta \otimes \boldsymbol{e}_r)\,.$$

(c) Construct the complementary potential energy $\mathcal{U}^*(\boldsymbol{\sigma})$ and determine its minimum.

(d) Construct the strain field corresponding to the approximate stress field determined in the minimization process, and verify its compatibility.

(e) Construct the displacement field.

Hint: Use the **IntegrateStrain** command and do not forget to add an infinitesimal rigid displacement field for completeness of solution: $\boldsymbol{a} + \boldsymbol{b} \times \boldsymbol{x}$.

(f) Compute the linear and angular momenta applied by the shaft and the effective torsional stiffness of the joint.

(g) What happens to the solution as $R_e \longrightarrow \infty$?

Hint: See notebook **C07_torsion_sphere.nb**

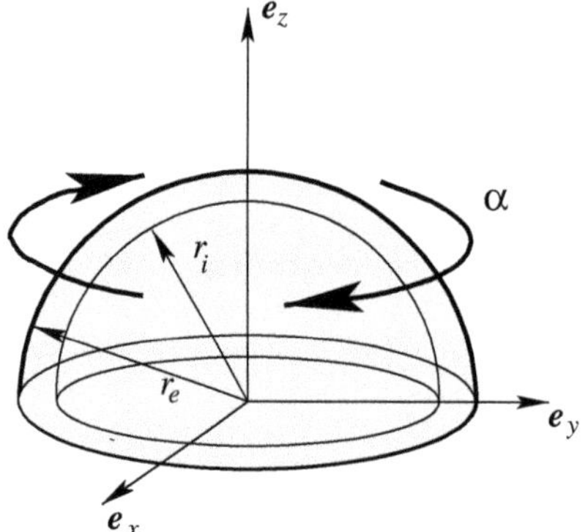

Figure 7.12. A hemispherical bond submitted to shear.

8. Elastic deformation of a hemi spherical joint (Dumontet et al., 1998)

A connexion between two shafts is constructed using a spherical bond layer as illustrated in Figure 7.12. The bond occupies a hollow hemispherical domain, with internal and external radii denoted by R_i and R_e, respectively, and is made from an isotropic elastic material. The material of both shafts being joined is much stiffer than that of the bond and can be considered to be rigid in this analysis. The shafts have a common axis along Oz and are rotated with respect to each other by an angle α that is accommodated by the bond. The joint is therefore subjected to shear traction loading that can be thought to be similar to that applied to the skin of an orange when juice is extracted from it.

The solution will be constructed within the spherical coordinate system (r, θ, φ) oriented in a standard way with respect to the cartesian system of coordinates.

(a) Describe the well-posed problem for the loading of the joint.

(b) Determine the *statically admissible stress fields* of the form

$$\boldsymbol{\sigma}(r, \theta, \phi) = \sigma_{r\varphi}(r, \theta)\,(\boldsymbol{e}_r \otimes \boldsymbol{e}_\varphi + \boldsymbol{e}_\varphi \otimes \boldsymbol{e}_r)\,.$$

(c) Construct the complementary potential energy $\mathcal{U}^*(\boldsymbol{\sigma})$ and determine its minimum.

Hint: The mimisation process leads to the Euler equation of the complementary potential energy. Use the standard package **`Calculus`VariationalMethods`** for your calculations.

(d) Construct the strain field corresponding to the approximate stress field determined in the minimisation process, and show its compatibility.

(e) Construct the displacement field.

Hint: Use the **`IntegrateStrain`** command and do not forget to add an infinitesimal rigid displacement field for completnees of the solution: $\boldsymbol{a} + \boldsymbol{b} \times \boldsymbol{x}$.

(f) Compute the linear and angular momenta applied by the shafts and the effective torsional stiffness of the bond.

(g) What happens to the solution as $R_i \longrightarrow 0$?

Hint: See notebook **`C07_torsion_sphere.nb`.**

APPENDIX 1

Differential operators

The solution of physical problems can often be simplified if it is expressed in a particular curvilinear coordinate system instead of the classical cartesian coordinate system.

The mathematical presentation of this subject follows the evolution of ideas as presented in (Malvern, 1969; Soos and Teodosiu, 1983). Our development of programming is built upon the kernel of an existing standard MATHEMATICA package, **`VectorAnalysis`**. Our additions and modifications are grouped together in the form of a new package called **`Tensor2Analaysis`**.

A.1.1 THE MATHEMATICAL DEFINITIONS

Orthogonal curvilinear coordinate systems

Let us introduce a new system of coordinates $\theta_1, \theta_2, \theta_3$ related to the cartesian coordinates by the functions

$$x_i = x_i(\theta_\alpha) \quad k = 1, 2, 3; \quad \alpha = 1, 2, 3. \tag{1.1}$$

We shall assume that the functions can be inverted and possess sufficient smoothness properties. The inverse transformations will be denoted by

$$\theta_\alpha = \theta_\alpha(x_k) \quad \alpha = 1, 2, 3 \quad k = 1, 2, 3. \tag{1.2}$$

The above smoothness and inversibility hypothesis implies that the *Jacobian* matrix has a nonvanishing determinant in the domain considered:

$$J = \det \begin{bmatrix} \dfrac{\partial x_1}{\partial \theta_1} & \dfrac{\partial x_1}{\partial \theta_2} & \dfrac{\partial x_1}{\partial \theta_3} \\ \dfrac{\partial x_2}{\partial \theta_1} & \dfrac{\partial x_2}{\partial \theta_2} & \dfrac{\partial x_2}{\partial \theta_3} \\ \dfrac{\partial x_3}{\partial \theta_1} & \dfrac{\partial x_3}{\partial \theta_2} & \dfrac{\partial x_3}{\partial \theta_3} \end{bmatrix}. \tag{1.3}$$

A point $\boldsymbol{P}$ in the Euclidean space can be identified by its position vector, denoted by

$$\boldsymbol{OP} = \boldsymbol{x} = x_k \boldsymbol{i}_k = x_1 \boldsymbol{i}_1 + x_2 \boldsymbol{i}_2 + x_3 \boldsymbol{i}_3, \tag{1.4}$$

where $\boldsymbol{i}_k$, $k = 1, 2, 3$ denote the basis vectors of the cartesian system. The point $\boldsymbol{P}$ can be identified either by the cartesian coordinates (x_1, x_2, x_3) or by the curvilinear coordinates $(\theta_1, \theta_2, \theta_3)$.

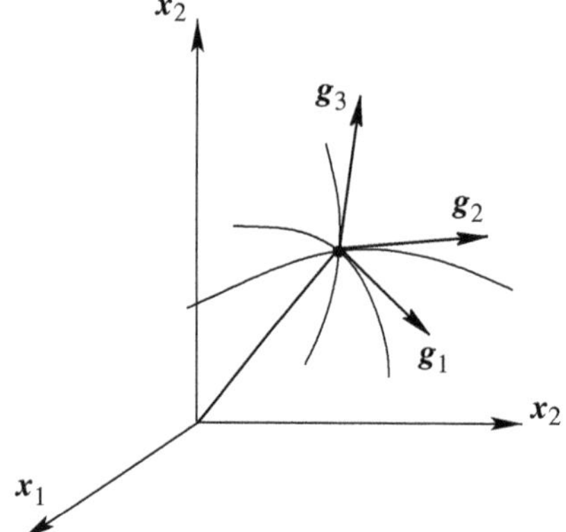

Figure A.1.1. An orthogonal system of coordinates defined by the functions $x_i(\theta_\alpha)$.

The operators **JacobianMatrix** and **JacobianDeterminant** are already defined in the **VectorAnalysis** package and provide the tools for calculating the mathematical objects introduced above.

The vectors that are defined for each $\alpha = 1, 2, 3$ by

$$\boldsymbol{g}_\alpha = \frac{\partial \boldsymbol{x}}{\partial \theta_\alpha} = \frac{\partial x_k}{\partial \theta_\alpha} \boldsymbol{i}_k \tag{1.5}$$

are tangent to the coordinate lines defined by varying θ_α while keeping the other two coordinates fixed. These vectors therefore can be associated in the physical sense with the rate of change of a particular coordinate along a line while other coordinates remain unchanged.

In a general curvilinear coordinate system, $\theta_1, \theta_2, \theta_3$ may have different physical dimensions. For example, in the spherical or cylindrical coordinate systems the coordinates have dimensions either of length (L) or angle (dimensionless).

We now define

$$\boldsymbol{e}_\alpha = \frac{1}{|\boldsymbol{g}_\alpha|} \boldsymbol{g}_\alpha \quad \forall \alpha = 1, 2, 3. \tag{1.6}$$

The norms of tangent vectors have the significance of scale factors and are also referred to as *Lamé coefficients*, expressed as

$$h_\alpha = |\boldsymbol{g}_\alpha| = \sqrt{\left(\frac{\partial x_1}{\partial \theta_\alpha}\right)^2 + \left(\frac{\partial x_2}{\partial \theta_\alpha}\right)^2 + \left(\frac{\partial x_3}{\partial \theta_\alpha}\right)^2} \quad \forall \alpha = 1, 2, 3. \tag{1.7}$$

The vectors $(\boldsymbol{e}_1, \boldsymbol{e}_2, \boldsymbol{e}_3)$ are now normalized to form a vector basis.

In the following we will *only* consider *orthogonal* curvilinear coordinate systems, for which the vectors $(\boldsymbol{e}_1, \boldsymbol{e}_2, \boldsymbol{e}_3)$ are mutually orthogonal:

$$\boldsymbol{e}_\alpha \cdot \boldsymbol{e}_\beta = \delta_{\alpha\beta} \quad \alpha, \beta = 1, 2, 3. \tag{1.8}$$

The spatial orientation of the curvilinear basis $(\boldsymbol{e}_1, \boldsymbol{e}_2, \boldsymbol{e}_3)$ is different for each point $\boldsymbol{P}$ of the Euclidean space, whereas the orientation of the cartesian basis $(\boldsymbol{e}_1, \boldsymbol{e}_2, \boldsymbol{e}_3)$ remains the same.

The components of the vectors in the curvilinear basis can be expressed in the cartesian basis in the form

$$\boldsymbol{e}_\alpha = q_{k\alpha} \cdot \boldsymbol{i}_k, \tag{1.9}$$

where

$$q_{k\alpha} = \frac{1}{h_\alpha}\frac{\partial x_k}{\partial \theta_\alpha} \quad \alpha = 1, 2, 3, \quad k = 1, 2, 3. \tag{1.10}$$

The columns of the matrix $\boldsymbol{Q} = (q_{\alpha k})$ are the normalised columns of the Jacobian matrix of the transformation. Moreover, due to the fact that both vector bases are orthogonal, we also have

$$q_{k\alpha}q_{k\beta} = \delta_{\alpha\beta} \qquad q_{k\alpha}q_{m\alpha} = \delta_{km}. \tag{1.11}$$

The operations defined before and the computation of the matrix $\boldsymbol{Q} = (q_{\alpha k})$ can be coded as follows:

```
cs = ExpandCoordSys[coordsys]
uvw = List @@ Take[cs, 3]
sf = ScaleFactors[cs];
q = Inner[ Divide, JacobianMatrix[uvw] , sf, List];
```

Programming remark

The above MATHEMATICA operations show that it is possible to define the scale factors h_α from the definition of the coordinate system as a transformation (see equation (1.1)). However, we retain the structure of the **VectorAnalysis** package in which the **ScaleFactors** are defined 'by hand' by enumerating each of the cases for individual coordinate systems in a special **Module**.

Length, surface, and volume elements

The length, surface, and volume elements are used to compute the curvilinear, surface, and volume integrals in the curvilinear coordinate system.

The length element of the α coordinate curve is given by

$$dl_\alpha = h_\alpha d\theta_\alpha. \tag{1.12}$$

The surface element on the (α, β) coordinate surface is given by

$$ds_{\alpha\beta} = h_\alpha h_\beta d\theta_\alpha d\theta_\beta. \tag{1.13}$$

The volume element is defined as

$$dv = h_1 h_2 h_3 d\theta_1 d\theta_2 d\theta_3. \tag{1.14}$$

We remark that the product $h_1h_2h_3$ is equal to the Jacobian determinant J, defined at the beginning of this chapter. We have obtained the formula

$$dv = J\ d\theta_1 d\theta_2 d\theta_3. \tag{1.15}$$

Tensors

Higher rank tensors can be understood as generalisations of vectors involving bilinear operations such as the direct product $\otimes$. For example, a tensor $\boldsymbol{T}$ of rank n can be created by

$$\boldsymbol{T} = T_{\alpha_1,\alpha_2,\ldots,\alpha_n}\boldsymbol{e}_{\alpha_1} \otimes \boldsymbol{e}_{\alpha_2} \otimes \ldots \otimes \boldsymbol{e}_{\alpha_n}. \tag{1.16}$$

Orthogonal curvilinear coordinate systems differ from general curvilinear coordinate systems in that the covariant and contravariant components of the tensors are equal, and thus the manipulation of these systems is simplified. From the physical viewpoint the orthogonality of a coordinate systems simplifies many expressions appearing in the equations. This is due to the fact that the normal to the coordinate surface defined by any two coordinates is tangent to the third coordinate line.

Defining tensors in MATHEMATICA

From the MATHEMATICA viewpoint tensors are represented by **List**'s of **List**'s. To construct a vector we can proceed using the **Table** command in the following way:

```
myvector = Table[ToExpression["v"<>ToString[i]], {i,3}]
tensor2  = Table[ToExpression["t"<>ToString[i]<>ToString[j]], {i,3},{j,3}]
tensor3  = Table[ToExpression["t"<>ToString[i]<>ToString[j]<>ToString[k]],
                 {i,3},{j,3},{k,3}].
```

The dimensions of these objects can be computed by

```
Dimensions[myvector]
Dimensions[tensor3].
```

A useful tool for visualising tensors is the command **MatrixForm**,

```
tensor2 // MatrixForm
```

which displays second- (and higher-) order tensors and matrices or arrays of matrices.

To view the resulting form for larger order tensors, execute the commands

```
tensor4 = Table[
    ToExpression["t"<>ToString[i]<>ToString[j]<>ToString[k]<>ToString[l]],
    {i,3},{j,3},{k,3},{l,3}]
tensor4 // MatrixForm.
```

For tensors of higher rank some experience may be required to locate any given component in a display of this type.

One has to be careful when using **MatrixForm**. It is not just an *Output* command, as it also changes the structure of the object:

```
Head[ tensor4 ]
Head[ tensor4 // MatrixForm ].
```

A way to display an object in matrix form without changing its internal storage format is to use brackets as follows:

```
(tensor5= 2 tensor4) // MatrixForm
```

Tensor index operations in the `Tensor2Analysis` pacage

Tensor index operators

`$ScaQ[f]`	True only if field `f` is a scalar
`$VecQ[f]`	True only if field `f` is a 3-vector (tensor of rank 1)
`$TenQ[f]`	True only if field `f` if a 3-tensor of rank 2
`Twirl`	Generalised transpose of tensor
`GTr`	Generalised trace of tensor
`GDot`	Generalised dot product of tensors

In order to perform operations with indices correctly (e.g., transposition), we need to distinguish between scalars and tensors of different ranks. Furthermore, our analysis is specialised to three-dimensional space.

Logical test functions **`$ScaQ`**, **`$VecQ`**, and **`$TenQ`** are created as follows:

```
$ScaQ[v_] := Not[ListQ[v]] || (VectorQ[v] && Length[v] == 1);
$VecQ[v_] := VectorQ[v] && Length[v] == 3;
$TenQ[v_] := MatrixQ[v] && Dimensions[v] == {3,3};
```

Next, we define a new operator **`Twirl`** as a generalisation and correction of the MATHEMATICA operator **`Transpose`** for interchanging two indices of a matrix. This new definition is necessary because the existing operator **`Transpose`** does not perform transposition of arbitrary indices correctly when applied to higher rank objects. Operator **`Twirl`** is used in order to obtain from tensor $\boldsymbol{T}$,

$$\boldsymbol{T} = T_{\alpha_1,\alpha_2,\ldots,\alpha_i,\ldots,\alpha_j,\ldots,\alpha_p}\boldsymbol{e}_{\alpha_1} \otimes \ldots \otimes \boldsymbol{e}_{\alpha_i} \otimes \ldots \otimes \boldsymbol{e}_{\alpha_j} \otimes \ldots \otimes \boldsymbol{e}_{\alpha_n}, \tag{1.17}$$

another tensor transposed in the indices α_i, α_j,

$$\boldsymbol{T} = T_{\alpha_1,\alpha_2,\ldots,\alpha_j,\ldots,\alpha_i,\ldots,\alpha_p}\boldsymbol{e}_{\alpha_1} \otimes \ldots \otimes \boldsymbol{e}_{\alpha_i} \otimes \ldots \otimes \boldsymbol{e}_{\alpha_j} \otimes \ldots \otimes \boldsymbol{e}_{\alpha_n}, \tag{1.18}$$

by the application of

$$\texttt{Twirl}[\boldsymbol{T}, \{i, j\}]. \tag{1.19}$$

Therefore the operator **`Twirl`** should be defined for a general permutation of indices as

```
Twirl[t_?ListQ, ord_?ListQ] :=
  Module[ {rank = TensorRank[t], d = Dimensions[t], ind, perm},
    (ind = Map[ToExpression["i" <> ToString[#]] &, Range[rank]];
       perm = Map[{ToExpression["i" <> ToString[#]], d[[#]]} &, ord];
       Table[Part[t, Evaluate[Sequence @@ind]],
         Evaluate[Sequence @@perm]] ) /;
     ( Signature[ord] =!= 0 && Length[ord] === rank)
    ]

Twirl[t_?ListQ, ord_?IntegerQ] :=
```

```
    Twirl[t, Prepend[Drop[Range[TensorRank[t]], {ord} ], ord]]

Twirl[t_?ListQ] := Twirl[t, 2]
```

The trace operator `Tr` is the convolution over any two repeated indices of a given tensor object, that is, summation of the terms on the generalized diagonal. Trace can only be defined for tensor fields of at least second rank.

For a second rank tensor field $\boldsymbol{T}$,

$$\operatorname{tr} \boldsymbol{T} = T_{\alpha\alpha} = T_{11} + T_{22} + T_{33}. \tag{1.20}$$

For higher rank tensor fields the summation can be performed over any pair of indices.

We define the generalised trace

$$\boldsymbol{A}_{\alpha_1,\alpha_2,\ldots,\gamma_k,\ldots\gamma_l,\ldots\alpha_n}, \tag{1.21}$$

where the summation index is denoted by γ placed in the k and l index positions. This can be programmed as follows,

```
GTr[f_?ListQ,n1_?IntegerQ,n2_?IntegerQ]:= Tr[Twirl[Twirl[f,n1],n2],Plus,2];
```

The operation of taking a `Dot` product between any two tensors can also be applied to any pair of indices. For example, consider the expression

$$\boldsymbol{A}_{\alpha_1,\alpha_2,\ldots,\gamma_k,\ldots,\alpha_n} \boldsymbol{B}_{\beta_1,\beta_2,\ldots,\gamma_l,\ldots,\beta_m}, \tag{1.22}$$

where the summation index is indicated by γ placed once in the k and l index positions. We implement this operator in MATHEMATICA with the help of `Twirl` and the existing operator `Inner`:

```
GDot[A_?ListQ,B_?ListQ,kA_?IntegerQ,lB_?IntegerQ] :=
        Inner[Times,A,Twirl[B,lB],Plus,kA];
```

Differential operators in curvilinear coordinate systems

Differential operators

`Grad[f]`	Gradient of field `f`
`Div[f]`	Divergence of field `f`
`Curl[f]`	Curl of field `f`
`Laplacian[f]`	Laplacian of field `f`
`Biharmonic[f]`	Biharmonic of field `f`, that is, twice the Laplacian of field `f`
`Inc[f]`	Incompatibility of field `f`

The gradient operator is defined in cartesian coordinates as

$$\nabla = \boldsymbol{i}_k \frac{\partial}{\partial x_k}.$$

Using definitions of parameters already introduced for curvilinear coordinate systems and chain differentiation rule for composite functions, we obtain

$$\nabla = \boldsymbol{i}_k \frac{\partial}{\partial x_k} = \boldsymbol{e}_\alpha q_{k\alpha} \frac{\partial}{\partial x_k} = \boldsymbol{e}_\alpha \frac{1}{h_\alpha} \frac{\partial x_k}{\partial \theta_\alpha} \frac{\partial}{\partial x_k} = \boldsymbol{e}_\alpha \frac{1}{h_\alpha} \frac{\partial}{\partial \theta_\alpha} = \boldsymbol{e}_\alpha \nabla_\alpha. \tag{1.23}$$

Here

$$\nabla_\alpha = \frac{1}{h_\alpha} \frac{\partial}{\partial \theta_\alpha}$$

represents differentiation along the coordinate line θ_α.

There are two difficulties with the application of differential operators to vector and tensor fields:

- The first one comes from indexing the differentiation operation with respect to existing indices. In other words, is the (i, j) component of the gradient of the vector field $\boldsymbol{v} = v_k \boldsymbol{i}_k$

$$\frac{\partial v_i}{\partial x_j} \quad \text{or} \quad \frac{\partial v_j}{\partial x_i}$$

The same question arises for the (i, j, k) componenent of second-order tensors $\boldsymbol{T} = T_{nm} \boldsymbol{i}_n \otimes \boldsymbol{i}_m$:

$$\frac{\partial T_{ij}}{\partial x_k} \quad \text{or} \quad \frac{\partial T_{kj}}{\partial x_i}$$

Therefore we shall define two operators:

the *pregradient* $\overrightarrow{\nabla}$, where differentiation appears as the *first* index of the gradient

the *postgradient* $\overleftarrow{\nabla}$, where differentiation appears as the *last* index of the gradient.

- The second stems from the basic fact that basis vectors and hence components in curvilinear coordinate systems depend on the coordinates $\boldsymbol{e}_\alpha = \boldsymbol{e}_\alpha(\theta_1, \theta_2, \theta_3)$.

Let us consider the differentiation of $\boldsymbol{T}$ a tensor of rank n in the direction of fixed α:

$$\nabla_\alpha \boldsymbol{T} = \frac{1}{h_\alpha} \frac{\partial}{\partial \theta_\alpha} \left(T_{\alpha_1 \alpha_2 \ldots \alpha_p} \boldsymbol{e}_{\alpha_1} \otimes \boldsymbol{e}_{\alpha_2} \otimes \ldots \otimes \boldsymbol{e}_{\alpha_n} \right) \tag{1.24}$$

$$= \frac{1}{h_\alpha} \frac{\partial}{\partial \theta_\alpha} \left(T_{\alpha_1 \alpha_2 \ldots \alpha_p} \right) \boldsymbol{e}_{\alpha_1} \otimes \boldsymbol{e}_{\alpha_2} \otimes \ldots \otimes \boldsymbol{e}_{\alpha_n} \tag{1.25}$$

$$+ T_{\alpha_1 \alpha_2 \ldots \alpha_p} \frac{1}{h_\alpha} \frac{\partial}{\partial \theta_\alpha} \left(\boldsymbol{e}_{\alpha_1} \otimes \boldsymbol{e}_{\alpha_2} \otimes \ldots \otimes \boldsymbol{e}_{\alpha_n} \right). \tag{1.26}$$

In order to understand the differentiation of basis vectors, let us write the following series of equalities:

$$\nabla_\alpha \boldsymbol{e}_\beta = \nabla_\alpha \left(q_{k\beta} \boldsymbol{i}_k \right) = \left(\nabla_\alpha q_{k\beta} \right) \boldsymbol{i}_k = \left(\nabla_\alpha q_{k\beta} \right) q_{k\gamma} \boldsymbol{e}_\gamma = \langle \alpha\beta\gamma \rangle \boldsymbol{e}_\gamma. \tag{1.27}$$

In the above expression we encounter the *Hessian* tensor given by the following expression:

$$\langle \alpha\beta\gamma \rangle = \left(\nabla_\alpha q_{k\beta} \right) q_{k\gamma} = \frac{1}{h_\alpha} \frac{\partial q_{k\beta}}{\partial \theta_\alpha} q_{k\gamma}.$$

The following series of properties can be established and are left as an exercise for the reader:

$$\langle \alpha\beta\gamma \rangle = -\langle \alpha\gamma\beta \rangle \quad \forall \alpha, \beta, \gamma \tag{1.28}$$

and

$$\langle \alpha\rho\alpha \rangle = -\langle \alpha\alpha\rho \rangle \tag{1.29}$$

$$= \frac{1}{h_\alpha} \nabla_\rho h_\alpha \quad \forall \alpha \neq \rho \text{ with no summation in } \alpha. \tag{1.30}$$

The computation of the $\langle \alpha\beta\gamma \rangle$ symbol can be achieved through a simple combination of MATHEMATICA commands:

```
uvw = List @@ Take[cs, 3];
sf = ScaleFactors[cs];
q = Inner[ Divide, JacobianMatrix[uvw] , sf, List];
dq = Inner[ Divide, Outer[ D, Transpose[ q], uvw ], sf, List ]
hessian = Transpose[ Simplify[ Inner[Times, dq, q, Plus, 2]]].
```

We can combine these commands in a module defined as

```
Hessian3Tensor[arg_:$CoordinateSystem] :=
    Module[{pt, cs},
        Hessian3Tensor[pt, cs] /; (If[$VecQ[arg],
                                          cs = $CoordinateSystem;
                                          pt = arg,
                                          cs = $ExpandCoordSys[arg];
                                          If[cs =!= $Failed,
                                              pt = List @@ Take[cs, 3]]];
                                          cs =!= $Failed)];

Hessian3Tensor[pt_?$VecQ, coordsys_:$CoordinateSystem] :=
    Module[{cs = $ExpandCoordSys[coordsys], sf, q, dq, uvw},
        (uvw = List  Take[cs, 3];
         sf = ScaleFactors[cs];
         q = Inner[ Divide, JacobianMatrix[uvw] , sf, List];
         dq = Inner[ Divide, Outer[ D, Transpose[ q], uvw ], sf, List ] /.
                             $DAbsSign /. $PointRule[cs, pt];
         Transpose[ Simplify[ Inner[Times, dq, q, Plus, 2]]] ) /;
                             (cs =!= $Failed)].
```

The definitions for the operator ∇ for scalar, vector-valued, or tensor-valued fields follow in a logical way as

```
Nabla[f_?$ScaQ, coordsys_: $CoordinateSystem] :=
    Module[{cs = $ExpandCoordSys[coordsys]},
        Outer[D,{f}, List @@ Take[cs, 3]][[1]]/ScaleFactors[cs] /;
                (cs =!= $Failed)]
```

```
Nabla[f_?$VecQ, coordsys_:$CoordinateSystem] :=
    Module[{ cs=$ExpandCoordSys[coordsys],sf,ht },
        (sf=ScaleFactors[cs]; ht=Hessian3Tensor[cs];
        Transpose[Inner[ Divide, Outer[D, f, List @@Take[cs, 3]],
        sf,List ]  ] ) /;
                (cs =!=  $Failed)]

Nabla[f_?$TenQ, coordsys_:$CoordinateSystem] :=
     Module[ { cs =$ExpandCoordSys[coordsys],sf,ht },
        (sf=ScaleFactors[cs]; ht=Hessian3Tensor[cs];
         Twirl[ Inner[ Divide, Outer[D, f, List @@ Take[cs, 3]],/;
              sf,List ] ,3 ]   ) (cs =xs!=$Failed)].
```

The gradient operator

Next we shall carefully distinguish between the pre- and post- versions of the differential operators.

Consider a scalar function,

$$f : \Omega \longrightarrow \mathbb{R},$$

and define the gradient in a curvilinear coordinate system as the following vector field:

$$\mathrm{grad} f = \overrightarrow{\nabla} f = \overleftarrow{\nabla} f = (\nabla_\alpha f)\, \boldsymbol{e}_\alpha. \tag{1.31}$$

The pregradient and postgradient of the vector field $\boldsymbol{v} : \Omega \longrightarrow \mathbb{R}^3$ are second-order tensors defined as

$$\overrightarrow{\nabla}\ \boldsymbol{v} = (\nabla_\alpha v_\beta + \langle \alpha\gamma\beta \rangle v_\gamma)\, \boldsymbol{e}_\alpha \otimes \boldsymbol{e}_\beta \tag{1.32}$$

$$\overleftarrow{\nabla}\ \boldsymbol{v} = (\nabla_\beta v_\alpha + \langle \beta\gamma\alpha \rangle v_\gamma)\, \boldsymbol{e}_\alpha \otimes \boldsymbol{e}_\beta. \tag{1.33}$$

We remark that in the case of a vector field the difference between the two operators is just a transposition:

$$\overleftarrow{\nabla}\ \boldsymbol{v} = (\overrightarrow{\nabla}\ \boldsymbol{v})^T.$$

We shall identify the gradient of $\boldsymbol{v}$ with the postgradient of $\boldsymbol{v}$, so that

$$\mathrm{grad}\, \boldsymbol{v} = \nabla \boldsymbol{v} = \overleftarrow{\nabla}\ \boldsymbol{v} = (\overrightarrow{\nabla}\ \boldsymbol{v})^T.$$

The pre- and postgradients of the second-order tensor field $\boldsymbol{T} : \Omega \longrightarrow \mathbb{R}^3 \times \mathbb{R}^3$ are third-order tensors, defined as :

$$\overrightarrow{\nabla}\ \boldsymbol{T} = (\nabla_\alpha T_{\beta\gamma} + \langle \alpha\lambda\beta \rangle T_{\lambda\gamma} + \langle \alpha\lambda\gamma \rangle T_{\beta\lambda})\, \boldsymbol{e}_\alpha \otimes \boldsymbol{e}_\beta \otimes \boldsymbol{e}_\gamma \tag{1.34}$$

$$\overleftarrow{\nabla}\ \boldsymbol{T} = (\nabla_\gamma T_{\alpha\beta} + \langle \gamma\lambda\alpha \rangle T_{\lambda\beta} + \langle \gamma\lambda\beta \rangle T_{\alpha\lambda})\, \boldsymbol{e}_\alpha \otimes \boldsymbol{e}_\beta \otimes \boldsymbol{e}_\gamma. \tag{1.35}$$

We shall identify the gradient of the tensor field $\boldsymbol{T}$ with the postgradient of $\boldsymbol{T}$, so that

$$\mathrm{grad}\, \boldsymbol{T} = \nabla \boldsymbol{T} = \overleftarrow{\nabla}\ \boldsymbol{T}.$$

In MATHEMATICA we write

```
Grad[f_?$ScaQ, coordsys_:$CoordinateSystem] :=
    Module[{cs = $ExpandCoordSys[coordsys]},
        Outer[D, {f}, List @@ Take[cs, 3]][[1]]/ScaleFactors[cs] /;
                (cs =!= $Failed)]

Grad[f_?$VecQ, coordsys_:$CoordinateSystem] :=
    Module[{cs =$ExpandCoordSys[coordsys],sf,ht},
        (sf=ScaleFactors[cs]; ht=Hessian3Tensor[cs];
        Nabla[f,coordsys]   +   GDot[ ht, f, 2, 1] ) /;
                (cs =!= $Failed)]


Grad[f_?$TenQ, coordsys_:$CoordinateSystem] :=
    Module[{cs = $ExpandCoordSys[coordsys],sf,ht},
       (sf=ScaleFactors[cs]; ht=Hessian3Tensor[cs];
        Nabla[f,coordsys]
        +  GDot [ht, f, 2, 1]
        +  Twirl[ GDot [ht, f, 2, 2], { 1,3,2 } ]   ) /;
                (cs =!= $Failed)].
```

The divergence operator

The divergence of a field can be defined as the *trace* of the gradient of this field.

For a vector field $\boldsymbol{v}$ we have

$$\operatorname{div} \boldsymbol{v} = \operatorname{tr}(\operatorname{grad} \boldsymbol{v}) = (\nabla_\alpha v_\alpha + \langle \alpha\gamma\beta \rangle v_\gamma)\, \boldsymbol{e}_\alpha \otimes \boldsymbol{e}_\beta. \tag{1.36}$$

The divergence of the second-order tensor field $\boldsymbol{T}$ is a first-order tensor defined as

$$\operatorname{div} \boldsymbol{T} = \operatorname{tr}(\operatorname{grad} \boldsymbol{T}) = \boldsymbol{T} \cdot \boldsymbol{e}_\alpha \nabla_\alpha = (\nabla_\alpha T_{\alpha\beta} + \langle \lambda\alpha\lambda \rangle T_{\alpha\beta} + \langle \alpha\lambda\beta \rangle T_{\alpha\lambda})\, \boldsymbol{e}_\beta. \tag{1.37}$$

The preceeding definitions are now translated into MATHEMATICA as

```
Div[f_?$VecQ, coordsys_:$CoordinateSystem] :=
  Module[cs = $ExpandCoordSys[coordsys],sf,ht,
      (sf=ScaleFactors[cs]; ht=Hessian3Tensor[cs];
      Tr[ Grad[f,coordsys] ] ) /;
              (cs =!= $Failed)]

Div[f_?$ TenQ, coordsys_:$CoordinateSystem] :=
  Module[{cs = $ExpandCoordSys[coordsys],sf,ht,grad},
      (sf=ScaleFactors[cs]; ht=Hessian3Tensor[cs];
      GTr[Grad[f,coordsys], 1,2 ] )/;
              (cs =!= $Failed)].
```

The curl operator

In the definition of the **curl** operator we use the Levy-Civita fully antisymmetric third rank symbol $\epsilon_{\alpha\beta\lambda}$ which can be readily implemented in MATHEMATICA using the **Signature** function as follows:

```
epst= Table[ Signature[{i, j, k}], {i, 3}, {j, 3}, {k, 3}]
```

For a vector field $\boldsymbol{v}$ define the curl operator as

$$\operatorname{curl} \boldsymbol{v} = \boldsymbol{e}_\alpha \nabla_\alpha \times \boldsymbol{v} = \epsilon_{\alpha\beta\lambda}(\nabla_\alpha v_\beta - \langle\alpha\beta\gamma\rangle v_\gamma)\boldsymbol{e}_\lambda. \tag{1.38}$$

In a similar way we have for a second-order tensor field $\boldsymbol{T}$

$$\operatorname{curl} \boldsymbol{T} = \boldsymbol{e}_\alpha \nabla_\alpha \times \boldsymbol{T} = \boldsymbol{e}_\alpha \nabla_\alpha \times (T_{\beta\gamma}\, \boldsymbol{e}_\beta \otimes \boldsymbol{e}_\gamma) \tag{1.39}$$

$$= \epsilon_{\gamma\lambda\alpha}(\nabla_\gamma T_{\lambda\beta} + \langle\gamma\rho\lambda\rangle\, T_{\rho\beta} + \langle\gamma\rho\beta\rangle\, T_{\lambda\rho})\, \boldsymbol{e}_\alpha \otimes \boldsymbol{e}_\beta. \tag{1.40}$$

This can be programmed as follows:

```
Curl[f_?$VecQ, coordsys_:$CoordinateSystem] :=
    Module[{cs = $ExpandCoordSys[coordsys], sf,ht,epst,df,br},
        ( sf=ScaleFactors[cs]; ht=Hessian3Tensor[cs];
          epst= Table[ Signature[{i, j, k}], {i, 3}, {j, 3}, {k, 3}];
          df=Nabla[ f, coordsys];
          br=df - GDot[ht,f,3,1];
        GTr[ GDot[ epst, br,1,1] ,1,3]
              ) /; (cs =!= $Failed)]

Curl[f_?$TenQ, coordsys_:$CoordinateSystem] :=
    Module[{cs = $ExpandCoordSys[coordsys], sf,ht,epst,df,br},
        ( sf=ScaleFactors[cs]; ht=Hessian3Tensor[cs];
          epst= Table[ Signature[{i, j, k}], {i, 3}, {j, 3}, {k, 3}];
          df=Nabla[ f, coordsys];
          br=df + GDot[ht,f,2,1] + Twirl[ GDot[ht,f,2,2], {1,3,2} ];
        GTr[ GDot[epst,br,1,1], 1,3 ]
              ) /; (cs =!= $Failed)].
```

Laplacian, biharmonic, and inc operators

The Laplacian operator is a second-order differential operator defined as

$$\triangle = \operatorname{div} \operatorname{grad} = \nabla \cdot \nabla. \tag{1.41}$$

Because the Laplacian operator is obtained simply as a combination of the previously defined operators, we code it as follows:

```
Laplacian[f_, coordsys_:$CoordinateSystem] :=
    Module[{cs = $ExpandCoordSys[coordsys]},
        Div[Grad[f, coordsys], coordsys] /; (cs =!= $Failed)].
```

The **Grad** and **Div** operators take into account the scalar, vector, or tensor character of the field f and apply the appropriate definition.

We shall also mention the biharmonic operator, which plays a particularly important role in elasticity. This operator is simply defined as

$$\triangle\triangle = (\nabla \cdot \nabla)(\nabla \cdot \nabla). \tag{1.42}$$

It has been implemented as

```
Biharmonic[f_, coordsys_:$CoordinateSystem] :=
    Module[{cs = $ExpandCoordSys[coordsys]},
        Laplacian[Laplacian[f, coordsys], coordsys] /; (cs =!= $Failed)].
```

Finally, we define the incompatibility operator **Inc**, which can be applied to second rank tensors. It is defined as

$$\operatorname{inc} \boldsymbol{T} = (\operatorname{curl} (\operatorname{curl} \boldsymbol{T})^T)^T. \tag{1.43}$$

It is implemented in MATHEMATICA as follows:

```
Inc[f_?$TenQ, coordsys_:$CoordinateSystem] :=
    Module[{cs = $ExpandCoordSys[coordsys]},
Simplify[Transpose[Curl[Transpose[Curl[f, coordsys]]]]] /;
(cs =!= $Failed)].
```

For a compatible tensor field (i.e., one that can be integrated to a vector), the result of the application of the incompatibility operator is identically zero.

Classical results for differential operators

The following results are not directly related to the implementation of the **Tensor2Analysis** package. However, because they are intimately related to the definitions of differential operators, we state them here. These results have direct relevance to the properties of potential functions in elasticity.

Potential fields and Stokes' theorem

Let there be a second-order tensor field. Then there exists a vector field $\boldsymbol{f}$ such that

$$\boldsymbol{g} = \operatorname{grad} \boldsymbol{f}$$

if and only if

$$(\operatorname{curl} \boldsymbol{g}^T)^T = 0.$$

This also ensures that the integral of $\boldsymbol{g}$ on any closed path $\gamma \subset \Omega$ vanishes:

$$\int_\gamma \boldsymbol{g} \cdot dx = 0.$$

In a general tensor field framework one can replace g with a tensor of rank p and $\boldsymbol{f}$ with a tensor field of rank $p+1$.

For mathematically interested readers we shall recall that the preceeding compatibility conditions are a particular case of the *Poincaré Lemma* (Spivak, 1965) which states as follows

Given Ω and an open star-shaped domain and the 1-differential form

$$\boldsymbol{\omega} = g_i dx_i,$$

the following statements are equivalent:

- $\boldsymbol{\omega}$ is *exact*, that is, there exists f such that $\boldsymbol{\omega} = df$, which implies $\boldsymbol{g} = \operatorname{grad} f$.
- $\boldsymbol{\omega}$ is *closed*, that is, $d\boldsymbol{\omega} = 0$, which implies equation (2.30).

Differential forms

With a vector field $\boldsymbol{v} = v_1\boldsymbol{e}_1 + v_2\boldsymbol{e}_2 + v_3\boldsymbol{e}_3$, one can associate the following differential forms Spivak (1965):

$$\begin{aligned}
\boldsymbol{\omega}_{\boldsymbol{v}}^1 &= v_1 dx_1 + v_2 dx_2 + v_3 dx_3 \\
\boldsymbol{\omega}_{\boldsymbol{v}}^1 &= v_1 dx_2 \wedge dx_3 + v_2 dx_3 \wedge dx_1 + v_3 dx_1 \wedge dx_2.
\end{aligned}$$

Then the following relations are valid:

$$\begin{aligned}
df \; &= \boldsymbol{\omega}_{\text{grad}f}^1 \\
d(\boldsymbol{\omega}_{\boldsymbol{v}}^1) &= \boldsymbol{\omega}_{\text{rot}\boldsymbol{v}}^2 \\
d(\boldsymbol{\omega}_{\boldsymbol{v}}^2) &= (\text{div}\boldsymbol{v}) dx_1 \wedge dx_2 \wedge dx_3.
\end{aligned}$$

Using the preceding equalities and applying Poincaré's Lemma implies that

- if $\text{rot}\boldsymbol{v} = 0$ then there exists f such that $\boldsymbol{v} = \text{grad}f$
- if $\text{div}\boldsymbol{v} = 0$ then there exists $\boldsymbol{f}$ such that $\boldsymbol{v} = \text{grad}\boldsymbol{f}$.

Stokes' Theorem can be reformulated in terms of differential forms as follows (see Figure A.1.2).

If $\boldsymbol{\omega}$ is a differential form and c is an n-cube, that is, a regular domain or surface, then

$$\int_c d\boldsymbol{\omega} = \int_{\partial c} \boldsymbol{\omega}.$$

If we now apply the preceeding result to the vector field $\boldsymbol{v}$ and tensor field $\boldsymbol{T}$, then

$$\int_\Omega \operatorname{div} \boldsymbol{v}\, dv = \int_{\partial\Omega} \boldsymbol{v} \cdot \boldsymbol{n}\, ds, \qquad \int_\Omega \operatorname{div} \boldsymbol{T}\, dv = \int_{\partial\Omega} \boldsymbol{T} \cdot \boldsymbol{n}\, ds$$

$$\int_\Sigma (\operatorname{curl} \boldsymbol{v}) \cdot \boldsymbol{n}\, ds = \int_{\partial\Omega} \boldsymbol{v} \cdot \boldsymbol{t}\, dc, \qquad \int_\Sigma (\operatorname{curl} \boldsymbol{T}^T)^T \cdot \boldsymbol{n}\, ds = \int_{\partial\Omega} \boldsymbol{T} \cdot \boldsymbol{t}\, dc.$$

Solutions of the Poisson equation

Let ψ be a continuous differentiable scalar field defined on a closed set $\overline{\Omega}$.

Then

$$\varphi(\boldsymbol{x}) = -\frac{1}{4\pi} \int_\Omega \frac{\psi(\boldsymbol{y})}{|\,\boldsymbol{x} - \boldsymbol{y}\,|}\, dv \tag{1.44}$$

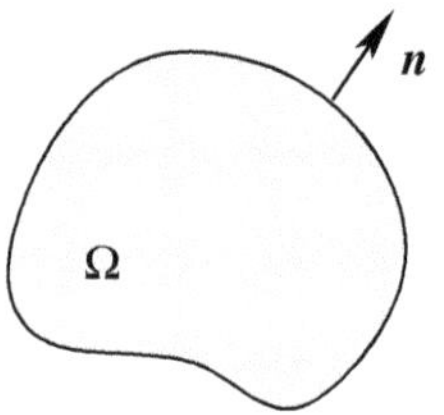

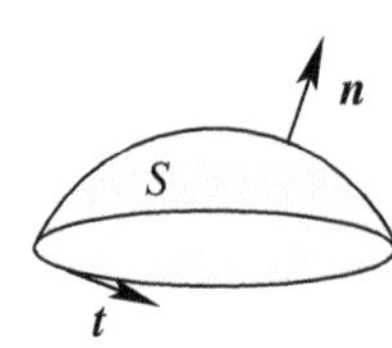

Figure A.1.2. Two forms of domains which can be considered as n-cubes embedded in $\mathbb{R}^3$: a three-dimensional domain Ω, and a two-dimensional surface S.

is twice continuously differentiable and satisfies the following equation:

$$\triangle\varphi = \psi. \tag{1.45}$$

Similar results can be established for vector and tensor fields.

Representation of curl curl

The following identity holds:

$$\text{curl}\,\text{curl} = \text{grad}\,\text{div} - \triangle. \tag{1.46}$$

For example, for the vector field $\boldsymbol{v}$ the above equation can be written as

$$\text{curl}\,\text{curl}\boldsymbol{v} = \text{grad}\,\text{div}\boldsymbol{v} - \triangle\boldsymbol{v}. \tag{1.47}$$

A direct consequence of the last formula is the Helmholtz representation of vector fields.

Helmholtz representation of a vector field

Let $\boldsymbol{v}$ be a continuous vector field.

Then there exist a scalar field φ and a vector field $\boldsymbol{\psi}$ such that

$$\boldsymbol{v} = \nabla\phi + \text{curl}\,\boldsymbol{\psi} \quad \text{div}\,\boldsymbol{\psi} = 0. \tag{1.48}$$

Similar results can be established for tensor fields.

Gradient integration

The integration of a gradient in order to recover the original function is a more delicate operation than differentiation. Recall first that this operation is possible only if certain compatibility relations are verified, as explained in the text of Chapter 2 and in preceeding sections.

IntegrateGrad functions provided with the book package and given below proceed on the assumption that the user has successfully verified compatibility before applying the procedure. The algorithm then simply performs integration along coordinate lines using **DSolve** and appropiate **ScaleFactors**.

```
IntegrateGrad[nf_?$VecQ, coordsys_:$CoordinateSystem] :=
  Module[{cs = $ExpandCoordSys[coordsys], sf, ht},
  (

    thef[v_] := f[v] ;
    C2frule[v_] =  C[1] ->  thef[v] ;
```

```
    var = cs[[1]];
    func = thef[var];

    eq = Thread[ Grad[func][[1]] == nf [[1]]];
    sol = DSolve[ eq, thef[var], var , GeneratedParameters -> C][[1]];

    var = cs[[2]];
    func = func /. sol /. C2frule[var] ;
    eq = Thread[ Grad[func][[2]] == nf [[2]]];

    sol = DSolve[ eq, thef[var], var , GeneratedParameters -> C][[1]];

    var = cs[[3]];
    func = func /. sol /. C2frule[var];
    eq = Thread[ Grad[func][[3]] == nf [[3]]];

    sol = DSolve[ eq, thef[var], var , GeneratedParameters -> C][[1]];

    func = func /. sol ; func


  )

      /; (cs =!= $Failed  || Curl[nf] =!= {0,0,0})]

IntegrateGrad[nf_? $TenQ, coordsys_:$CoordinateSystem] :=
  Module[ {cs = $ExpandCoordSys[coordsys ]},
  (

    thef[v_] := {f1[v], f2[v], f3[v]} ;
    C2frule[v_] = Thread[ Table[ C[i], {i, 3}] ->  thef[v] ];

    var = cs[[1]];
    func = thef[var];

    eq = Thread[ Grad[func][[All, 1]] == nf [[All, 1]]];
    sol = DSolve[ eq, thef[var], var , GeneratedParameters -> C][[1]];

    var = cs[[2]];
    func = func /. sol /. C2frule[var] ;
    eq = Thread[ Grad[func][[All, 2]] == nf [[All, 2]]];

    sol = DSolve[ eq, thef[var], var , GeneratedParameters -> C][[1]];

    var = cs[[3]];
    func = func /. sol /. C2frule[var];
```

```
    eq = Thread[ Grad[func][[All, 3]] == nf [[All, 3]]];
    sol = DSolve[ eq, thef[var], var , GeneratedParameters -> C][[1]];

    func = func /. sol ; func

  )
     /; (cs = ! = $Failed    ||  Curl[Transpose[nf]] =!=
                                 { {0,0,0}, {0,0,0}, {0,0,0}})]
```

APPENDIX 2

MATHEMATICA tricks

OUTLINE

This appendix is devoted to a brief discussion of the somewhat less than common MATHEMATICA constructs that are used in the main text. Because the presentation below does not systematically cover the range of MATHEMATICA techniques available, we recommend that, in case of need, readers seek details through other MATHEMATICA resources: notebook help files and the MATHEMATICA book (Wolfram, 1999). Within the large literature on this subject we refer to an entry-level introduction by Nancy Blachman (1992), or else to one of the highly structured monographs of Roman Maeder (1997).

The subject matter of this appendix is split into four sections devoted to list manipulation, definition of functions, algebraic handling of expressions, and graphics.

A.2.1 LIST MANIPULATION

Lists are important items within MATHEMATICA, because most objects are internally represented by general lists,

```
mylist1 = { item1, item2 , .... , itemn}
mylist2 = { item1 , {item21, item22, {item231, item232, item233}} ,
  item3 ... , itemn}
```

where the items represent different MATHEMATICA objects. Note, for example, that **mylist2** consists of a series of sublists.

A command that allows one to get rid of the sublists and to bring all elements within the list to one level is **Flatten**:

```
In[.]:= Flatten[ mylist2 ]
Out[.]= { item1 , {item21, item22, ,item231, item232, item233 ,
  item3 ... , itemn}.
```

A finer level of manipulation is afforded by using options; for example, **Flatten[mylist, n]** permits 'flattening' of the list up to level **n** of the sublist. The reverse operation of splitting a list into sublists is performed by the **Partition** operator.

Different items within a list can be extracted using commands **[[.]]** or **Part**; for example,

```
theitem231 = mylist2[[2,3,1]]
theitem231 = Part[ mylist2 , 2 , 3 , 1].
```

While discussing generalised lists, we note that the principles also apply to algebraic expressions of the form

```
a + b^2.
```

Two important operators must be mentioned here. **Fullform** reveals the internal form of the expression in a generalised form,

```
In[.]:= FullForm[a + b^2 ]
Out[.]//FullForm=  Plus[a, Power[b, 2]]
```

and **Head** recoveres the zero-order element of the generalised list, which defines its outermost type,

```
In[.]:= Head[ a + b^2 ]
Out[.]=  Plus

In[.]:= Head[ mylist2 ]
Out[.]=  List.
```

Construction of lists using Table and other operators

Lists can be constructed directly by using operators such as **Table, Array, and Range** or may be obtained by the application of more complex operators such as like **NestList**.

Let us briefly illustrate the mode of using **Table** to create a two-dimensional list:

```
In[.]:= mytable= Table[ a[i] b[j] , {i,3}, {j,3} ]
Out[.]= {{a[1] b[1], a[1] b[2], a[1] b[3]},
        {a[2] b[1], a[2] b[2], a[2] b[3]},
        {a[3] b[1], a[3] b[2], a[3] b[3]}}
```

To display the table in the form of an array, one has to use the **MatrixForm** operator, as follows:

```
In[.]:= mytable // MatrixForm.
```

Assigning element names in lists

When handling multidimensional lists or tables, it is sometimes necessary to carry out assignment of element names to reveal the position of this element within the list. The situation is similar to that encountered in hand manipulation, when in vector or tensor notation we write

$$\boldsymbol{v} = v_i\,\boldsymbol{e}_i, \qquad \boldsymbol{t} = t_{ij}\,\boldsymbol{e}_i \otimes \boldsymbol{e}_j,$$

so that the notation for vector or tensor components is clearly associated with their meaning. We present one particular way of creating this type of element names:

```
mylist = Table[ ToExpression["mylist"<>ToString[i]<>ToString[j] ],
              {i,6}, {j,3} ].
```

Here **ToString** transforms the numerical value of the iterator into a string. This string is then concatenated with the tensor name (e.g., the string **"mylist"**) using **<>**, and finally transformed back into the name of a MATHEMATICA expression using **ToExpression**. This series of operations has been used in examples to create formal expressions for elasticity and stress tensors.

Other operations on lists: Map, Apply, Thread

It is a common feature of many programming languages that in order to apply a function to all elements of a list, it is necessary to create a *do* loop. MATHEMATICA provides several commands that can perform this type of operation, such as **Do** and **While**. However, there exist also a range of commands that are specifically designed to act on all members of the list. Among these commands **Map** is of special interest, because it does not require knowledge of the length of the list in advance. It applies to general expressions and distributes the action of its first argument onto the first level elements of the list, as illustrated by the following examples:

```
In[.]:= v = {v1, v2, v3, v4};

In[.]:= Map[ f , v]
Out[.]= {f[v1], f[v2], f[v3], f[v4]}

In[.]:= Map[ f ,  a + b^2 + c]
Out[.]= f[a] + f[b^2] + f[c].
```

In the same family of operators one finds the operator **Apply**, which makes its second parameter (an expression) into an argument of the function that appears as its first parameter, as shown next:

```
In[.]:= Apply[ f , v ]
Out[.]= f[v1, v2, v3, v4]

In[.]:= Apply[ f,  a + b^2 + c]
Out[.] = f[a, b^2, c].
```

Another important operator is **Thread**, which permits switching the head between two expressions. In some respects its action is similar to that of **Map**:

```
In[.]:= Thread[ f[ a + b^2 + c ], Plus]
Out[.]= f[a] + f[b^2] + f[c].
```

However, it is also particularly convenient to use it in other situations when functions or operations need to be distributed 'through' the list, as happens in the construction of a system of equations below:

```
In[.]:= v == w
Out[.]= {v1, v2, v3, v4} == {w1, w2, w3, w4}

In[.]:= Thread[ v == w ]
Out[.]= {v1 == w1, v2 == w2, v3 == w3, v4 == w4}.
```

Vector-type operations on lists: `Inner, Outer`

In vector and tensor calculus two operations are of special interest whenever combinations of two tensors are to be formed, namely, inner and outer multiplication. As a particular example, consider the two vectors below:

$$\boldsymbol{v} = v_i\, \boldsymbol{e}_i, \qquad \boldsymbol{w} = w_i\, \boldsymbol{e}_i.$$

Forming inner and outer products between these two vectors produces a scalar and a second-order tensor, respectively, which are defined using the Einstein summation convention as

$$\boldsymbol{v} \cdot \boldsymbol{w} = v_i\, w_i \qquad \boldsymbol{v} \otimes \boldsymbol{w} = v_i\, w_j \boldsymbol{e}_i \otimes \boldsymbol{e}_j.$$

The MATHEMATICA operators for this are `Inner` and `Outer`. The operator `Dot` is a direct implementation of the usual dot product definition:

```
In[.]:= v = {v1, v2, v3, v4}; w = {w1, w2, w3, w4};
```

```
In[.]:= Dot[ v , w ]
Out[.]= v1 w1 + v2 w2 + v3 w3 + v4 w4.
```

The same result can also be obtained using `Inner`. However, `Inner` can be used in a more general way, as can be seen in the next example, where the inner product is completed with respect to the functions `f` and `g`:

```
In[.]:= Inner[ f, v , w, g ]
Out[.]=  g[f[v1, w1], f[v2, w2], f[v3, w3], f[v4, w4]].
```

Here the outermost operation is the application of `g`, and the innermost operation is the application of `f`. To obtain a dot product using `Inner` we must set `f = Times` and `g = Plus`, so that

```
In[.]:= Inner[ Times , v , w, Plus ]
Out[.]= v1 w1 + v2 w2 + v3 w3 + v4 w4.
```

`Outer` works in a similar way; that is, it creates a second-order tensor from the elements of two vectors using the function `f`:

```
In[.]:= Outer[ f , v , w  ]
Out[.]= {{f[v1, w1], f[v1, w2], f[v1, w3], f[v1, w4]},
         {f[v2, w1], f[v2, w2], f[v2, w3], f[v2, w4]},
         {f[v3, w1], f[v3, w2], f[v3, w3], f[v3, w4]},
         {f[v4, w1], f[v4, w2], f[v4, w3], f[v4, w4]}}.
```

To obtain the outer product one sets `f = Times`, so that

```
In[.]:= Outer[ Times , v , w  ]
Out[.]= {{v1 w1, v1 w2, v1 w3, v1 w4},
         {v2 w1, v2 w2, v2 w3, v2 w4},
         {v3 w1, v3 w2, v3 w3, v3 w4},
         {v4 w1, v4 w2, v4 w3, v4 w4}}.
```

List manipulation

`Table, Array, Range`	Creation of lists
`Flatten, Partition`	Rearrangement of sublist structure
`Inner, Outer`	Inner and outer operations and products
`Thread, Apply, Map`	Application of functions to lists
`ToString, ToExpression`	Manipulation of expression names and strings

A.2.2 FUNCTIONS

A mathematical function is defined as the application of a transformation that turns a set A into a set B:

$$f : A \longrightarrow B, \qquad \text{so that } f(a) = b, \qquad a \in A, \quad b \in B.$$

The definition of function is understood and used differently in common computations, depending on the context. Within MATHEMATICA there exist different ways to define and manipulate objects that possess properties of mathematical functions.

Start with an example showing the difference between `=` (a.k.a. **`Set`**) and `:=` (a.k.a. **`SetDelayed`**) operators:

```
In[.]:= a = 5; b = 7; c = a + b;
        c
Out[.]= 12
```

```
In[.]:= a = 5; b = 7; c := a + b;
        c
Out[.]= 12
```

```
In[.]:= b = 9;
        c
Out[.]= 14.
```

The illustration is meant to show that *lhs* `=` *rhs* and *lhs* `:=` *rhs* differ insofar that they assign either the *evaluated* or the *unevaluated* right-hand-side value *rhs* to the left-hand-side *lhs*. Assigning an unevaluted value implies that the value will be evaluated whenever the value of the left-hand-side *lhs* is required in the calculation, and hence the result will depend on the values of other variables that are current at the time of evaluation.

In this respect, evaluating **`c`** using **`c`** `=`... simply represents a short form of writing out **`Evaluate[c]`**.

A simple assignment operation for a function such as $f[x]$ simply establishes that every occurrence of $f[x]$ must be replaced with the right-hand side:

```
In[.]:= f[x] := Sin[ x^2]
Out[.]= Sin[x^2 ].
```

But this has no effect for $f[y]$:

```
In[.]:= f[y]
Out[.]= f[y].
```

In practice we are more frequently interested in assigning expressions to functions in a way that would allow argument(s) such as **x** to be substituted into the expression, that is, replaced by another value (e.g., this might be y). This is done using a special underscore sign, _, that is an example of a **Pattern**:

```
In[.]:= f[x_] := Sin[ x^2]
Out[.]= Sin[x^2 ]

In[.]:= f[y + z]
Out[.]= Sin[(y + z)^2 ].
```

Patterns are used in the more general sense to define entire ranges of parameters that pass certain conditions. For example, for functions requiring tensor-valued arguments, we perform a check of their type using patterns in **TensQ** in the **Tensor2Analysis** package.

The derivative of a function is computed using **D** or **Derivative** commands. The example below shows that one has to be careful in defining the function and its arguments, as well as the variable with respect to which the differentiation takes place.

```
In[.]:= f[x_] := Sin[ x^2];

In[.]:= D[ f, x]
Out[.]= 0

In[.]:= D[f[x] , x]
Out[.]= 2 x Cos[x^2 ]

In[.]:= Derivative[1][f[x]]
Out[.]=(Sin[x^2])'

In[.]:= Derivative[1][f]
Out[.]= 2 Cos[#1^2 ] #1 &.
```

The first and the third attempt to differentiate were unsuccessful due to the reasons explained previously, whereas the second and fourth attempts gave the correct answer. In the last attempt the answer appears to be expressed in a strange format called *pure function*, which is another way of expressing functions. The **#1** and the **&** denote the argument and the fact that the expression represents a function. Let us compute the vaule of this function for **x**:

```
In[30]:= (2 Cos[#1^2 ] #1 &)[x]
Out[30]= 2 x Cos[x ].
```

The main utility of pure functions reveals itself in operations where the name of the function is not known or is being modified, such as when the application of the commands **Map, Apply, Select,...** to arbitrary functions is being considered.

Functions

`f[x_] := ...`	typical function definition with a general argument
`= , :=`	**`Set`** and **`SetDelayed`** commands for associating values to symbols
`_`	**`Pattern`** command that helps define types of argument(s)
`( ...) &`	Pure function
`#, #1, ...#n`	Arguments in pure functions
`Function`	Construction of an abstract function
`D, Derivative`	Computes derivative(s) of a function

A.2.3 ALGEBRAIC HANDLING OF EXPRESSIONS

The main rule of thumb when manipulating algebraic expressions is to proceed in small steps, similarly to the way one would perform operations by hand. This will permit MATHEMATICA operators to perform most efficiently even on enormous expressions, which can be very time-consuming to transform. We do not aim to present here the art and the techniques of this manipulation, although some use is made in the main text of various convenient tricks. We discuss briefly, however, a question regarding simplification of expressions, that is, how to obtain reductions

$$\sqrt{a^2+a^4} = a\sqrt{a}, \qquad \frac{1}{\sqrt{a^2+a^4}} = \frac{1}{a}\frac{1}{\sqrt{1+a^2}}, \qquad \text{if } a > 0.$$

It is easy to discover that the application of **`Simplify`**, or even of the **`FullSimplify`** operator with the *Assumptions* option available in version 5 of MATHEMATICA, does not provide the desired answer:

```
In[.]:= aa = Sqrt[a^2 + a^4]
Out[.]= Sqrt[a^2  + a^4 ]

In[.]:= FullSimplify[aa]
Out[.]= Sqrt[a^2  + a^4 ]

In[.]:= FullSimplify[aa, a [Element] Reals && a > 0]
Out[.]= Sqrt[a^2  + a^4 ]
```

One of the possible workarounds is to define directly a replacement rule for the expressions to be processed and apply it directly, as follows:

```
In[.]:= f[x_] := Sin[x]

In[.]:= f[ aa ] /. Sqrt[a^2 + a^4] -> a Sqrt[1 + a^2]
Out[.]= Sin[a Sqrt[1 + a ]]
```

In certain cases it is important to check the internal form of the expression within MATHEMATICA using the **`FullForm`** operator in order to understand the functions involved; for example, $1/\sqrt{a}$ becomes

```
In[65]:= FullForm[ 1 / Sqrt[a]]
Out[65]//FullForm= Power[a, Rational[-1, 2]]
```

and not `Power[Power[ a, Rational[1,2]], -1]`, as one might have expected. This also explains which replacement rules will and will not work.

This technique has been used in the computation of strain, stretch, and rotation tensors for simple shear in Chapter 2.

Handling algebraic expressions

`Simplify, FullSimplify`	Simplifies an expression
`a -> b`	Assigns a rule
`expr /. rule`	ReplaceAll, replaces suitable atoms using `rules` in `expr`

A.2.4 GRAPHICS

Standard routines within MATHEMATICA already provide a series of well-documented graphics functions, `Plot, Plot3D, ContourPlot, ListPlot,...`, which present practical solutions to most questions. In additions, we briefly highlight an interesting option, `DisplayFunction`, that permits one to choose where the graphics information is channeled. In superposing different graphics, it is useful to keep intermediate images hidden and display only the final result. This can be achieved using `DisplayFunction` in the following way:

```
In[.]:=
p = Plot[ x^2 , {x,0, 5}, DisplayFunction -> Identity ]
q = Plot[ x^5 , {x,0, 5}, DisplayFunction -> Identity ]
Show[ p, q, DisplayFunction -> $DisplayFunction]
```

Examples in Chapter 2 illustrate how one can build different types of graphics starting with primitive objects `Line,Disk,...`, assigning colours with `Hue` and `GrayLevel`. As a further illustration of graphics manipulation, Appendix 3 introduces a series of commands that allow plotting contour maps of a given function on a deformed mesh.

APPENDIX 3

Plotting parametric meshes

OUTLINE

The need to plot contour maps over complex-shaped domains arises frequently in the course of analysis of stress and strain distributions with respect to different coordinate systems. This appendix is dedicated to the construction of a series of graphics tools to build parametric meshes and to plot functions over new domains.

A mesh is defined in cartesian coordinates by the function

$$\boldsymbol{f}(u, v, w) = f_x(u, v, w)\,\boldsymbol{e}_x + f_y(u, v, w)\,\boldsymbol{e}_y + f_z(u, v, w)\,\boldsymbol{e}_z,$$

$$(u, v, z) \in [u_0, u_1] \times [v_0, v_1] \times [w_0, w_1]$$

and a series of three increments *du*, *dv*, *dz* on each of the coordinate lines. We propose to colour each mesh element using a colour function defined as

$$g(u, v, z), \qquad (u, v, z) \in [u_0, u_1] \times [v_0, v_1] \times [w_0, w_1].$$

The tools constructed here stand in a close relationship with the already existing commands of the **`ParametricPlot`** type or the **`ComplexMap`** package, which are standard with the MATHEMATICA distribution.

The main ingredients of the parametric plot package are presented next. The collection of commands and their assembly can be further understood by looking at the **`ParametricMesh`** package that is provided together with this book. Further insight into the techniques of command and package building can be found in Maeder (1997).

Building a 2D mesh

The sequence of commands introduced below shows how the mesh is first built up from rectangular elements, and then colour is associated with each element.

We start by defining a rectangular domain of the parameters

$$(u, v) \in [u_0, u_1] \times [v_0, v_1]$$

and associating a standard spacing with each of the parametric axes, **`ndu`** and **`ndv`**.

Then the function defining the deformation of the domain is introduced:

$$\boldsymbol{f}\,(u, v) = f_x\,\boldsymbol{e}_x + f_y\,\boldsymbol{e}_y$$
$$\boldsymbol{f}\,: [u_0, u_1] \times [v_0, v_1] \longrightarrow \mathbb{R}^2.$$

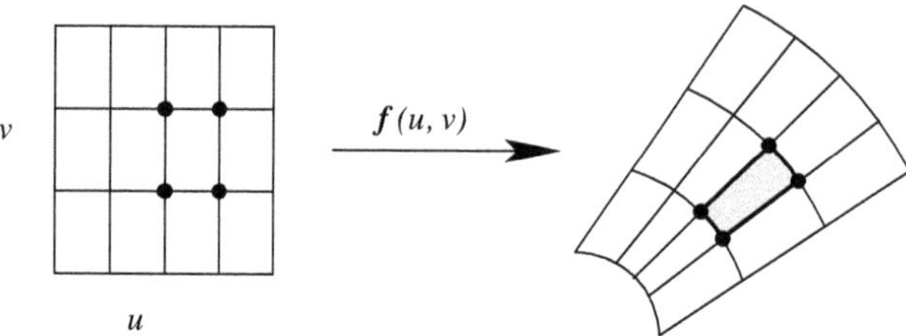

Figure A.3.1. A schematic illustration of the nodal positions and a rectangular mesh element in the (u, v) parameter space and the corresponding curvilinear domain spanned by the mesh and a distorted mesh element in the real space.

```
Thread[{u0, u1, ndu} = {0, 1, 0.1}];
Thread[{v0, v1, ndv} = {0, 1, 0.1}];
fu = u + 2 v;
fv = v;
```

Lists of nodal coordinates in the parameter space (u, v) are created by **Apply**-ing the **Range** command to the initial data.

The outer product of the two lists creates mesh nodes in the parameter space of (u, v). See Fig. A.3.1.

```
uu = Range @@ {u0, u1, ndu};
vv = Range @@ {v0, v1, ndv};

thenodes = Outer[ List, uu, vv] ;
```

The projection of mesh nodes from the parameter space into the real space is created by mapping f over the coordinate pairs (u, v).

The nodes are then paired into groups of four in order to define individual rectangular elements of the mesh (patches) that are going to be associated with particular colours.

```
meshnodes = Map[
    {fu , fv}  /. Thread[{u,v} -> #]&,
    thenodes, {2}];

thequads = Map[
    Flatten[#,1][[{1,2,4,3}]] &,
    Flatten[
       Partition[ meshnodes, {2,2}, {1,1}],
       1]];
```

The final mesh will be defined as a collection of closed rectangular lines.

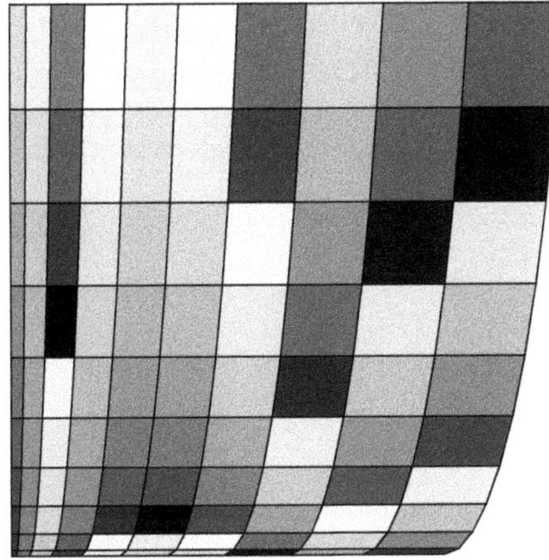

Figure A.3.2. Deformation of the $[0, 2] \times [0, 1]$ domain under the application of the function $\boldsymbol{f}(u, v) = 3u^2 \boldsymbol{e}_x + 2u \sin v \boldsymbol{e}_y$ using the value of the colour function at the average deformed node position as the piecewise constant filling colour.

ClosedLine appends the first node of each four-node set that defines a rectangular domain to the list of nodes in order to close the line. This function is mapped over all rectangular sets to obtain the mesh.

The mesh can be plotted by using **Graphics** to create the graphics objects and using **Show** to display it.

```
ClosedLine[ptlist_]:=Line[ Append[ptlist, First[ptlist]] ]

themesh = Map[ ClosedLine, thequads];

Show[ Graphics[themesh] ]
```

The filling of the mesh is created by a set of coloured rectangular patches. We first define a piecewise constant colour function by associating the value of this function at the average node position within each rectangular element with the entire element.

The colour function and the **Polygon** command are then mapped over the collection of elements to obtain the filling, which can be plotted using a syntax similar to that used for the mesh. (See Figure A.3.2.)

```
MeanNode[ nl_] := 0.25 Plus @@  Drop[nl, -1]
fillcolor[ ptlist_] :=
   GrayLevel[ FractionalPart[Norm[MeanNode[ ptlist ]]]]

thefill = Map[ {fillcolor[#] , Polygon[#]} &, thequads ];

Show[ Graphics[thefill] ]
```

Building a 2D surface in 3D

The image of a 2D surface within a 3D space can now be built in a way similar to that presented before, starting from a mesh of rectangular elements and associating a patch of colour with each element. Coding of the command follows the same steps as before and adjoins new space coordinates to geometrical objects.

The main change in the procedure with respect to that used for a 2D mesh is that the mapping transformation

$$f(u,v) = f_x \boldsymbol{e}_x + f_y \boldsymbol{e}_y + f_z \boldsymbol{e}_z$$

$$f : [u_0, u_1] \times [v_0, v_1] \longrightarrow \mathbb{R}^3$$

now contains the additional third coordinate.

```
Thread[{u0, u1, ndu} = {0, 1, 0.1}];
Thread[{v0, v1, ndv} = {0, 1, 0.1}];
fu = u + 2 v;
fv = v;
fz = u^2;
```

Nodes and meshes are obtained in the same way as in 2D, with the only difference being that the resulting geometric objects have an additional coordinate.

```
uu = Range @@ {u0, u1, ndu};
vv = Range @@ {v0, v1, ndv};

thenodes = Outer[ List, uu, vv] ;
meshnodes = Map[  {fu , fv, fz} /.Thread[{u,v} -> #]&, thenodes, {2}];
thequads = Map[   Flatten[#,1][[{1,2,4,3} ]]  & ,
            Flatten[ Partition[ meshnodes, {2,2}, {1,1}], 1]];
themesh = Map[ ClosedLine, thequads];
thefill = Map[ {fillcolor[#] , Polygon[ # ]} &, thequads ];
```

Displaying the objects is achieved in a similar way using **Graphics3D** and **Show**. We note in passing the use of the option **Lighting -> False**, which ensures that ambient lighting that affects the colour of the underlying object is switched off. This option is a characteristic feature of 3D display routines and did not appear in the discussion of 2D meshes.

```
Show[ Graphics3D[themesh] ]
Show[ Graphics3D[thefill] , Lighting -> False]
```

Building a 3D box as a collection 2D surfaces

If both the parameter space and the range of the map are three-dimensional, that is, f is defined as

$$f(u,v,w) = f_x \boldsymbol{e}_x + f_y \boldsymbol{e}_y + f_z \boldsymbol{e}_z \qquad (u,v,z) \in [u_0,u_1] \times [v_0,v_1] \times [w_0,w_1],$$

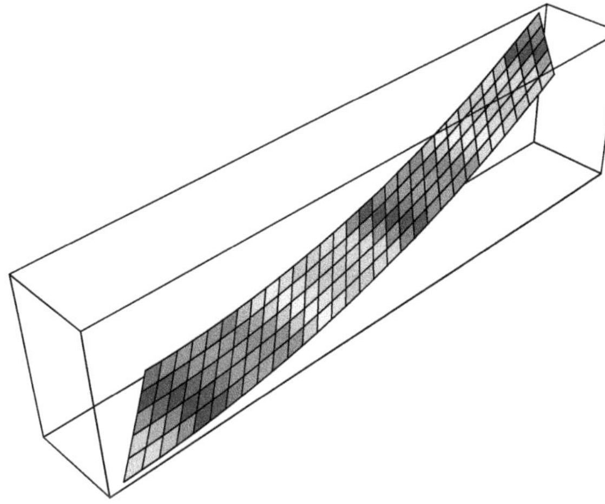

Figure A.3.3. Usage of **ParametricMesh3D**: The map of the $[0,3] \times [0,0.5]$ domain under the application of the function $\boldsymbol{f}(u,v) = (u+v)\boldsymbol{e}_x + v\boldsymbol{e}_y + (0.7 + 0.1(u^2+v^2))\boldsymbol{e}_z$.

then one must consider only external sufaces of a 3D parametric box. These are created by the maps

$$\boldsymbol{f}_{w_i}(u,v) = \boldsymbol{f}(u,v,w_i) \qquad i = 0,1 \qquad (u,v) \in [u_0,u_1] \times [v_0,v_1]$$

and $\boldsymbol{f}_{v_i}(u,w) = \boldsymbol{f}(u,v_i,w)$ and $\boldsymbol{f}_{u_i}(v,w) = \boldsymbol{f}(u_i,v,w)$ are defined in a similar way.

The six external surfaces can be displayed using a combination of the previously defined commands to plot 2D surfaces in 3D space. Further details of these commands can be found in the package **ParametricMesh** provided with this book. See Figures A.3.3 and A.3.4

The ParametricMesh package

Mathematica commands and their assemblies discussed above have been organised into three principal modules, **ParametricMesh, ParametricMesh3D, ParametricBox** which together allow one to create surfaces in 2D and 3D and to colour exterior surfaces

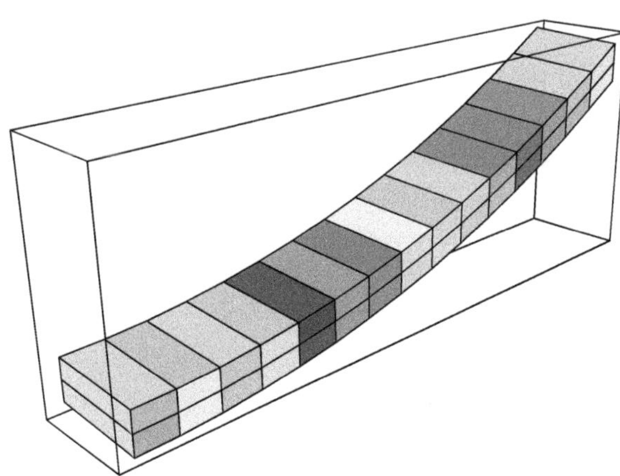

Figure A.3.4. Usage of **ParametricBox3D**: the map of the $[0,3] \times [0,0.5] \times [-0.1,0.1]$ domain under the application of the function $\boldsymbol{f}(u,v,w) = u\boldsymbol{e}_x + v\boldsymbol{e}_y + (w + 0.1(u^2+v^2))\boldsymbol{e}_z$.

of a boxed domain in 3D, respectively. Examples of command call formats are given in the notebook **CA3_package_examples.nb**.

Parametric mesh operators

ParametricMesh[*f_args, domain_args, opts*]	Builds a 2D mesh in 2D
ParametricMesh3D[*f_args, domain_args, opts*]	Builds a 2D mesh in 3D
ParametricBox[*f_args, domain_args, opts*]	Builds a boxed domain in 3D

Options

Fill -> True / False	Defines if the elements of the mesh are filled or not
FillColor -> g	Defines the colour function
Mesh -> True / False	Defines if the mesh is displayed or not
Plotpoints -> n	Defines the number of intermediate points in the mesh
Lighting -> False	Standard **Graphics3D** option that turns the ambient lighting off and makes colours of the object visible

Bibliography

Bahder, T. B. *Mathematica for scientists and engineers*. Addison–Wesley, 1994.

Ballard, P., and Millard, A. *Modélisation et calcul des structures élancées*. Ecole Polytechnique, Département de Mécanique, 2005.

Bamberger, Y. *Mécanique de l'ingénieur*. Hermann, 1997.

Barber, J. R. *Elasticity*, second edition. Kluwer, 2002.

Blachman, N. *Mathematica: A practical approach*. Prentice Hall, 1992.

Boussinesq, M. J. *Application des potentiels*. Gauthier–Villars, 1885.

Bonnet, M., and Constantinescu, A. "Inverse problems in elasticity," *Inverse problems* 21 No. 2 (April 2005) R1–R50.

Chadwick, P., Vianello, M., and Cowin, S. C. A new proof that the number of linear anisotropic elastic symmetries is eight. *J. Mech. Phys. Solids* 49, 2001.

Coirier, J. *Mécanique des mileux continus*. Dunod, 1997.

Cowin, S. C., and Mehrabadi, M. Eigentensors of linear elastic materials. *Q. J. Mech. Appl. Math*. 43, 1990.

Cowin, S. C., and Mehrabadi, M. The structure of the linear anisotropic symmetries. *J. Mech. Phys. Solids* 40(7):1459–1471, 1992.

Dumontet, H., Duvaut, G., Léne, F., Muller, P., and Turbé, N. *Exercices corrigés de mécanique des mileux continus*. Dunod, 1998.

Eason, J., and Ogden, R. W., eds. *Elasticity, mathematical models and applications*. Ellis Horwood, Chichester, 1964.

Galerkin, B. G. Contribution à la solution générale du problème de la théorie de l'elasticité dans la cas de trois dimensions. *Comptes Rendus Acad. Sci. Paris*, 190: 1047–1048, 1930.

Germain, P. *Mécanique des milieux continus*. Masson, Paris, 1983.

Griffith, A. A. The phenomena of rupture and flow in solids. *Phil. Trans. R. soc London Ser.* A. 221:163–180, 1921.

Gurtin, M. E. *An introduction to continuum mechanics*. Academic Press, London, 1982.

Halmos, P. R. *Finite dimensional vector spaces*. Princeton University Press, 1959.

Hoff, N. J. The applicability of Saint-Venant's principle to airplane structures. *J.Aero.Sci.* 12:455, 1952.

Huerre, P. *Mécanique des fluides*. Editions de l'Ecole Polytechnique, Palaiseau, France, 2001.

Inglis, C. E. Stresses in a plate due to the presence of cracks and sharp corners. *Trans. Inst. Naval Arch.* 219–230, 1913.

Kestelman, H. *Modern theories of integration*. Dover, 1960.

Lehnitski, S. G *Theory of elasticity of an anisotropic body*. Mir, Moscow, 1981.

Love, A. E. H. *A treatise on the mathematical theory of elasticity*. Dover, 1944.

Maeder, R. *Programming in Mathematica*. Addison–Wesley, 1997.

Malvern, L. E. *Introduction the mechanics of continuous medium*. Prentice Hall, 1969.

Marsden, J. E., and Hughes, T. J. R. *Foundations of elasticity*. Prentice Hall, 1982; Dover, 1994.

Muskhelishvili, N. I. *Some basic problems of mathematical theory of elasticity*. Noordhoff, Groningen, 1953.

Nye, J. F. *Physical properties of crystals*. Clarendon, Oxford, 1985.

Obala, J. *Exercices et problémes de mécanique des mileux continus*. Masson, 1997.

Ogden, R. W. *Non-linear elastic deformations*. Dover, 1997.

Polytechnique Collective. *Recueil: Texte des contrôles des connaissances*. Département de Mécanique, Ecole Polytechnique, 1990–2005.

Saint Venant, Barré de Mémoire sur la torsion des prismes. *Mémoires Savants étrangers*, 1855.

Salençon, J. *Handbook of continuum mechanics : General concepts : Thermoelasticity*. Springer-Verlag, 2001.

Soos, E., and Teodosiu, C. *Calcul tensorial cu aplicaţii in mecanica solidelor*. Editura Ştiinţifica şi enciclopedica, 1983.

Soutas-Little, R. W. *Elasticity*. Dover, 1973.

Spivak, M. *Calculus on manifolds*. Addison–Wesley, 1965.

Sternberg, E. On Saint Venant's principle. *Q. J. Appl. Math.* 11:393–402, 1954.

Timoshenko, S., and Goodier, J. N. *Theory of elasticity*. McGraw–Hill, 1951.

Truesdell, C. A. *Essays in the history of mechanics*. Springer-Verlag, 1968.

Westergaard, H. M. *Theory of elasticity and plasticity*. Harvard University Press, 1952.

Williams, M. L. Stress singularities resulting from various boundary conditions in angular corners of plates in extension. *J. Appl. Mech.*, 19:526–528, 1952.

Wolfram, S. *The Mathematica book*. Wolfram Media, Cambridge Univ. Press, 1999.

Index

For EU product safety concerns, contact us at Calle de José Abascal, 56–1°,
28003 Madrid, Spain or eugpsr@cambridge.org.

www.ingramcontent.com/pod-product-compliance
Ingram Content Group UK Ltd.
Pitfield, Milton Keynes, MK11 3LW, UK
UKHW050730090726
473066UK00013B/1098

* 9 7 8 1 1 0 7 4 0 6 1 3 1 *